Correspondence Analysis in Practice

Third Edition

CHAPMAN & HALL/CRC
Interdisciplinary Statistics Series

Seriese ditors: N.K eiding,B .J.T.M organ,C .K. Wikle,P .v ande rH eijden

Publishedtitle s

AGE-PERIOD-COHORT ANALYSIS: NEW MODELS, METHODS, AND EMPIRICAL APPLICATIONS Y. Yang and K. C. Land

ANALYSIS OF CAPTURE-RECAPTURE DATA R. S. McCrea and B. J.T. Morgan

AN INVARIANT APPROACH TO STATISTICAL ANALYSIS OF SHAPES S. Lele and J. Richtsmeier

ASTROSTATISTICS G. Babu and E. Feigelson

BAYESIAN ANALYSIS FOR POPULATION ECOLOGY R. King, B. J.T. Morgan, O. Gimenez, and S. P. Brooks

BAYESIAN DISEASE MAPPING: HIERARCHICAL MODELING IN SPATIAL EPIDEMIOLOGY, SECOND EDITION A. B. Lawson

BIOEQUIVALENCE AND STATISTICS IN CLINICAL PHARMACOLOGY S. Patterson and B. Jones

CLINICAL TRIALS IN ONCOLOGY, THIRD EDITION S. Green, J. Benedetti, A. Smith, and J. Crowley

CLUSTER RANDOMISED TRIALS R. J. Hayes and L. H. Moulton

CORRESPONDENCE ANALYSIS IN PRACTICE, THIRD EDITION M. Greenacre

DESIGN AND ANALYSIS OF QUALITY OF LIFE STUDIES IN CLINICAL TRIALS, SECOND EDITION D.L. Fairclough

DYNAMICAL SEARCH L. Pronzato, H. Wynn, and A. Zhigljavsky

FLEXIBLE IMPUTATION OF MISSING DATA S. van Buuren

GENERALIZED LATENT VARIABLE MODELING: MULTILEVEL, LONGITUDI-NAL, AND STRUCTURAL EQUATION MODELS A. Skrondal and S. Rabe-Hesketh

GRAPHICAL ANALYSIS OF MULTI-RESPONSE DATA K. Basford and J. Tukey

INTRODUCTION TO COMPUTATIONAL BIOLOGY: MAPS, SEQUENCES, AND GENOMES M. Waterman

MARKOV CHAIN MONTE CARLO IN PRACTICE W. Gilks, S. Richardson, and D. Spiegelhalter

MEASUREMENT ERROR ANDMISCLASSIFICATION IN STATISTICS AND EPIDE-MIOLOGY: IMPACTS AND BAYESIAN ADJUSTMENTS P. Gustafson

MEASUREMENT ERROR: MODELS, METHODS, AND APPLICATIONS J. P. Buonaccorsi

Published titles

MEASUREMENT ERROR: MODELS, METHODS, AND APPLICATIONS
J. P. Buonaccorsi

MENDELIAN RANDOMIZATION: METHODS FOR USING GENETIC VARIANTS IN CAUSAL ESTIMATION S. Burgess and S.G. Thompson

META-ANALYSIS OF BINARY DATA USING PROFILE LIKELIHOOD D. Böhning, R. Kuhnert, and S. Rattanasiri

MISSING DATA ANALYSIS IN PRACTICE T. Raghunathan

POWER ANALYSIS OF TRIALS WITH MULTILEVEL DATA M. Moerbeek and S. Teerenstra

SPATIAL POINT PATTERNS: METHODOLOGY AND APPLICATIONS WITH R A. Baddeley, E Rubak, and R. Turner

STATISTICAL ANALYSIS OF GENE EXPRESSION MICROARRAY DATA T. Speed

STATISTICAL ANALYSIS OF QUESTIONNAIRES: A UNIFIED APPROACH BASED ON R AND STATA F. Bartolucci, S. Bacci, and M. Gnaldi

STATISTICAL AND COMPUTATIONAL PHARMACOGENOMICS R. Wu and M. Lin

STATISTICS IN MUSICOLOGY J. Beran

STATISTICS OF MEDICAL IMAGING T. Lei

STATISTICAL CONCEPTS AND APPLICATIONS IN CLINICAL MEDICINE J. Aitchison, J.W. Kay, and I.J. Lauder

STATISTICAL AND PROBABILISTIC METHODS IN ACTUARIAL SCIENCE P.J. Boland

STATISTICAL DETECTION AND SURVEILLANCE OF GEOGRAPHIC CLUSTERS P. Rogerson and I. Yamada

STATISTICS FOR ENVIRONMENTAL BIOLOGY AND TOXICOLOGY A. Bailer and W. Piegorsch

STATISTICS FOR FISSION TRACK ANALYSIS R.F. Galbraith

VISUALIZING DATA PATTERNS WITH MICROMAPS D.B. Carr and L.W. Pickle

Chapman & Hall/CRC
Interdisciplinary Statistics Series

Correspondence Analysis in Practice

Third Edition

Michael Greenacre

Universitat Pompeu Fabra
Barcelona, Spain

CRC Press
Taylor & Francis Group
Boca Raton London New York

CRC Press is an imprint of the
Taylor & Francis Group, an **informa** business

A CHAPMAN & HALL BOOK

CRC Press
Taylor & Francis Group
6000 Broken Sound Parkway NW, Suite 300
Boca Raton, FL 33487-2742

First issued in paperback 2021

ISBN 13: 978-0-367-78251-1 (pbk)
ISBN 13: 978-1-4987-3177-5 (hbk)

Library of Congress Cataloging-in-Publication Data

Names: Greenacre, Michael J.
Title: Correspondence analysis in practice / Michael Greenacre.
Description: Third edition. | Boca Raton, Florida : CRC Press, [2017] |
Series: Interdisciplinary statistics | Includes bibliographical references
and index.
Identifiers: LCCN 2016036265| ISBN 9781498731775 (hardback : alk. paper) |
ISBN 9781498731782 (e-book)
Subjects: LCSH: Correspondence analysis (Statistics)
Classification: LCC QA278.5 .G74 2017 | DDC 519.5/37--dc23
    LC record available at https://lccn.loc.gov/2016036265

Visit the Taylor & Francis Web site at
http://www.taylorandfrancis.com

and the CRC Press Web site at
http://www.crcpress.com

To Françoise, Karolien and Gloudina

Contents

Preface

This book is a revised and extended third edition of the second edition of *Correspondence Analysis in Practice* (Chapman & Hall/CRC, 2007), first published in 1993. In the original first edition I wrote the following in the Preface, which is still relevant today:

"Correspondence analysis is a statistical technique that is useful to all students, researchers and professionals who collect categorical data, for example data collected in social surveys. The method is particularly helpful in analysing crosstabular data in the form of numerical frequencies, and results in an elegant but simple graphical display which permits more rapid interpretation and understanding of the data. Although the theoretical origins of the technique can be traced back over 50 years, the real impetus to the modern application of correspondence analysis was given by the French linguist and data analyst Jean-Paul Benzécri and his colleagues and students, working initially at the University of Rennes in the early 1960s and subsequently at the Jussieu campus of the University of Paris. Parallel developments of correspondence analysis have taken place in the Netherlands and Japan, centred around such pioneering researchers as Jan de Leeuw and Chikio Hayashi. My own involvement with correspondence analysis commenced in 1973 when I started my doctoral studies in Benzécri's Data Analysis Laboratory in Paris. The publication of my first book *Theory and Applications of Correspondence Analysis* in 1984 coincided with the beginning of a wider dissemination of correspondence analysis outside of France. At that time I expressed the hope that my book would serve as a springboard for a much wider and more routine application of correspondence analysis in the future. The subsequent evolution and growing popularity of the method could not have been more gratifying, as hundreds of researchers were introduced to the method and became familiar with its ability to communicate complex tables of numerical data to non-specialists through the medium of graphics. Researchers with whom I have collaborated come from such varying backgrounds as sociology, ecology, palaeontology, archaeology, geology, education, medicine, biochemistry, microbiology, linguistics, marketing research, advertising, religious studies, philosophy, art and music. In 1989 I was invited by Jay Magidson of Statistical Innovations Inc. to collaborate with Leo Goodman and Clifford Clogg in the presentation of a two-day short course in New York, entitled "Correspondence Analysis and Association Models: Geometric Representation and Beyond". The participants were mostly marketing professionals from major American companies. For this course I prepared a set of notes which reinforced the practical, user-oriented approach to correspondence analysis. ... The positive reaction of the audience was infectious and inspired me subsequently to present short courses on correspondence analysis in South Africa, England and Germany. It is from the notes prepared for these courses that this book has grown."

Extract from preface of first edition of Correspondence Analysis in Practice (1993)

In 1991 Prof. Walter Kristof of Hamburg University proposed that we organize a conference on correspondence analysis, with the assistance of Dr. Jörg Blasius of the Zentralarchiv für Empirische Sozialforschung (Central Archive

The CARME conferences

for Empirical Social Research) at the University of Cologne. This conference was the first international one of its kind and drew a large audience to Cologne from Germany and neighbouring European countries. This initial meeting developed into a series of quadrennial conferences, repeated in 1995 and 1999 in Cologne, at the Pompeu Fabra University in Barcelona (2003), the Erasmus University in Rotterdam (2007), Agrocampus Rennes (2011) and the University of Naples Federico II (2015). The 1991 conference led to the publication of the book *Correspondence Analysis in the Social Sciences*, while the 1995 conference gave birth to another book, *Visualization of Categorical Data*, both of which received excellent reviews. For the 1999 conference on *Large Scale Data Analysis*, participants had to present analyses of data from the multinational International Social Survey Programme (ISSP — see `www.issp.org`). This interdisciplinary meeting included presentations not only on the latest methodological developments in survey data analysis but also topics as diverse as religion, the environment and social inequality. In 2003 we returned to the original theme for the Barcelona conference, which was baptized with the Catalan girl's name CARME, standing for Correspondence Analysis and Related MEthods; hence the formation of the CARME network (`www.carme-n.org`). This led once more to Jörg Blasius and myself editing a third book, *Multiple Correspondence Analysis and Related Methods*, which was published in 2006. As with the two previous volumes, our idea was to produce a multi-authored book, inviting experts in the field to contribute. Our editing task was to write the introductory and linking material, unifying the notation and compiling a common reference list and index. As a result of the Rennes conference in 2011, which celebrated 50 years of correspondence analysis, the book *Visualization and Verbalization of Data* was published, half of which is devoted to the history of multivariate analysis. These books mark the pace of development of correspondence analysis, at least in the social sciences, and are highly recommended to anyone interested in deepening their knowledge of this versatile statistical method as well as methods related to it.

New material in third edition I have been very gratified to be invited to prepare a new edition of *Correspondence Analysis in Practice*, having accumulated considerably more experience in social and environmental research in the nine years since the publication of the second edition. Apart from revising the existing chapters, five new chapters have been added, on "Compositional Data Analysis" (an area highly related to correspondence analysis), "Analysis of Matched Matrices" (joint analysis of data tables with the same rows and columns), "Correspondence Analysis of Networks" (applying correspondence analysis to graphs), "Co-Inertia and Co-Correspondence Analysis" (analysis of relationships between two tables with common rows), and "Permutation Tests" (performing statistical inference in the context of correspondence analysis and related methods). All in all, I can say that this third edition contains almost all my practical knowledge of the subject, after more than 40 years working in this area.

At a conference I attended in the 1980s, I was given this lapel button with its nicely ambiguous maxim, which could well be the motto of correspondence analysts all over the world:

"Statisticians count!"

To illustrate the more technical meaning of this motto, and to give an initial example of correspondence analysis, I made a count of the most frequent words in each of the 30 chapters of this new edition. I had to aggregate variations of the same word, e.g. "coordinate" and "coordinates", "plot" and "plotting", a process called *lemmatization* in textual data analysis. The top 10 most frequent words were, in descending order of frequency: "row/s", "profile/s", "inertia" (which is the way correspondence analysis measures variance in a table), "point/s", "column/s", "data", "CA" (abbreviation for correspondence analysis), "variable/s", "value/s" and "average". I omitted words that occur in one chapter only, such as "fuzzy" and "degree", which are specific to a single chapter, and removed words that described particular applications. This left an eventual total of 167 words, which can be regarded as reflecting the methodological content of the book.

Textual analysis of third edition

	analyse/sis	association/s	asymmetric	average	axis/es	...
Chap 1	10	0	0	0	15	...
Chap 2	0	0	0	29	22	...
Chap 3	0	0	0	55	0	...
Chap 4	0	6	0	22	0	...
Chap 5	0	0	0	22	13	...
Chap 6	0	0	0	8	0	...
Chap 7	0	0	0	29	0	...
Chap 8	47	0	0	14	20	...
Chap 9	0	0	14	6	32	...
Chap 10	0	0	17	0	14	...
⋮	⋮	⋮	⋮	⋮	⋮	...
Total	369	12	39	370	277	...

Exhibit 0.1:
First few rows and columns of the table of counts of the 167 most frequent words in the 30 chapters of Correspondence Analysis in Practice, Third Edition, visualized in Exhibit 0.2 using correspondence analysis.

Exhibit 0.2:
Correspondence analysis display of 30 chapters of the present book in terms of the most frequent words in each chapter. Numbers in boldface indicate the positions of the chapters, and their proximity signifies relative similarity of word distribution. Directions of the words give the interpretation for the positioning of the chapters. Technically, this is a so-called "contribution biplot" (see Chapter 13).

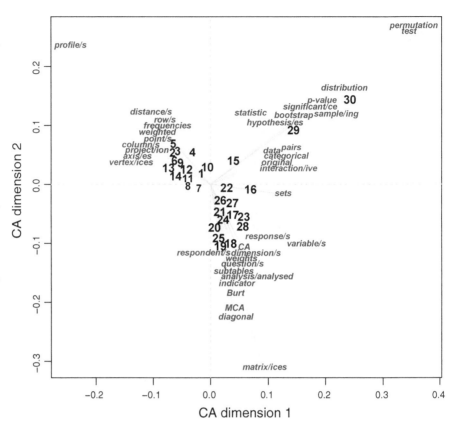

The final table of word counts was composed of the 30 chapters as rows, and the 167 words as columns (see Exhibit 0.1). This table is very *sparse*, i.e. it has many zeros. In fact, 80% of the cells of the table have no counts. Correspondence analysis copes quite well with such data, which has made it a popular method in research areas such as linguistics, archaeology and ecology, where data sets of frequency counts occur that are very sparse.

Exhibit 0.2 shows the "map" of the table, resulting from applying correspondence analysis. The first thing to notice is that the rows (chapters, indicated by their numbers) and the columns (words, connected to the centre by lines) are displayed with respect to two "dimensions". These dimensions are determined by the analysis with the objective of exposing the most important features of the associations between chapters and words. An alternative way of thinking about this is that the chapters are mapped according to the similarity in their distributions of words, with closer chapters being more similar and distant chapters more different. Then the directions of the words explain the differences between the chapters. Not all the words are shown, because about two-thirds of the words turn out to be not so important for the interpretation of the result, so only those words are shown that contribute highly to the positioning of the chapters. Without further explanation of the concepts underlying correspondence analysis (after all, this is the aim of the book that follows!) the map clearly shows three sets of words emanating from the centre. The words out to the top right clearly distinguish Chapters 29 and 30 from all the others — these are the chapters that concentrate on the sampling, distributional and inferential properties of correspondence analysis, with main keywords "permutation" and "test". Chapter 15 on clustering also tends in that direction because it contains some hypothesis testing. Out in the upper left direction are all the words describing basic concepts and terminology of correspondence analysis associated with Chapters 1–14 that introduce the method and develop it, exemplified by the most prominent keyword "profile/s". Towards the bottom of the map are the words associated with a generalization of correspondence analysis, called multiple correspondence analysis, usually applied to questionnaire data, described in later chapters. This method involves various coding schemes in different types of matrices, hence the important keyword "matrix/ices" down below.

Format of third edition

Like the second edition, the book maintains its didactic format, with exactly eight pages per chapter to provide a constant amount of material in each chapter for self-learning or teaching (a feature that has been commented on favourably in book reviews of the second edition). One of my colleagues remarked that it was like writing 14-line sonnets with strict rules for metre and rhyming, which was certainly true in this case: the format definitely contributed to the creative process. The margins are reserved for section headings as well as captions of the tables and figures — these captions tend to be more informative than conventional one-liner ones. Each chapter has a short introduction and its own "Contents" list on the first page, and the chapter always ends with a summary in the form of a bulleted list.

Appendices

As in the first and second editions, the book's main thrust is towards the practice of correspondence analysis, so most technical issues and mathematical aspects are gathered in a theoretical appendix at the end of the book. It is followed by a computational appendix, which describes some features of the R language relevant to the methods in the book, including the **ca** package for correspondence analysis. R scripts are placed on the website www.carme-n.org, along with several of the data sets. No references at all are given in the 30 chapters — instead, a brief bibliographical appendix is given to point readers towards further readings and more complete literature sources. A glossary of the most important terms in the book is also provided and the book concludes with some personal thoughts in the form of an epilogue.

Acknowledgements

The first edition of this book was written in South Africa, and the second and present third editions in Catalonia, Spain. Many people and institutions have contributed in one way or another to this project. I would like to thank the BBVA Foundation in Madrid and its director Prof. Rafael Pardo, for support and encouragement in my work on correspondence analysis. The BBVA Foundation has published a Spanish translation of the second edition of *Correspondence Analysis in Practice*, called *La Práctica del Análisis de Correspondéncias*, available for free download at www.multivariatestatistics.org.

I owe a similar debt of gratitude to my colleagues and the institution of the *Universitat Pompeu Fabra* in Barcelona, where I have been working since 1994, one of the most innovative universities in Europe, recently ranked the most productive Spanish university after only 25 years of its existence.

I would like to thank all my friends and colleagues in many countries for moral and intellectual support, especially my wife Zerrin Aşan Greenacre, Jörg & Beate Blasius, Trevor & Lynda Hastie, Carles Cuadras, John Gower, Jean-Paul Benzécri, Angelos Markos, Alfonso Iodice d'Enza, Patrick Groenen, Pieter Kroonenberg, Cajo ter Braak, Jan de Leeuw, Ludovic Lebart, Jean-Marie & Annick Monget, Pierre & Martine Teillard, Michael Meimaris, Michael Friendly, Antoine de Falguerolles, John Aitchison, Michael Browne, Cas Crouse, Fred Lombard, June Juritz, Francesca Little, Karl Jöreskog, Lesley Andres, Barbara Cottrell, Raul Primicerio, Michaela Aschan, Salve Dahle, Stig Falk-Petersen, Reinhold Fieler, Sabine Cochrane, Paul Renaud, Haakon Hop, Tor & Danielle Korneliussen, Yasemin El-Menouar, Ekkehard & Ingvill Mochmann, Maria Rohlinger, Antonella Curci, Gianna Mastrorilli, Paola Bordandini, Oleg Nenadić, Walter Zucchini, Thierry Fahmy, Kimmo Vehkalahti, Öztaş Ayhan, Simo Puntanen, George Styan, Juha Alho, François Theron, Volker Hooyberg, Gurdeep Stephens & Pascal Courty, Antoni Bosch & Helena Trias, Teresa Garcia-Milà, Tamara Djermanovic, Guillem Lopez, Xavier Calsamiglia, Andreu Mas-Colell, Xavier Freixas, Frederic Udina, Albert Satorra, Jan Graffelman & Nuria Satorra, Rosemarie Nagel, Anna Espinal, Carolina Chaya, Carlos Pérez, Moya Berry, Alan Griffiths, Dianne Fortescue, Tasos & Androula Ladikos, Jerry & Mary Ann Reedy, Bodo & Bärbel Bilinski, Andries & Gaby Claassens, Judy Twycross, Romà Revelles & Carme Clotet of *Niu Nou* restaurant, Santi Careta, Marta Andreu, Rita Lugli & Danilo Guaitoli, José Penalva & Nuria Serrano, Gabor Lugosi and the whole community of Gréixer in the Pyrenees — you have all played a part in this story!

I also fondly remember dear friends and colleagues who have influenced my career, but who have sadly passed away in recent years: Paul Lewi, Cas Troskie, Dan Bradu, Reg & Kay Griffiths, Leo & Wendy Theron, Tony Brink, Jan Visser, Victor Thiessen, Al McCutcheon and Ingram Olkin, as well as Ruben Gabriel, from whom I learnt such a lot about statistics and life.

Particular thanks go to Angelos Markos and Antoine de Falguerolles for valuable comments on parts of the manuscript, and to Oleg Nenadić for his continuing collaboration in the development of our **ca** package in R.

Like the first and second editions, I have dedicated this book to my three daughters, Françoise, Karolien and Gloudina, who never cease to amaze me by their joy, sense of humour and diversity.

Finally, I thank the commissioning editor, Rob Calver, as well as Rebecca Davies and Karen Simon of Chapman & Hall/CRC Press, for placing their trust in me and for their constant cooperation in making this third edition of *Correspondence Analysis in Practice* become a reality.

Michael Greenacre
Barcelona

Scatterplots and Maps

Correspondence analysis is a method of data analysis for representing tabular data graphically. Correspondence analysis is a generalization of a simple graphical concept with which we are all familiar, namely the *scatterplot*. The scatterplot is the representation of data as a set of points with respect to two perpendicular coordinate axes: the horizontal axis often referred to as the x-axis and the vertical one as the y-axis. As a gentle introduction to the subject of correspondence analysis, it is convenient to reflect for a short time on our perception of scatterplots and how we interpret them in relation to the data they represent graphically. Particular emphasis will be placed on how we interpret distances between points in a scatterplot and when scatterplots can be seen as a spatial map of the data.

Contents

During the original writing of this book, I was reflecting on the journeys I had made during the year to Norway, Canada, Greece, France and Germany. According to my diary I spent periods of 18 days in Norway, 15 days in Canada and 29 days in Greece. Apart from these longer trips I also made several short trips to France and Germany, totalling 24 days. This numerical description of my time spent in foreign countries can be visualized in the graphs of Exhibit 1.1. This seemingly trivial example conceals several issues that are relevant to our perception of graphs of this type that represent data with respect to two coordinate axes, and which will eventually help us to

Data set 1: My travels

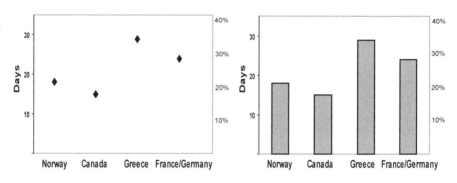

Exhibit 1.1:
Graphs of number of days spent in foreign countries in one year, in scatterplot and bar-chart formats respectively. A percentage scale, expressing days relative to the total of 86 days, is given on the right-hand side of each graph.

understand correspondence analysis. Let me highlight these issues one at a time.

Continuous variables

The left-hand vertical axis labelled Days represents the scale of a numeric piece of information often referred to as a *continuous*, or *numerical*, variable. The scale on this axis is the number of days spent in some foreign country, and the ordering from zero days at the bottom end of the scale to 30 days at the top end is clearly defined. In the bar-chart form of this display, given in the right-hand graph of Exhibit 1.1, bars are drawn with lengths proportional to the values of the variable. Of course, the number of days is a rounded approximation of the time actually spent in each country, but we call this variable continuous because the underlying time variable is indeed truly continuous.

Expressing data in relative amounts

The right-hand vertical axis of each plot in Exhibit 1.1 can be used to read the corresponding percentage of days relative to the total of 86 days. For example, the 18 days in Norway account for 21% of the total time. The total of 86 is often called the *base* relative to which the data are expressed. In this case there is only one set of data and therefore just one base, so in these plots the original absolute scale on the left and the relative scale on the right can be depicted on the same graph.

Categorical variables

In contrast to the vertical *y*-axis, the horizontal *x*-axis is clearly not a numerical variable. The four points along this axis are just positions where we have placed labels denoting the countries visited. The horizontal scale represents a *discrete*, or *categorical*, variable. There are two features of this horizontal axis that have no substantive meaning in the graph: the ordering of the categories and the distances between them.

Firstly, there is no strong reason why Norway has been placed first, Canada second and Greece third, except perhaps that I visited these countries in that order. Because the France/Germany label refers to a collection of shorter trips scattered throughout the year, it was placed after the others. By the way, in this type of representation where order is essentially irrelevant, it is usually a good idea to re-order the categories in a way that has some substantive meaning, for example in terms of the values of the variable. In this example we could order the countries in descending order of days, in which case we would position the countries in the order Greece, France/Germany, Norway and Canada, from most visited to least. This simple re-arrangement assists in the interpretation of data, especially if the data set is much larger: for example, if I had visited 20 different countries, then the order would contain relevant information that is not quickly deduced from the data in their original ordering.

Ordering of categories

Secondly, there is no reason why the four points are at equal intervals apart on the axis. There is also no immediate reason to put them at different intervals apart, so it is purely for convenience and aesthetics that they have been equally spaced. Using correspondence analysis we will show that there are substantively interesting ways to define intervals between the categories of a variable such as this one, when it is related to other variables. In fact, correspondence analysis will be shown to yield values for the categories where both the distances between the categories and their ordering have substantive meaning.

Distances between categories

Since the ordering of the countries is arbitrary on the horizontal axis of Exhibit 1.1, as well as the distances between them, there would be no sense in measuring and interpreting distances between the displayed points in the left-hand graph. The only distance measurement that has meaning is in the strictly vertical direction, because of the numerical nature of the vertical axis that indicates frequency (left-hand scale) or relative frequency (right-hand scale).

Distance interpretation of scatterplots

In some special cases, the two variables that define the axes of the scatterplot are of the same numerical nature and have comparable scales. For example, suppose that 20 students have written a mathematics examination consisting of two parts, algebra and geometry, each part counting 50% towards the final grade. The 20 students can be plotted according to their pair of grades, shown in Exhibit 1.2. It is important that the two axes representing the respective grades have scales with unit intervals of identical lengths. Because of the similar nature of the two variables and their scales, it is possible to judge distances in any direction of the display, not only horizontally or vertically. Two points that are close to each other will have similar results in the examination, just like two neighbouring towns having a small geographical distance between

Scatterplots as maps

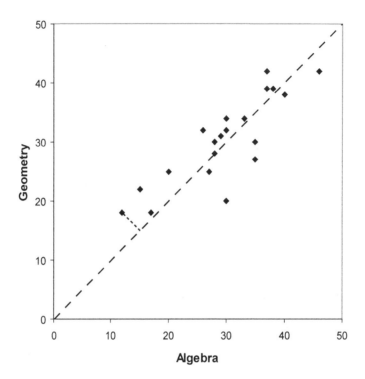

them. Thus, one can comment here on the shape of the scatter of points and the fact that there is a small cluster of four students with high grades and a single student with very high grades. Exhibit 1.2 can be regarded as a *map*, because the position of each student can be regarded as a two-dimensional position, similar to a geographical location in a region defined by latitude and longitude co-ordinates.

Calibration of a   Maps have interesting geometric properties. For example, in Exhibit 1.2 the
direction in the   45° dashed line actually defines an axis for the final grades of the students,
map   combining the algebra and geometry grades. If this line is calibrated from 0
(bottom left) to 100 (top right), then each student's final grade can be read
from the map by projecting each point perpendicularly onto this line. An
example is shown of a student who received 12 out of 50 and 18 out of 50
for the two sections, respectively, and whose position projects onto the line at
coordinates 15 and 15, corresponding to a total grade of 30.

Information-   The scatterplots in Exhibit 1.1 and Exhibit 1.2 are different ways of expressing
transforming   in graphical form the numerical information in the two sets of travel and
nature of the   examination data respectively. In each case there is no loss of information
display   between the data and the graph. Given the graph it is easy to recover the data

exactly. We say that the scatterplot or map is an "information-transforming instrument" — it does not process the data at all; it simply expresses the data in a visual format that communicates the same information in an alternative way.

In my travel example, the categorical variable "country" has four categories, and, since there is no inherent ordering of the categories, we refer to this variable more specifically as a *nominal* variable. If the categories are ordered, the categorical variable is called an *ordinal* variable. For example, a day could be classified into three categories according to how much time is spent working: (i) less than one hour (which I would call a "holiday"), (ii) more than one but less than six hours (a "half day", say) and (iii) more than six hours (a "full day"). These categories, which are based on the continuous variable "time spent daily working" divided up into intervals, are ordered and this ordering is usually taken into account in any graphical display of the categories. In many social surveys, questions are answered on an ordinal scale of response, for example, an ordinal scale of importance: "not important"/"somewhat important"/"very important". Another typical example is a scale of agreement/disagreement: "strongly agree"/"somewhat agree"/"neither agree nor disagree"/"somewhat disagree"/"strongly disagree". Here the ordinal position of the category "neither agree nor disagree" might not lie between "somewhat agree" and "somewhat disagree"; for example, it might be a category used by some respondents instead of a "don't know" response when they do not understand the question or when they are confused by it. We shall treat this topic later in this book (Chapter 21) once we have developed the tools that allow us to study patterns of responses in multivariate questionnaire data.

Nominal and ordinal variables

COUNTRY	Holidays	Half days	Full days	TOTAL
Norway	6	1	11	18
Canada	1	3	11	15
Greece	4	25	0	29
France/Germany	2	2	20	24
TOTAL	13	31	42	86

Exhibit 1.3: *Frequencies of different types of day in four sets of trips.*

Let us suppose now that the 86 days of my foreign trips were classified into one of the three categories *holidays*, *half days* and *full days*. The *cross-tabulation* of country by type of day is given in Exhibit 1.3. This table can be considered in two different ways: as a set of rows or a set of columns. For example, each column is a set of frequencies characterizing the respective type of day, while each row characterizes the respective country. Exhibit 1.4(a) shows the latter way, namely a plot of the frequencies for each country (row), where the horizontal axis now represents the type of day (column). Notice that, because the categories of the variable "type of day" are ordered, it makes sense to connect

Plotting more than one set of data

the categories by lines. Clearly, if we want to make a substantive comparison between the countries, then we should take into account the fact that different numbers of days in total were spent in each country. Each country total forms a different base for the re-expression of the corresponding row in Exhibit 1.3 as a set of percentages (Exhibit 1.5). These percentages are visualized in Exhibit 1.4(b) in a plot that expresses better the different *compositions* of days in the respective trips.

Exhibit 1.4:
Plots of (a) frequencies in Exhibit 1.3 and (b) relative frequencies in each row expressed as percentages.

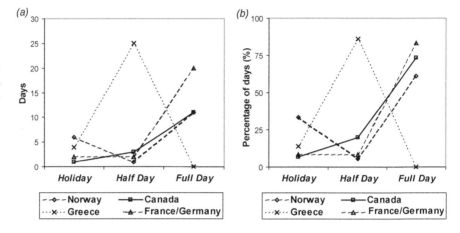

Exhibit 1.5:
Percentages of types of day in each country, as well as the percentages overall for all countries combined; rows add up to 100%.

COUNTRY	Holidays	Half days	Full days
Norway	33%	6%	61%
Canada	7%	20%	73%
Greece	14%	86%	0%
France/Germany	8%	8%	83%
Overall	*15%*	*36%*	*49%*

Interpreting absolute or relative frequencies

There is a lesson to be learnt from these displays that is fundamental to the analysis of frequency data. Each trip has involved a different number of days and so corresponds to a different base as far as the frequencies of the types of days are concerned. The 6 *holidays* in Norway, compared to the 4 in Greece, can be judged only in relation to the total number of days spent in these respective countries. As percentages they turn out to be quite different: 6 out of 18 is 33%, while 4 out of 29 is 14%. It is the visualization of the relative frequencies in Exhibit 1.4(b) that gives a more accurate comparison of how I spent my time in the different countries. The "marginal" frequencies (18, 15, 29, 24 for the countries, and 13, 31, 42 for the day types) are also interpreted relative to their respective totals — for example, the last row of Exhibit 1.5 shows the percentages of day types for all countries combined, and could have been plotted similarly in Exhibit 1.4(b).

Any conclusion drawn from the points' positions in Exhibit 1.4(b) is purely an interpretation of the data and not a statement of the statistical significance of the observed feature. In this book we shall address the statistical aspects of graphical displays only towards the end of the book (Chapters 29 and 30); for the most part we shall be concerned only with the question of data visualization and interpretation. The deduction that I had proportionally more holidays in Norway than in the other countries is certainly true in the data and can be seen strikingly in Exhibit 1.4(b). It is an entirely different question whether this phenomenon is statistically compatible with a model or hypothesis of my behaviour that postulates that the proportion of holidays was generally intended to be the same for all countries, in which case any observed differences are purely random. Most of statistical methodology concentrates on problems where data are fitted and compared to a theoretical model or preconceived hypothesis, with little attention being paid to enlightening ways for describing data, interpreting data and generating hypotheses. A typical example in the social sciences is the use of the ubiquitous chi-square statistic to test for association in a cross-tabulation. Often statistically significant association is found but there are no simple tools for detecting which parts of the table are responsible for this association. Correspondence analysis is one tool that can fill this gap, allowing the data analyst to see the pattern of association in the data and to generate hypotheses that can be tested in a subsequent stage of research. In most situations data description, interpretation and modelling can work hand-in-hand with one other. But there are situations where data description and interpretation assume supreme importance, for example when the data represent the whole population of interest.

Describing and interpreting data, vs. modelling and statistical inference

As data tables increase in size, it becomes more difficult to make simple graphical displays such as Exhibit 1.4, owing to the overabundance of points. For example, suppose I had visited 20 countries during the year and had a breakdown of time spent in each one of them, leading to a table with many more rows. I could also have recorded other data about each day in order to study possible relationships with the type of day I had; for example, the weather on each day — "fair weather", "partly cloudy" or "rainy". So the table of data might have many more columns as well as rows. In this case, to draw graphs such as Exhibit 1.4, involving many more categories and with 20 sets of points traversing the plot, would result in such a confusion of points and symbols that it would be difficult to see any patterns at all. It would then become clear that the descriptive instrument being used, the scatterplot, is inadequate in bringing out the essential features of the data. This is a convenient point to introduce the basic concepts of correspondence analysis, which is also a method for visualizing tabular data, but which can easily accommodate larger data sets in a natural and intuitive way.

Large data sets

1. Scatterplots involve plotting two variables, with respect to a horizontal axis and a vertical axis, often called the "x-axis" and "y-axis" respectively.

2. Usually the x variable is a completely different entity to the y variable. We can often interpret distances along at least one of the axes in the specific sense of measuring the distance according to the scale that is calibrated on the axis. It is usually meaningless to measure or interpret oblique distances in the plot.

3. In a few cases the x and y variables are similar entities with comparable scales, in which case interpoint distances can be interpreted as a measure of difference, or dissimilarity, between the plotted points. In this special case we call the scatterplot a *map*. For such maps it is important that the horizontal and vertical scales have physically equal units, i.e. the *aspect ratio* of the axes is equal to 1.

4. When plotting positive quantities (usually frequencies in our context), both the absolute and relative values of these quantities are of interest.

5. The more complex the data are, the less convenient it is to represent these data in a scatterplot.

6. This book is concerned with visually describing and interpreting complex information, rather than modelling it.

Profiles and the Profile Space

## 2

The concept of a set of relative frequencies, or a *profile*, is fundamental to correspondence analysis (referred to from now on by its abbreviation CA). Such sets, or *vectors*, of relative frequencies have special geometric features because the elements of each set add up to 1 (or 100%). In analysing a frequency table, relative frequencies can be computed for rows or for columns — these are called row or column profiles respectively. In this chapter we shall show how profiles can be depicted as points in a profile space, illustrating the concept in the special case when each profile consists of only three elements.

Contents

Let us look again at the data in Exhibit 1.3, a table of frequencies with four rows (the countries) and three columns (the type of day). The first and most basic concept in CA is that of a *profile*, which is a set of frequencies divided by their total. Exhibit 2.1 shows the row profiles for these data: for example, the profile of Norway is [0.33 0.06 0.61], where $0.33 = 6/18$, $0.06 = 1/18$, $0.61 = 11/18$. We say that this is the "profile of Norway across the types of day". The profile may also be expressed in percentage form, i.e. [33% 6% 61%] in this case, as in Exhibit 1.5. In a similar way, the profile of Canada across the day types is [0.07 0.20 0.73], concentrated mostly in the *full day* category, as is Norway. In contrast, Greece has a profile of [0.14 0.86 0.00], concentrated mostly in the *half day* category, and so on. The percentages are plotted in Exhibit 1.4(b) on page 6.

Profiles

COUNTRY	*Holidays*	*Half days*	*Full days*
Norway	0.33	0.06	0.61
Canada	0.07	0.20	0.73
Greece	0.14	0.86	0.00
France/Germany	0.08	0.08	0.83
Average	*0.15*	*0.36*	*0.49*

Average profile   In addition to the four country profiles, there is an additional row in Exhibit 2.1 labelled *Average*. This is the profile of the final row [13 31 42] of Exhibit 1.3, which contains the column sums of the table; in other words this is the profile of all the trips aggregated together. In Chapter 3 we shall explain more specifically why this is called the *average profile*. For the moment, it is only necessary to realize that, out of the total of 86 days travelled, irrespective of country visited, 15% were *holidays*, 36% were *half days* and 49% were *full days* of work. When comparing profiles we can compare one country's profile with another, or we can compare a country's profile with the average profile. For example, eyeballing the figures in Exhibit 2.1, we can see that of all the countries, the profiles of Canada and France/Germany are the most similar. Compared to the average profile, these two profiles have a higher percentage of *full days* and are below average on *holidays* and *half days*.

Row profiles   In the above we looked at the row profiles in order to compare the different
and column   countries. We could also consider Exhibit 1.3 as a set of columns and compare
profiles   how the different types of days are distributed across the countries. Exhibit 2.2 shows the column profiles as well as the average column profile. For example, of the 13 *holidays* 46% were in Norway, 8% in Canada, 31% in Greece and 15% in France/Germany, and so on for the other columns. Since I spent different numbers of days in each country, these figures should be checked against those of the average column profile to see whether they are lower or higher than the average pattern. For example, 46% of all *holidays* were spent in Norway, whereas the number of days spent in Norway was just 21% of the total of 86 — in this sense there is a high number of *holidays* there compared to the average.

COUNTRY	*Holidays*	*Half days*	*Full days*	*Average*
Norway	0.46	0.03	0.26	*0.21*
Canada	0.08	0.10	0.26	*0.17*
Greece	0.31	0.81	0.00	*0.34*
France/Germany	0.15	0.07	0.48	*0.28*

Looking again at the proportion 0.46 (= 6/13) of *holidays* spent in Norway (Exhibit 2.2) and comparing it to the proportion 0.21 (= 18/86) of all days spent in that country, we can calculate the ratio 0.46/0.21 = 2.2, and conclude that *holidays* in Norway were just over twice the average. Exactly the same conclusion is reached if a similar calculation is made on the row profiles. In Exhibit 2.1 the proportion of *holidays* in Norway was 0.33 (= 6/18) whereas for all countries the proportion was 0.15 (= 13/86). Thus, there are 0.33/0.15 = 2.2 times as many holidays compared to the average, the same ratio as was obtained when arguing from the point of view of the column profiles (this ratio is called the *contingency ratio* and will re-appear in future chapters). Whether we argue via the row profiles or column profiles we arrive at the same conclusion. In Chapter 8 it will be shown that CA treats the rows and columns of a table in an equivalent fashion or, as we say, in a *symmetric* way.

Symmetric treatment of rows and columns

Nevertheless, it is true in practice that a table of data is often thought of and interpreted in a non-symmetric, or *asymmetric*, fashion, either as a set of rows or as a set of columns. For example, since each row of Exhibit 1.3 constitutes a different country (or pair of countries in the case of France/Germany), it might be more natural to think of the table row-wise, as in Exhibit 2.1. Deciding which way is more appropriate depends on the nature of the data and the researcher's objective, and the decision is often not a conscious one. One concrete manifestation of the actual choice is whether the researcher refers to row or column percentages when interpreting the data. Whatever the decision, the results of CA will be invariant to this choice, but the interpretation will adapt to the researcher's viewpoint.

Asymmetric consideration of the data table

Let us consider the four row profiles and average profile in Exhibit 2.1 and a completely different way to plot them. Rather than the display of Exhibit 1.4(b), where the horizontal axis serves only as labels for the type of day and the vertical axis represents the percentages, we now propose using three axes corresponding to the three types of day, which is a scatterplot in three dimensions. To imagine three perpendicular axes is not difficult: merely look down into an empty corner of the room you are sitting in and you will see three axes as shown in Exhibit 2.3. Each of the three edges of the room serves as an axis for plotting the three elements of the profile. These three values are now considered to be coordinates of a single point that represents the whole profile — this is quite different from the graph in Exhibit 1.4(b) where there is a point for each of the three profile elements. The three axes are labelled *holidays*, *half days* and *full days*, and are calibrated in fractional profile units from 0 to 1. To plot the four profiles is now a simple exercise. Norway's profile of [0.33 0.06 0.61] (see Exhibit 2.1) is 0.33 of a unit along axis *holidays*, 0.06 along axis *half days* and 0.61 along axis *full days*. To take another example, Greece's profile of [0.14 0.86 0.00] has a zero coordinate in the *full days* direction, so its position is on the "wall", as it were, on the left-hand side of the display, with coordinates 0.14 and 0.86 on the two axes

Plotting the profiles in the profile space

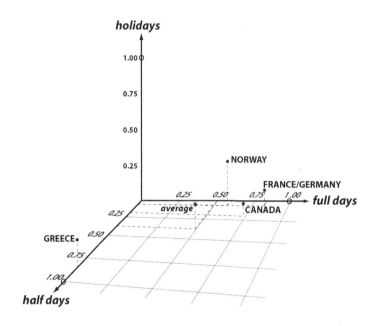

holidays and half days that define the "wall". All other row profile points in this example, including the average row profile [0.15 0.36 0.49], can be plotted in this three-dimensional space.

*Vertex points
define the
extremes of the
profile space*

With a bit of imagination it might not be surprising to discover that the profile points in Exhibit 2.3 all lie exactly in the plane defined by the triangle that joins the extreme *unit points* [1 0 0], [0 1 0] and [0 0 1] on the three respective axes, as shown in Exhibit 2.4. This triangle is equilateral and its three corners are called *vertex points* or *vertices*. The vertices coincide with extreme profiles that are totally concentrated into one of the day types. For example, the vertex point [1 0 0] corresponds to a trip to a country consisting only of *holidays* (fictional in my case, unfortunately). Likewise, the vertex point [0 0 1] corresponds to a trip consisting only of *full days* of work.

*Triangular (or
ternary)
coordinate system*

Having realized that all profile points in three-dimensional space actually lie exactly on a flat (two-dimensional) triangle, it is possible to lay this triangle flat, as in Exhibit 2.5. Looking at the profile points in a flat space is clearly better than trying to imagine their three-dimensional positions in the corner of a room! This particular type of display is often referred to as the *triangular* (or *ternary*) *coordinate system* and may be used in any situation where we have sets of data consisting of three elements that add up to 1, as in the case of the row profiles in this example. Such data are common in geology and chemistry, for example where samples are decomposed into three constituents, by weight

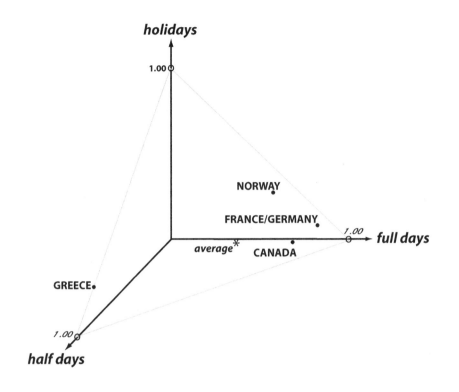

Exhibit 2.4:
The profile points in Exhibit 2.3 lie exactly on an equilateral triangle joining the vertex points of the profile space. Thus the three-dimensional profiles are actually two-dimensional. The profile of Greece lies on the edge of the triangle because it has zero full days.

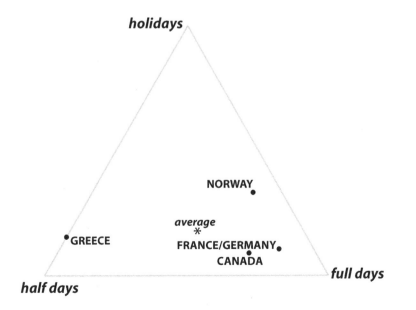

Exhibit 2.5:
The triangle in Exhibit 2.4 that contains the row (country) profiles. The three corners, or vertices, of the triangle represent the columns (day types).

Exhibit 2.6:
*Norway's profile
[0.33 0.06 0.61] is
positioned using
triangular
coordinates as
shown, using the
sides of the triangle
as axes. Each side is
calibrated in profile
units from 0 to 1.*

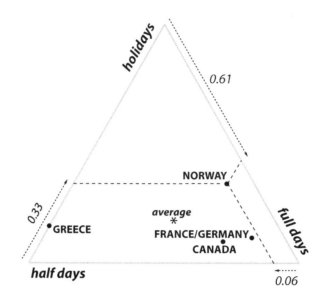

or by volume. A particular sample is characterized by the three proportions of the constituents and can thus be displayed as a single point with respect to triangular (or ternary) coordinates.

*Positioning a
point in a
triangular
coordinate system*

Given a blank equal-sided triangle and the profile values, how can we find the position of a profile point in the triangle, without passing via the underlying three-dimensional space of Exhibits 2.3 and 2.4? In the triangular coordinate system the sides of the triangle define three axes. Each side is considered to have a length of 1 and can be calibrated accordingly on a linear scale from 0 to 1. In order to position a profile in the triangle, its three values on these axes determine three lines drawn from these values parallel to the respective sides of the triangle. For example, to position Norway, as illustrated in Exhibit 2.5, we take a value of 0.33 on the *holidays* axis (see Exhibit 2.6), 0.06 on the *half days* axis and 0.61 on the *full days* axis. Lines from these coordinate values drawn parallel to the sides of the triangle all meet at the point representing Norway. In fact, any two of the three profile coordinates are sufficient to situate a profile in this way, and the remaining coordinate is always superfluous, which is another way of demonstrating that the profiles are inherently two-dimensional.

*Geometry of
profiles with more
than three
elements*

The triangular coordinate system may be used only for profiles with three elements. But the idea can easily be generalized to profiles with any number of elements, in which case the coordinate system is known as the *barycentric coordinate system* ("barycentre" is synonymous with "weighted average", to be explained in the next chapter, page 19). The dimensionality of this coordinate system is always one less than the number of elements in the profile. For example, we have just seen that three-element profiles are contained exactly in a two-dimensional triangular profile space. For profiles with four elements the dimensionality is three and the profiles lie in a four-pointed tetrahedron in three-dimensional space. The two-dimensional triangle and the three-dimensional tetrahedron are examples of what is known in mathematics as a *regular simplex*. R code for visualizing an example in three dimensions is given in the Computing Appendix, pages 257–258, so you can get a feeling for three-dimensional profile space. For higher-dimensional profiles some strong imagination would be needed to be able to "see" the profile points spaces of dimension greater than three, but fortunately CA will be of great help to us in visualizing such multidimensional profiles.

We have illustrated the concept of a profile using frequency data, which is the prime example of data suitable for CA. But CA is applicable to a much wider class of data types; in fact it can be used whenever it makes sense to express the data in relative amounts, i.e. data on a so-called *ratio scale*. For example, suppose we have data on monetary amounts invested by countries in different areas of research — the relative amounts would be of interest, e.g. the percentage invested in environmental research, biomedecine, etc. Another example is of morphometric measurements on a living organism, for example measurements in centimeters on a fish, its length and width, length of fins, etc. Again all these measurements can be expressed relative to the total, where the total is a surrogate measure for the size of the fish, so that we would be analysing and comparing the shapes of different fish in the form of profiles rather than the original values.

Data on a ratio scale

A necessary condition of the data for CA is that all observations are on the same scale: for example, counts of particular individuals in a frequency table, a common monetary unit in the table of research investments, centimeters in the morphometric study. It would make no sense in CA to analyse data with mixed scales of measurement, unless a pre-transformation is conducted to homogenize the scales of the whole table. Most of the data sets in this book are frequency data, but in Chapter 26 we shall look at a wide variety of other types of data and ways of recoding them to be suitable for CA.

Data on a common scale

SUMMARY:
Profiles and the
Profile Space

1. The *profile* of a set of frequencies (or any other amounts that are positive or zero) is the set of frequencies divided by their total, i.e. the set of relative frequencies.

2. In the case of a cross-tabulation, the rows or columns define sets of frequencies which can be expressed relative to their respective totals to give row profiles or column profiles.

3. The marginal frequencies of the cross-tabulation can also be expressed relative to their common total (i.e. the grand total of the table) to give the average row profile and average column profile.

4. Comparing row profiles to their average leads to the same conclusions as comparing column profiles to their average.

5. Profiles consisting of m elements can be plotted as points in an m-dimensional space. Because their m elements add up to 1, these profile points occupy a restricted region of this space, an $(m-1)$-dimensional subspace known as a *simplex*. This simplex is enclosed within the edges joining all pairs of the m unit vectors on the m perpendicular axes. These unit points are also called the *vertices* of the simplex or profile space. The coordinate system within this simplex is known as the *barycentric coordinate system*.

6. A special case that is easy to visualize is when the profiles have three elements, so that the simplex is simply a triangle that joins the three vertices. This special case of the barycentric coordinate system is known as the *triangular* (or *ternary*) *coordinate system*.

7. The idea of a profile can be extended to data on a *ratio scale* where it is of interest to study relative values. In this case the set of numbers being profiled should all have the same scale of measurement.

## 3

Masses and Centroids

There is an equivalent way of thinking about the positions of the profile points in the profile space, and this will be useful to our eventual understanding and interpretation of correspondence analysis (CA). This is based on the notion of a weighted average, or centroid, of a set of points. In the calculation of an ordinary (unweighted) average, each point receives equal weight, whereas a weighted average allows different weights to be associated with each point. When the points are weighted differently, then the centroid does not lie exactly at the "geographical" centre of the cloud of points, but tends to lie in a position closer to the points with higher weight.

Contents

We now use a typical set of data in social science research, a cross-tabulation (or "cross-classification") of two variables from a survey. The table, given in Exhibit 3.1, concerns 312 readers of a certain newspaper, in particular their level of thoroughness in reading the newspaper. Based on data collected in the survey, each respondent was classified into one of three groups: *glance* readers, *fairly thorough* readers and *very thorough* readers. These reading classes have been cross-tabulated against education, an ordinal variable with five categories ranging from some primary education to some tertiary education. Exhibit 3.1 shows the raw frequencies and the education group profiles in parentheses, i.e. the row profiles. The triangular coordinate plot of the row profiles, in the style described in Chapter 2, is given in Exhibit 3.2. In this display the corner points, or vertices, of the triangle represent the three readership groups — remember that each vertex is at the position of a "pure" row profile totally concentrated into that category; for example, the *very thorough* vertex *C3* is representing a fictitious row profile of [0 0 1] that contains 100% *very thorough* readers.

Data set 2: Readership and education groups

EDUCATION GROUP	Glance C1	Fairly thorough C2	Very thorough C3	Total	Row masses
Some primary E1	5 (0.357)	7 (0.500)	2 (0.143)	*14*	*0.045*
Primary completed E2	18 (0.214)	46 (0.548)	20 (0.238)	*84*	*0.269*
Some secondary E3	19 (0.218)	29 (0.333)	39 (0.448)	*87*	*0.279*
Secondary completed E4	12 (0.119)	40 (0.396)	49 (0.485)	*101*	*0.324*
Some tertiary E5	3 (0.115)	7 (0.269)	16 (0.615)	*26*	*0.083*
Total	*57*	*129*	*126*	*312*	
Average row profile	*(0.183)*	*(0.413)*	*(0.404)*		

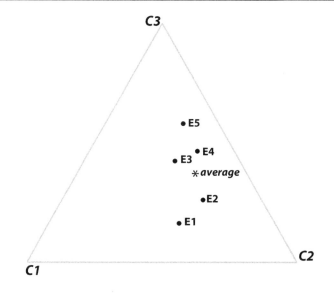

*Points as
weighted averages*   Another way to think of the positions of the education groups in the triangle is as weighted averages. Assigning weights to the values of a variable is a well-known concept in statistics. For example, in a class of 26 students, suppose that the average grade turns out to be 7.5, calculated by summing the 26 grades and dividing by 26. In fact, 3 students obtain the grade of 9, 7 students obtain an 8, and 16 students obtain a 7, so that the average grade can be determined equivalently by assigning weights of 3/26 to the grade of 9, 7/26 to the grade of 8 and 16/26 to the grade of 7 and then calculating the weighted average. Here the weights are the relative frequencies of each grade, and because the grade of 7 has more weight than the others, the weighted av-

erage of 7.5 is "closer" to this grade, whereas the ordinary arithmetic average of the three values 7, 8 and 9 is clearly 8.

Looking at the last row of data in Exhibit 3.1, for education group E5 (some tertiary education), we see the same frequencies of 3, 7 and 16 for the three respective readership groups, and associated relative frequencies of 0.115, 0.269 and 0.615. The idea now is to imagine 3 cases situated at the *glance* vertex *C1* of the triangle, 7 cases at the *fairly thorough* vertex *C2* and 16 cases at the *very thorough* vertex *C3*, and then consider what would be the average *position* for these 26 cases. In other words, we do not associate the weights with values of a variable but with positions in the profile space, in this case the positions of the vertex points. There are more cases at the *very thorough* corner, so we would expect the average position of E5 to be closer to this vertex, as is indeed the case. For the same reason, row profile E1 lies far from the *very thorough* corner *C3* because it has a very low weight (2 out of 14, or 0.143) on this category. Hence each row profile point is positioned within the triangle as an average point, where the profile values, i.e. relative frequencies, serve as the weights allocated to the vertices. Thus, we can think of the profile values not only as coordinates in a multidimensional space, but also as weights assigned to the vertices of a simplex. This idea can be extended to higher-dimensional profiles: for example, a profile with four elements is also at an average position with respect to the four corners of a three-dimensional tetrahedron, weighted by the respective profile elements.

Profile values are weights assigned to the vertices

Alternative terms for weighted average are *centroid* or *barycentre*. Some particular examples of weighted averages in the profile space are given in Exhibit 3.3. For example, the profile point $[1/3 \; 1/3 \; 1/3]$, which gives equal weight to the three corners, is positioned exactly at the centre of the triangle, equidistant from the corners, in other words at the ordinary average position of the three vertices. The profile $[1/2 \; 1/2 \; 0]$ is at a position midway between the first and second vertices, since it has equal weight on these two vertices and zero weight on the third vertex. In general, we can write a formula for the position of a profile as the centroid of the three vertices as follows, for a profile $[a \; b \; c]$ where $a + b + c = 1$:

Each profile point is a weighted average, or centroid, of the vertices

$$centroid \; position = (a \times \text{vertex 1}) + (b \times \text{vertex 2}) + (c \times \text{vertex 3})$$

For example, the position of education group E5 in Exhibit 3.2 is obtained as follows:

E5 $= (0.115 \times glance) + (0.269 \times fairly \; thorough) + (0.615 \times very \; thorough)$

Similarly, the position of the average profile is also a weighted average of the vertex points:

average $= (0.183 \times glance) + (0.413 \times fairly \; thorough) + (0.404 \times very \; thorough)$

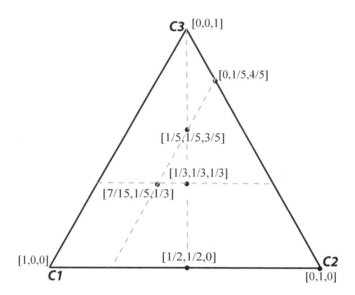

The average is farther from the *glance* corner since there is less weight on the *glance* vertex than on the other two, which have approximately the same weights (see Exhibit 3.2).

*Average profile
is also a weighted
average of the
profiles themselves*

The average profile is a rather special point — not only is it a centroid of the three vertices as we have just shown, just like any profile point, but it is also a centroid of the five row profiles themselves, where different weights are assigned to the profiles. Looking again at Exhibit 3.1, we notice that the row totals are different: education group E1 (some primary education) includes only 14 respondents whereas education group E4 (secondary education completed) has 101 respondents. In the last column of Exhibit 3.1, headed "row masses", we have these marginal row frequencies expressed relative to the total sample size 312. Just as we thought of row profiles as weighted averages of the vertices, we can think of each of the five row profile points in Exhibit 3.2 being assigned weights according to their marginal frequencies, as if there were 14 respondents (proportion 0.045 of the sample) at the position E1, 84 respondents (0.269 of the sample) at the position E2, and so on. With these weights assigned to the five profile points, the weighted average position is exactly at the average profile point:

$$Average\ row\ profile = (0.045 \times E1) + (0.269 \times E2) + (0.279 \times E3)$$
$$+ (0.324 \times E4) + (0.083 \times E5)$$

This average row profile is at a central position amongst the row profiles but more attracted to the profiles observed with higher frequency.

*Row and
column masses*

The weights assigned to the profiles are so important in CA that they are given a specific name: *masses*. The last column of Exhibit 3.1 shows the row masses:

0.045, 0.269, 0.279, 0.324 and 0.083. The word "mass" is the preferred term in CA although it is entirely equivalent for our purpose to the term "weight". An alternative term such as mass is convenient here to differentiate this geometric concept of weighting from other forms of weighting that occur in practice, such as weights assigned to population subgroups in a sample survey.

All that has been said about row profiles and row masses can be repeated in a similar fashion for the columns. Exhibit 3.4 shows the same contingency table as Exhibit 3.1 from the column point of view. That is, the three columns have

EDUCATION GROUP	Glance C1	Fairly thorough C2	Very thorough C3	Total	Average column profile
Some primary E1	5 (0.088)	7 (0.054)	2 (0.016)	14	(0.045)
Primary completed E2	18 (0.316)	46 (0.357)	20 (0.159)	84	(0.269)
Some secondary E3	19 (0.333)	29 (0.225)	39 (0.310)	87	(0.279)
Secondary completed E4	12 (0.211)	40 (0.310)	49 (0.389)	101	(0.324)
Some tertiary E5	3 (0.053)	7 (0.054)	16 (0.127)	26	(0.083)
Total	57	129	126	312	
Column masses	0.183	0.413	0.404		

Exhibit 3.4: *Cross-tabulation of education group by readership cluster, showing column profiles and average column profile in parentheses, and the column masses.*

been expressed as relative frequencies with respect to their column totals, giving three profiles with five values each. The column totals relative to the grand total are now column masses assigned to the column profiles, and the average column profile is the set of row totals divided by the grand total. Again, we could write the average column profile as a weighted average of the three column profiles *C1*, *C2* and *C3*:

Average column profile $= (0.183 \times C1) + (0.413 \times C2) + (0.404 \times C3)$

Notice how the row and column masses play two different roles, as weights and as averages: in Exhibit 3.4 the average column profile is the set of row masses in Exhibit 3.1, and the column masses in Exhibit 3.4 are the elements of what was previously the average row profile in Exhibit 3.1.

At this point, even though the final key concepts in CA still remain to be explained, it is possible to make a brief interpretation of Exhibit 3.2. The vertices of the triangle represent the "pure profiles" of readership categories *C1*, *C2* and *C3*, whereas the education groups are "mixtures" of these readership categories and find their positions within the triangle in terms of their respec-

Interpretation in the profile space

tive proportions of each of the three categories. Notice the following aspects
of the display:

- The degree of spread of the profile points within the triangle gives an idea
  of how much variation there is the contingency table. The closer the profile
  points lie to the centroid, the less variation there is, and the more they
  deviate from the centroid, the more variation. The profile space is bounded
  and the most extreme profiles will lie near the sides of the triangle, or in
  the most extreme case at one of the vertices (for example, an illiterate
  group with profile $[1 \ 0 \ 0]$ would lie on the vertex *C1*). In tables of social
  science data such as this one, profiles usually occupy a small region of the
  profile space close to the average because the variation in profile values
  for a particular category will be relatively small. For example, the range
  in the first element (i.e. readership category *C1*) across the profiles is only
  from 0.115 to 0.357 (Exhibit 3.1), in a potential range from 0 to 1. In
  contrast, for data in ecological research, as we shall see later, the range of
  profile values is much higher, usually because of many zero frequencies in
  the table — the profiles are then more spread out inside the profile space
  (see the second example in Chapter 10).

- The profile points are stretched out in what is called a "direction of spread"
  more or less from the bottom to the top of the display. Looking from the
  bottom upwards, the five education group profiles lie in their natural order of increasing educational qualifications, from E1 to E5. At the top,
  group E5 lies closest to the vertex C3, which represents the highest category of *very thorough* reading — we have already seen that this group
  has the highest proportion (0.615) of these readers. At the bottom, the
  lower educational group is not far from the edge of the triangle which we
  know displays profiles with zero *C3* readers (for example, see the point
  $[1/2 \ 1/2 \ 0]$ in Exhibit 3.3 as an illustration of a point on the edge). The
  interpretation of this pattern would be that as we move up from the bottom of this display to the top, from lower to higher education, the profiles
  are generally changing with respect to their relative frequency of type *C3*
  as opposed to that of *C1* and *C2* combined, while there is no particular
  tendency towards either *C1* or *C2*. In addition, the relative frequency of
  C1 is decreasing as the education points move away from *C1* towards the
  edge joining *C2* and *C3*.

Merging rows   Suppose we wanted to combine the two categories of primary education, E1
or columns   and E2, into a new row of Exhibit 3.1, denoted by E1&2. There are two ways
of thinking about this. First, add the two rows together to obtain the row
of frequencies $[23 \ 53 \ 22]$, with total 98 and profile $[.235 \ .541 \ .224]$. The
alternative way is to think of the profile of E1&2 as the weighted average of
the profiles of E1 and E2:

$$[.235 \ .541 \ .224] = \frac{.045}{.314} \times [.357 \ .500 \ .143] + \frac{.269}{.314} \times [.214 \ .548 \ .238]$$

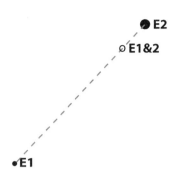

Exhibit 3.5:
Enlargement of positions of E1 and E2 in Exhibit 3.2, showing the position of the point E1&2 which merges the two categories; E2 has 6 times the mass of E1, hence E1&2 lies closer to E2 at a point which splits the line between the points in the ratio 84:14 = 6:1.

where the masses of E1 and E2 are .045 and .269, with sum .314 (notice that the weights in this weighted average are identical to 14/98 and 84/98, where 14 and 84 are the totals of rows E1 and E2). Geometrically, E1&2's profile lies on a line between E1 and E2, but closer to E2 as shown in Exhibit 3.5. The distances from E1 to E1&2 and E2 to E1&2 are in the same proportion as the totals 84 and 14 respectively; i.e. 6 to 1. E1&2 can be thought of as the balancing point of the two masses situated at E1 and E2, with the heavier mass at E2.

Distributionally equivalent rows or columns

Suppose that we had an additional row of data in Exhibit 3.1, a category of "no formal education" denoted by E0, with frequencies [10 14 4] across the reading categories. The profile of E0 is identical to E1's profile, because the frequencies in E0 are simply twice those of E1. The two sets of frequencies are said to be *distributionally equivalent*. Thus the profiles of E0 and E1 are at exactly the same point in the profile space, and can be merged into one point with mass equal to the combined masses of the two profiles, i.e. a single point with frequencies [15 21 6].

Changing the masses

The row and column masses are proportional to the marginal sums of the table. If the masses need to be modified for a substantive reason, this can be achieved by a simple transformation of the table. For example, suppose that we require the five education groups of Exhibit 3.1 to have masses proportional to their population sizes rather than their sample sizes. Then the table is rescaled by multiplying each education group profile by its respective population size. The row profiles of this new table are identical to the original row profiles, but the row masses are now proportional to the population sizes. Alternatively, suppose that the education groups are required to be weighted equally, rather than differentially as described up to now. If we regard the table of row profiles (or, equivalently, of row percentages) as the original table, then this table has row sums equal to 1 (or 100%), so that each education group is weighted equally. Hence analysing the table of profiles implies weighting each profile equally.

SUMMARY:
Masses and
Centroids

1. We assume that we are analysing a table of data and are concerned with the row problem, i.e. where the row profiles are plotted in the simplex space defined by the column vertices. Then each vertex point represents a column category in the sense that a row profile that is entirely concentrated into that category would lie exactly at that vertex point.

2. Each profile can be interpreted as the *centroid* (or weighted average) of the vertex points, where the weights are the individual elements of the profile. Thus a profile will tend to lie closer to those vertices for which it has higher values.

3. Each row profile in turn has a unique weight associated with it, called a *mass*, which is proportional to the row sum in the original table. The average row profile is then the centroid of the row profiles, where each profile is weighted by its mass in the averaging process.

4. Everything described above for row profiles applies equally to the columns of the table. In fact, the best way to make the jump from rows to columns is to re-express the table in its transposed form, where columns become rows, and vice versa — then everything applies exactly as before.

5. Rows (or columns) that are combined by aggregating their frequencies have a profile equal to the weighted average of the profiles of the component rows (or columns).

6. Rows (or columns) that have the same profile are said to be *distributionally equivalent* and can be combined into a single point with a mass equal to the sum of the masses of the combined rows (or columns).

7. Row (or column) masses can be modified to be proportional to prescribed values by a simple rescaling of the rows (or columns).

Chi-Square Distance and Inertia

4

In correspondence analysis (CA) the way distance is measured between profiles is a bit more complicated than the one that was used implicitly when we drew and interpreted the profile plots in Chapters 2 and 3. Distance in CA is measured using the so-called *chi-square distance* and this distance is the key to the many favourable properties of CA. There are several ways to justify the chi-square distance: some are more technical and beyond the scope of this book, while other explanations are more intuitive (see Appendix B, pages 270–271 for one theoretical justification). In this chapter we choose the latter approach, starting with a geometric explanation of the well-known chi-square statistic computed on a contingency table. All the ideas embodied in the chi-square statistic carry over to the chi-square distance in CA and to the related concept of inertia, which is the way CA measures variation in a data table.

Contents

Consider the data in Exhibit 3.1 again. Notice that, of the sample of 312 people, 57 (or 18.3%) are in readership category *C1* ("glance"), 129 (41.3%) in *C2* ("fairly thorough") and 126 (40.4%) in *C3* ("very thorough"); i.e. the average row profile is the set of proportions [0.183 0.413 0.404]. If there were no difference between the education groups as far as readership is concerned, we would expect that the profile of each row is more or less the same as the average profile, and would differ from it only because of random sampling fluctuations. Assuming no difference, or in other words assuming that the education groups are *homogeneous* with respect to their reading habits, what would we have expected the frequencies in row E5, for example, to be? There are 26 people in the E5 education group, and we would thus have expected 18.3% of them to be in category *C1*; i.e. $26 \times 0.183 = 4.76$ (although it is

Hypothesis of independence or homogeneity for a contingency table

Exhibit 4.1:
Observed frequencies, as given in Exhibit 3.1, along with expected frequencies (in parentheses) calculated assuming the homogeneity assumption to be true.

EDUCATION GROUP	Glance C1	Fairly thorough C2	Very thorough C3	Total	Row masses
Some primary E1	5 (2.56)	7 (5.78)	2 (5.66)	14	0.045
Primary completed E2	18 (15.37)	46 (34.69)	20 (33.94)	84	0.269
Some secondary E3	19 (15.92)	29 (35.93)	39 (35.15)	87	0.279
Secondary completed E4	12 (18.48)	40 (41.71)	49 (40.80)	101	0.324
Some tertiary E5	3 (4.76)	7 (10.74)	16 (10.50)	26	0.083
Total	*57*	*129*	*126*	*312*	
Average row profile	*0.183*	*0.413*	*0.404*		

ridiculous to talk of 0.76 of a person, it is necessary to maintain such fractions in these calculations). Likewise, we would have expected $26 \times 0.413 = 10.74$ of the E5 subjects to be in category *C2*, and $26 \times 0.404 = 10.50$ in category *C3*. There are various names in the literature given to this "assumption of no difference" between the rows of a contingency table (or, similarly, between the columns) — the "hypothesis of independence" is one of them, or perhaps more aptly for our purpose here, the "homogeneity assumption". Under the homogeneity assumption, we would therefore have expected the frequencies for E5 to be [4.76 10.74 10.50], but in reality they are observed to be [3 7 16]. In a similar fashion we can compute what each row of frequencies would be if the assumption of homogeneity were exactly true. Exhibit 4.1 shows the expected values in each row underneath their corresponding observed values. Notice that exactly the same expected frequencies are calculated if we argue from the point of view of column profiles, i.e. assuming homogeneity of the readership groups.

Chi-square (χ^2) statistic to test the homogeneity hypothesis

It is clear that the observed frequencies are always going to be different from the expected frequencies. The question statisticians now ask is whether these differences are large enough to contradict the assumed hypothesis that the rows are homogeneous, in other words whether the discrepancies between observed and expected frequencies are so large that it is unlikely they could have arisen by chance alone. This question is answered by computing a measure of discrepancy between all the observed and expected frequencies, as follows. Each difference between an observed and expected frequency is computed, then this difference is squared and finally divided by the expected frequency. This calculation is repeated for all pairs of observed and expected frequencies and the results are accumulated into a single figure — the *chi-square statistic*, denoted by χ^2:

$$\chi^2 = \sum \frac{(\text{observed} - \text{expected})^2}{\text{expected}}$$

Because there are 15 cells in this 5-by-3 (or 5×3) table, there will be 15 terms in this computation. For purposes of illustration we show only the first three and last three terms corresponding to rows E1 and E5:

$$\chi^2 = \frac{(5 - 2.56)^2}{2.56} + \frac{(7 - 5.78)^2}{5.78} + \frac{(2 - 5.66)^2}{5.66} + \cdots$$

$$+ \frac{(3 - 4.76)^2}{4.76} + \frac{(7 - 10.74)^2}{10.74} + \frac{(16 - 10.50)^2}{10.50} \qquad (4.1)$$

Calculating the χ^2 statistic

The grand total of the 15 terms in this calculation turns out to be equal to 26.0. The larger this value, the more discrepant the observed and expected frequencies are, i.e. the less convinced we are that the assumption of homogeneity is correct. In order to judge whether this value of 26.0 is large or small, we use probabilities of the chi-square distribution corresponding to the "degrees of freedom" associated with the statistic. For a 5×3 table, the degrees of freedom are $4 \times 2 = 8$ (one less than the number of rows multiplied by one less than the number of columns), and the p-value associated with the value 26.0 of the χ^2 statistic with 8 degrees of freedom is $p = 0.001$. This result tells us that there is an extremely small probability — one in a thousand — that the observed frequencies in Exhibit 4.1 can be reconciled with the homogeneity assumption. In other words, we reject the homogeneity of the table and conclude that it is highly likely that real differences exist between the education groups in terms of their readership profiles.

The statistical test of homogeneity described above is relevant to statistical inference, but we are more interested here in the ability of the χ^2 statistic to measure discrepancy from homogeneity, in other words to measure heterogeneity of the profiles. We shall now re-express the χ^2 statistic in a different form by dividing the numerator and denominator of each set of three terms for a particular row by the square of the corresponding row total. For example, looking just at the last three terms of the χ^2 calculation given in (4.1) above, we divide the numerator and denominator of each term by the square of E5's total, i.e. 26^2, in order to obtain observed and expected profiles in the numerators rather than the original raw frequencies:

Alternative expression of the χ^2 statistic in terms of profiles and masses

$$\chi^2 = 12 \text{ similar terms} \cdots + \frac{\left(\frac{3}{26} - \frac{4.76}{26}\right)^2}{\frac{4.76}{26^2}} + \frac{\left(\frac{7}{26} - \frac{10.74}{26}\right)^2}{\frac{10.74}{26^2}} + \frac{\left(\frac{16}{26} - \frac{10.50}{26}\right)^2}{\frac{10.50}{26^2}}$$

$$= 12 \text{ similar terms} \cdots$$

$$+ 26 \times \frac{(0.115 - 0.183)^2}{0.183} + 26 \times \frac{(0.269 - 0.413)^2}{0.413} + 26 \times \frac{(0.615 - 0.404)^2}{0.404} \qquad (4.2)$$

Notice in the last line above that one of the factors of 26 in each denominator has been taken out, so that the denominators are also profile values, equal to the average profile values. Each of the 15 terms in this calculation is thus of the form

$$\text{row total} \times \frac{(\text{observed row profile} - \text{expected row profile})^2}{\text{expected row profile}}$$

(Total) inertia is the χ^2 statistic divided by sample size We now make one more modification of the χ^2 calculation above to bring it into line with the CA concepts introduced so far: we divide both sides of Equation (4.2) by the total sample size so that each term involves an initial multiplying factor equal to the row mass rather than the row total:

$$\frac{\chi^2}{312} = 12 \text{ similar terms} \cdots$$
$$+ 0.083 \times \frac{(0.115 - 0.183)^2}{0.183} + 0.083 \times \frac{(0.269 - 0.413)^2}{0.413} + 0.083 \times \frac{(0.615 - 0.404)^2}{0.404}$$
$$(4.3)$$

where $0.083 = 26/312$ is the mass of row E5 (see Exhibit 4.1). The quantity χ^2/n on the left-hand side, where n is the grand total of the table, is called the *total inertia* in CA, or simply the *inertia*. It is a measure of how much variance there is in the table and does not depend on the sample size. In statistics this quantity has alternative names such as the mean-square contingency coefficient, and its square root is known as the phi coefficient (ϕ); hence we can denote the inertia by ϕ^2. If we gather together terms in (4.3) in groups of three corresponding to a particular row, we obtain the following form for the inertia:

$$\frac{\chi^2}{312} = \phi^2 = 4 \text{ similar groups of terms} \cdots$$
$$+ 0.083 \times \left[\frac{(0.115 - 0.183)^2}{0.183} + \frac{(0.269 - 0.413)^2}{0.413} + \frac{(0.615 - 0.404)^2}{0.404} \right] (4.4)$$

Each of the five groups of terms in this formula, one for each row of the table, is the row mass (e.g. 0.083 for row E5) multiplied by a quantity in square brackets which looks like a distance measure (or, to be precise, the square of a distance).

Euclidean, or Pythagorian, distance In (4.4) above, if it were not for the fact that each squared difference between observed and expected row profile elements is divided by the expected element, then the quantity in square brackets would be exactly the square of the "straight-line" regular distance between the row profile E5 and the average profile in three-dimensional physical space. This distance is also called the *Euclidean distance* or the *Pythagorian distance*. Let us state this in another way so that it is fully understood. Suppose we plot the two profile points $[0.115 \; 0.269 \; 0.615]$ and $[0.183 \; 0.413 \; 0.404]$ with respect to three perpendicular axes. Then the distance between them would be the square root of the sum of squared differences between the coordinates, as follows:

$$\text{Euclidean distance} = \sqrt{(0.115 - 0.183)^2 + (0.269 - 0.413)^2 + (0.615 - 0.404)^2}$$
$$(4.5)$$

This familiar distance, whose value in (4.5) is calculated as 0.264, is exactly the distance between the points E5 and their average in Exhibit 3.2.

However, the distance function in (4.4) is not the Euclidean distance — it involves an extra factor in the denominator of each squared term. Because this factor rescales or *reweights* each squared difference term, this variant of the Euclidean distance function is referred to in general as a *weighted Euclidean distance*. In this particular case where the scaling factors in the denominators are the expected profile elements, the distance is called the *chi-square distance*, or χ^2-distance for short. For example the χ^2-distance between row E5 and the centroid is:

$$\chi^2\text{-distance} = \sqrt{\frac{(0.115 - 0.183)^2}{0.183} + \frac{(0.269 - 0.413)^2}{0.413} + \frac{(0.615 - 0.404)^2}{0.404}} \quad (4.6)$$

and has value 0.431, higher than the Euclidean distance in (4.5) because each term under the square root sign has been increased in value. In the next chapter we will show how χ^2-distances can be visualized.

From (4.4) and (4.6) we can write the inertia in the following form:

$$\text{inertia} = \sum_i (i\text{-th mass}) \times (\chi^2\text{-distance from } i\text{-th profile to centroid})^2 \quad (4.7)$$

Geometric interpretation of inertia

where the sum is over the five rows of the table. Since the masses add up to 1, we can also express (4.7) in words by saying that the inertia is the weighted average of the squared χ^2-distances between the row profiles and their average profile. So the inertia will be high when the row profiles have large deviations from their average, and will be low when they are close to the average. Exhibit 4.2 shows a sequence of four small data matrices, each with five rows and three columns, as well as the display of the row profiles in triangular coordinates, going from low to high total inertia. The examples have been chosen especially to illustrate inertias in increasing order of magnitude. This sequence of maps also illustrates the concept of row–column association, or row–column correlation. When the inertia is low, the row profiles are not dispersed very much and lie close to their average profile. In this case we say that there is low association, or correlation, between the rows and columns. The higher the inertia, the more the row profiles lie closer to the column vertices, i.e. the higher is the row–column association. Later, in Chapter 8, we shall describe a correlation coefficient between the rows and columns which links up more formally to the inertia concept.

If all the profiles are identical and thus lie at the same point (their average), all chi-square distances are zero and the total inertia is zero. On the other hand, maximum inertia is attained when all the profiles lie exactly at the vertices of the profile space, in which case the maximum possible inertia can be shown to be equal to the dimensionality of the space (in the triangular examples of Exhibit 4.2, this maximum would be equal to 2).

Minimum and maximum inertia

Exhibit 4.2:
*A series of data
tables with
increasing total
inertia. The higher
the total inertia, the
greater is the
association between
the rows and
columns, displayed
by the higher
dispersion of the
profile points in the
profile space. The
values in these
tables have been
chosen specifically
so that the column
sums are all equal,
so the weights in the
χ^2-distance
formulation are the
same, and hence
distances we observe
in these maps are
true χ^2-distances.*

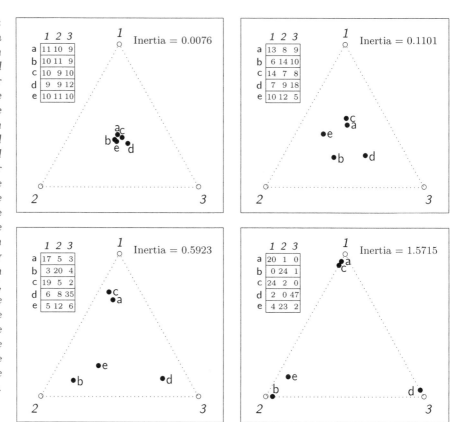

*Inertia of rows
is equal to inertia
of columns*

So far we have explained the concepts of profile, mass, χ^2-distance and inertia in terms of the rows of a data table. As we said in Chapter 3, everything described so far applies in an equivalent way to the columns of the table (see the column profiles, average profile and column masses in Exhibit 3.4). In particular, the calculation of inertia in (4.7) gives an identical result if it is calculated on the column profiles; i.e. the total inertia of a table is the weighted average of the squared χ^2-distance between the column profiles and their average profile, where the weights are now the column masses.

Some notation

This section is not essential to the understanding of the practical aspects of correspondence analysis and may be skipped. But for those who do want to understand the literature on correspondence analysis and its theory, this section will be useful (for example, we shall use these definitions in Chapter 14). We introduce some standard notation for the entities defined so far, using the data in Exhibit 3.1 as illustrations (the data have been repeated in Exhibit 4.1).

- n_{ij} — the element of the cross-tabulation (or contingency table) in the i-th row and j-th column, e.g. $n_{21} = 18$.

- n_{i+} — the total of the i-th row, e.g. $n_{3+} = 87$ (the $+$ in the subscript indicates summation over the corresponding index).

- n_{+j} — the total of the j-th column, e.g. $n_{+2} = 129$.

- n_{++}, or simply n, the grand total of the table, e.g. $n = 312$.

- p_{ij} — n_{ij} divided by the grand total of the table, e.g. $p_{21} = n_{21}/n = 18/312 = 0.0577$.

- r_i — the mass of the i-th row, i.e. $r_i = n_{i+}/n$ (which is the same as p_{i+}, the sum of the i-th row of relative frequencies p_{ij}); e.g. $r_3 = 87/312 = 0.279$; the vector of row masses is denoted by $\mathbf{r}$.

- c_j — the mass of the j-th column, i.e. $c_j = n_{+j}/n$ (which is the same as p_{+j}, the sum of the j-th column of relative frequencies p_{ij}); e.g. $c_2 = 129/312 = 0.414$; the vector of column masses is denoted by $\mathbf{c}$.

- a_{ij} — the j-th element of the profile of row i, i.e. $a_{ij} = n_{ij}/n_{i+}$; e.g. $a_{21} = 18/84 = 0.214$; the i-th row profile is denoted by vector $\mathbf{a}_i$.

- b_{ij} — the i-th element of the profile of column j, i.e. $b_{ij} = n_{ij}/n_{+j}$; e.g. $b_{21} = 18/57 = 0.316$; the j-th column profile is denoted by vector $\mathbf{b}_j$.

- $\sqrt{\sum_j (a_{ij} - a_{i'j})^2/c_j}$ — the χ^2-distance between the i-th and i'-th row profiles, denoted by $\|\mathbf{a}_i - \mathbf{a}_{i'}\|_c$; e.g. from Exhibit 3.1,

$$\|\mathbf{a}_1 - \mathbf{a}_2\|_c = \sqrt{\frac{(0.357-0.214)^2}{0.183} + \frac{(0.500-0.548)^2}{0.413} + \frac{(0.143-0.238)^2}{0.404}} = 0.374$$

- $\sqrt{\sum_i (b_{ij} - b_{ij'})^2/r_i}$ — the χ^2-distance between the j-th and j'-th column profiles, denoted by $\|\mathbf{b}_j - \mathbf{b}_{j'}\|_r$; e.g. from Exhibit 3.4,

$$\|\mathbf{b}_1 - \mathbf{b}_2\|_r = \sqrt{\frac{(0.088-0.054)^2}{0.045} + \frac{(0.316-0.357)^2}{0.269} + \ldots \text{etc.}} = 0.323$$

where $0.088 = 5/57$, $0.054 = 7/129$, $0.045 = 14/312$, etc.

- $\sqrt{\sum_j (a_{ij} - c_j)^2/c_j}$ — the χ^2-distance between the i-th row profile $\mathbf{a}_i$ and the average row profile $\mathbf{c}$ (the vector of column masses), denoted by $\|\mathbf{a}_i - \mathbf{c}\|_c$; e.g. from Exhibit 3.1,

$$\|\mathbf{a}_1 - \mathbf{c}\|_c = \sqrt{\frac{(0.357-0.183)^2}{0.183} + \frac{(0.500-0.413)^2}{0.413} + \frac{(0.143-0.404)^2}{0.404}} = 0.594$$

- $\sqrt{\sum_i (b_{ij} - r_i)^2/r_i}$ — the χ^2-distance between the j-th column profile $\mathbf{b}_j$ and the average column profile $\mathbf{r}$ (the vector of row masses), denoted by $\|\mathbf{b}_j - \mathbf{r}\|_r$; e.g. from Exhibit 3.4,

$$\|\mathbf{b}_1 - \mathbf{r}\|_r = \sqrt{\frac{(0.088-0.045)^2}{0.045} + \frac{(0.316-0.269)^2}{0.269} + \dots \text{etc.}} = 0.332$$

With this notation, the formula (4.7) for the total inertia is

$$\phi^2 = \frac{\chi^2}{n} = \sum_i r_i \|\mathbf{a}_i - \mathbf{c}\|_c^2 \qquad \text{(for the rows)}$$

$$= \sum_i r_i \sum_j \left(\frac{p_{ij}}{r_i} - c_j\right)^2 / c_j \qquad (4.8)$$

$$= \sum_j c_j \|\mathbf{b}_i - \mathbf{r}\|_r^2 \qquad \text{(for the columns)}$$

$$= \sum_j c_j \sum_i \left(\frac{p_{ij}}{c_j} - r_i\right)^2 / r_i \qquad (4.9)$$

and has the value 0.0833, hence $\chi^2 = 0.0833 \times 312 = 26.0$.

SUMMARY:
Chi-Square
Distance and
Inertia

1. The chi-square (χ^2) statistic is an overall measure of the difference between the observed frequencies in a contingency table and the expected frequencies, calculated under a hypothesis of homogeneity of the row profiles (or of the column profiles).

2. The *(total) inertia* of a contingency table is the χ^2 statistic divided by the total of the table.

3. Geometrically, the inertia measures how "far" the row profiles (or the column profiles) are from their average profile. The average profile can be considered to represent the hypothesis of homogeneity (i.e. equality) of profiles.

4. Distances between profiles are measured using the *chi-square distance* (χ^2-distance). This distance is similar to the *Euclidean* (or *Pythagorian*) distance between points in physical space, except that each squared difference between coordinates is divided by the corresponding element of the average profile. The χ^2-distance is an example of a weighted Euclidean distance.

5. The inertia can be rewritten in a form which can be interpreted as the weighted average of squared χ^2-distances between the row profiles and their average profile (similarly, between the column profiles and their average).

Plotting Chi-Square Distances

## 5

In Chapter 3 we interpreted the positions of two-dimensional profile points in a triangular coordinate system where distances were Euclidean distances. In Chapter 4 the chi-square distance (χ^2-distance) between profile points was defined, as well as its connection with the chi-square statistic and the inertia of a data matrix. The χ^2-distance is a weighted Euclidean distance, where each squared term corresponding to a coordinate is weighted inversely by the average profile value corresponding to that coordinate. So far we have not actually visualized the χ^2-distances between profiles, apart from Exhibit 4.2, where the average profile values were equal, so that the χ^2-distances were also Euclidean in that case. In this chapter we show that by a simple transformation of the profile space, the distances that we observe in our graphical display are actual χ^2-distances.

Contents

Exhibit 5.1 shows the row profiles of Exhibit 3.1 plotted according to perpendicular coordinate axes in the usual three-dimensional physical space. Here the distances between the profiles are not χ^2-distances, but rather (unweighted) Euclidean distances — see an example of the calculation in formula (4.5). In such a space distances between two profiles with elements x_j and y_j respectively (where $j = 1, \ldots, J$) are calculated by summing the squared differences between coordinates, of the form $(x_j - y_j)^2$, over all dimensions j and then taking the square root of the resultant sum. This is the usual "straight-line" Euclidean distance in physical space with which we are familiar. As we have seen, the χ^2-distance differs from this distance function by the division of each squared difference by the corresponding element of the average profile; i.e. each term is of the form $(x_j - y_j)^2/c_j$, where c_j is the corresponding element of the average profile. Since we can interpret and compare distances only in our familiar physical space, we need to be able to organize the points in the

Difference between χ^2-distance and ordinary Euclidean distance

Exhibit 5.1:
*The profile space
showing the profiles
of the education
groups on the
equilateral triangle
in three-dimensional
space; the distances
here are Euclidean
distances.*

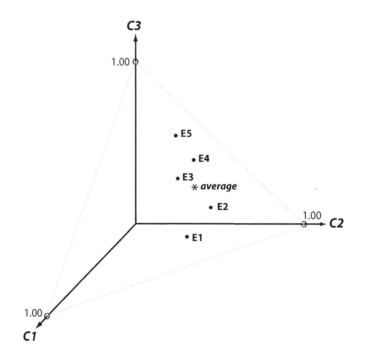

map in such a way that familiar "straight-line" distances turn out to be χ^2-distances. Luckily, this is possible thanks to a straightforward transformation of the profiles.

Transforming the coordinates before plotting

In the calculation of the χ^2-distance every term of the form $(x_j - y_j)^2/c_j$ can be rewritten as $(x_j/\sqrt{c_j} - y_j/\sqrt{c_j})^2$. This equivalent way of expressing the general term in the distance formula is identical in form to that of the ordinary Euclidean distance function; i.e. it is in the form of a squared difference. The only change is that the coordinates are not the original x_j and y_j values but the transformed $x_j/\sqrt{c_j}$ and $y_j/\sqrt{c_j}$ ones. This suggests that, instead of using the elements of the profiles as coordinates, we should rather use these elements divided by the square roots of the corresponding elements of the average profile. In that case the usual Euclidean distance between these transformed coordinates gives the χ^2-distance that we require.

Effect of the transformation in practice

The values of c_j are elements of the average profile and are thus all less than 1. Hence, dividing each profile element by its respective $\sqrt{c_j}$ will result in an increase in the values of all coordinates, but some will be increased more than others. If a particular c_j is relatively small compared to the others (i.e. the j-th column category has a relatively low frequency), then the corresponding coordinates $x_j/\sqrt{c_j}$ and $y_j/\sqrt{c_j}$ will be increased by a relatively large amount. Conversely, a large c_j corresponding to a more frequent category will lead to a relatively smaller increase in the transformed coordinates. Thus the effect of the transformation is to increase the values corresponding to low-frequency categories relatively more than the coordinates corresponding to high-frequency categories. We will see that this makes sense because the differences between low-frequency categories are generally smaller than those between high-frequency categories, so the transformation tends to balance out the contributions of the different categories.

In the untransformed space of Exhibit 5.1 the vertex points lie at one physical unit of measurement from the origin (or zero point) of the three coordinates axes. The first vertex, with coordinates $[1\ \ 0\ \ 0]$, is transformed to the position $[1/\sqrt{c_1}\ \ 0\ \ 0]$; i.e. its position on the first axis is stretched out to be $1/\sqrt{0.183} = 2.34$. Similarly, the second and third vertices are stretched out to values $1/\sqrt{c_2} = 1/\sqrt{0.413} = 1.56$ and $1/\sqrt{c_3} = 1/\sqrt{0.404} = 1.57$ respectively. These values are indicated at the vertices on the three axes in Exhibit 5.2. The profiles are plotted according to their transformed values and find new positions in the space, but are still in the triangle joining the transformed vertex points. Notice that the stretching is relatively more in the direction of $C1$, the category with the lowest marginal frequency.

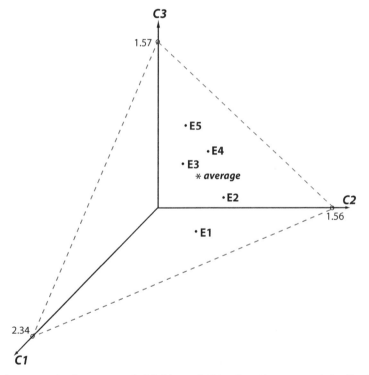

Exhibit 5.2:
The profile space, showing the vertex points of the triangle stretched by different amounts along each axis, so that distances between profiles are χ^2-distances.

There is an equivalent way of thinking of this situation geometrically. In the untransformed coordinate systems of Exhibits 2.4 and 5.1, the tic marks indicating the scales (for example, the values $0.1, 0.2, 0.3$, etc.) along the three axes were at equal intervals apart. The effect of the transformation is to stretch out the three vertices as shown in Exhibit 5.2. But we can still think of the three

Alternative interpretation in terms of recalibrated coordinate axes

vertex points as being one profile unit from the origin, but then the scales are different on the three axes. On the *C1* axis, an interval of 0.1 between two tic marks would be a physical length of 0.234, while on the *C2* and *C3* axes these intervals would be 0.156 and 0.157 respectively. Hence a unit interval on the *C1* axis is approximately 50% longer than the same interval on the other two axes. Along these recalibrated axes we would still use the original profile elements to situate a profile in three-dimensional space. Whichever way you prefer to think about the transformation, either as a transformation of the profile values or as a stretching and recalibrating of the axes, the outcome is the same: the profile points now lie in the stretched triangular space shown in Exhibit 5.2. In Exhibit 5.3 the stretched triangle has been laid flat and it is clear that vertex *C1*, corresponding to the rarest category of *glance* reading, has been stretched the most.

Exhibit 5.3:
The triangular space of the profiles in the stretched space of Exhibit 5.2 laid "flat" (compare with Exhibit 3.2). The triangle has been stretched most in the direction of C1, the category with the lowest frequency.

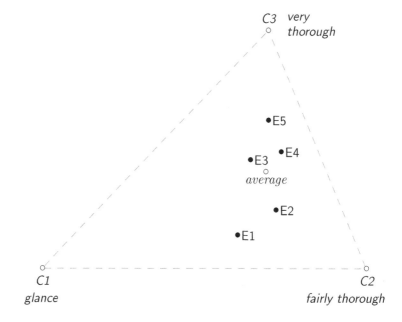

Geometric interpretation of the inertia and χ^2 statistic

Now that the observed straight-line distances in the transformed space are actual χ^2-distances, the profile points may be joined to the average point to show the χ^2-distances between the profiles and their average — see Exhibit 5.4. In the case of row profiles, if we associate each row mass with its respective profile, we know from formula (4.7) that the weighted sum of these squared distances is identical to the inertia of the table. If we associate the row totals with the profiles rather than the masses (where each row total is n times the respective row mass, n being the grand total of the whole table), then the weighted sum of these squared distances is equal to the χ^2 statistic. Equivalent results hold for the column profiles relative to their average point. Thus the inertia or χ^2 statistic may be interpreted geometrically as the degree of dispersion of the set of profile points (rows or columns) about their average, where the points are weighted by their relative frequencies (i.e. masses) or total frequencies respectively.

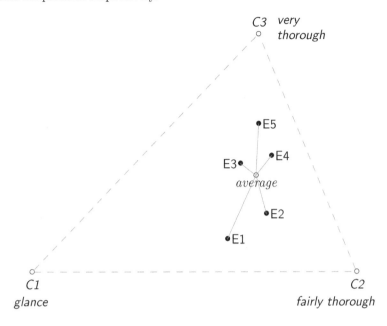

Exhibit 5.4:
The "stretched" profile space showing the χ^2-distances from the profiles to their centroid; the inertia is the weighted average of the sum of squares of these distances, and the χ^2 statistic is the inertia multiplied by the sample size ($n = 312$ in this example).

In Chapter 3 the concept was introduced of distributionally equivalent rows or columns of a table, i.e. those with the same profiles. Considering Exhibits 3.1 or 4.1 again, suppose that we could distinguish two types of *fairly thorough* readers, those who concentrated more on the political content of the newspaper and those who concentrated more on the cultural and sports sections; these categories could be denoted by *C2a* and *C2b*. Suppose further that in both these new columns, the relative frequencies (i.e. profiles) of education groups were the same; in other words, there was no difference between these two subdivisions of the *fairly thorough* reading group as far as education is concerned, i.e. they were distributionally equivalent. The subdivision of column *C2* into *C2a* and *C2b* brings no extra information about the differences between the education groups; hence any analysis of these data should give the same results whether *C2* is subdivided or left as a single category. An analysis that satisfies this property is said to obey the *principle of distributional equivalence*. If we used ordinary Euclidean distances between the education group profiles, this principle would not be obeyed because different results would be obtained if such a subdivision were made. The χ^2-distance, on the other hand, does obey the principle, remaining unaffected by such a partitioning of a category of the data matrix: if two distributionally equivalent columns are merged, the χ^2-distance between the rows does not change. In practice, this means that columns that have similar profiles can be aggregated with almost no effect on the geometry of the rows, and vice versa. This gives the researcher a certain assurance that introducing many categories into the anal-

Principle of distributional equivalence

ysis adds only substantive value and is not affected by some technical quirk that depends on the number of categories.

χ^2-distances make the contributions of categories more similar

We know now how to organize the display in order to visualize χ^2-distances, but why do we need to go to all this trouble to see χ^2-distances rather than Euclidean distances? There are many ways to defend the use of the χ^2-distance, some more technical than others, and the reason is more profound than simply being able to visualize the χ^2 statistic. One clear reason is that there are inherent differences in the variances of sets of frequencies. For example, in Exhibit 3.1 one can see that the range of profile values in the less frequent column *C1* (from 0.115 to 0.357) is less than the range in the more frequent column *C3* (from 0.143 to 0.615). This is a general rule for frequency data, namely that a set of smaller frequencies has less dispersion than a set of larger frequencies. The effect of this disparity in the spread of the profile values can be seen by measuring the contributions of each category to the distance function. For example, let us compare the squared values of the Euclidean distances and the χ^2-distances between the education group profiles and their centroid (average profile) in the data set of Exhibit 3.1. For example, for the fifth education group *E5*, the squared Euclidean distance between its profile and the centroid is

$$\text{Euclidean distance}^2 = (0.115 - 0.183)^2 + (0.269 - 0.413)^2 + (0.615 - 0.404)^2$$
$$= 0.00453 + 0.02080 + 0.04475$$
$$= 0.07008$$

whereas the squared χ^2-distance is

$$\chi^2\text{-distance}^2 = \frac{(0.115 - 0.183)^2}{0.183} + \frac{(0.269 - 0.413)^2}{0.413} + \frac{(0.615 - 0.404)^2}{0.404}$$
$$= 0.02480 + 0.05031 + 0.11081$$
$$= 0.18592$$

(see (4.5) and (4.6) on pages 28–29). Each of these squared distances is the sum of three values, one term for each column category, and can be expressed as percentages of the total to assess the contributions of each category of readership. For example, in the squared Euclidean distance category *C1* contributes 0.00453 out of 0.07008, which is 6.5%, whereas in the squared χ^2-distance its contribution is 0.02480 out of 0.18592, i.e. 13.3% (see the row *E5* in Exhibit 5.5). If all the terms for *C1* are summed over the five education groups and expressed as a percentage of the sum of squared distances we get the overall percentage contribution of 17.0% for the Euclidean distance, and 31.3% for the χ^2-distance (see the last row of Exhibit 5.5). This exercise illustrates the phenomenon that the lowest frequency category *C1* generally contributes less to the Euclidean distance compared to *C3*, for example, whereas in the χ^2-distance its contribution is boosted owing to the division by the average frequency.

Row	Euclidean			χ^2		
	C1	*C2*	*C3*	*C1*	*C2*	*C3*
E1	28.7	7.1	64.2	47.1	5.1	47.7
E2	2.1	38.7	59.1	4.7	37.2	58.1
E3	13.2	66.4	20.4	25.5	56.7	17.8
E4	37.1	2.8	60.1	56.6	1.9	41.5
E5	6.5	29.7	63.9	13.3	27.1	59.6
Overall	17.0	21.8	61.2	31.3	17.7	51.0

Exhibit 5.5:
Percentage contributions of each column category to the squared Euclidean and squared χ^2-distances from the row profiles to their centroid (data of Exhibit 3.1).

As described in Chapter 4, the χ^2-distance is an example of a weighted Euclidean distance, whose general definition is as follows:

Weighted Euclidean distance

$$\text{weighted Euclidean distance} = \sqrt{\sum_{j=1}^{p} w_j (x_j - y_j)^2} \qquad (5.1)$$

where w_j are nonnegative weights and x_j, $j = 1, \ldots, p$ and y_j, $j = 1, \ldots, p$ are two points in p-dimensional space. In principal component analysis (PCA), a method closely related to CA, the p dimensions are defined by continuous variables, often on different measurement scales. It is necessary to remove the effect of scale in some way, and this is usually done by dividing the data by the standard deviations s_j of the respective variables; i.e. by replacing observations x_j and y_j for variable j by x_j/s_j and y_j/s_j. This operation can be thought of as using a weighted Euclidean distance with weights $w_j = 1/s_j^2$, the inverse of the variances. In the definition of the χ^2-distance between profiles, the weights are equal to $w_j = 1/c_j$, i.e. the inverses of the average profile elements.

Although the profiles are on the same relative frequency scale, there is still a need to compensate for different variances, similar to the situation in PCA. The phenomenon that sets of frequencies with higher average have higher variance than those with a lower average is embodied in the *Poisson distribution* — one of the standard statistical distributions for count variables. A property of the Poisson distribution is that its variance is equal to its mean. Hence, transforming the frequencies by dividing by the square roots of the expected (mean) frequencies is one way of standardizing the data because the square root of the mean is a surrogate for the standard deviation. But it is not the only way to standardize, so why is the χ^2-distance so special? There are many advantages of the χ^2-distance, apart from its obeying the principle of distributional equivalence and giving CA the property of symmetry between the treatment of rows and columns. A more technical reason for using the χ^2-distance can be found in the properties of a multivariate statistical distri-

Theoretical justification of χ^2-distance

bution for count data, called the *multinomial distribution*. This subject will be discussed again in the Epilogue, Appendix E, pages 299–301.

SUMMARY:
Plotting
Chi-Square
Distances

1. χ^2-distances between profiles can be observed in ordinary physical (or Euclidean) space by transforming the profiles before plotting. This transformation consists of dividing each element of the profile by the square root of the corresponding element of the average profile.

2. Another way of thinking about χ^2-distances is not to transform the profile elements but to stretch the plotting axes by different amounts, so that a unit on each axis has a physical length inversely proportional to the square root of the corresponding element of the average profile.

3. The χ^2-distance is a special case of a weighted Euclidean distance where the weights are the inverses of the corresponding average profile values.

4. Assuming that we are plotting row profiles, the rescaling of the coordinates (or, equivalently, the stretching of the axes) can be regarded as a way of standardizing the columns of the table. This makes visual comparisons between the row profiles more equitable across the different columns.

5. The χ^2-distance obeys the *principle of distributional equivalence*, which guarantees stability in the distances between rows, say, when columns are disaggregated into columns with similar profiles, or when columns with similar profiles are aggregated.

Reduction of Dimensionality

## 6

Up to now, small data sets (Exhibits 2.1 and 3.1) were used specifically because they were low-dimensional and hence easy to visualize exactly. These tables with three columns involved three-dimensional profiles, which were actually two-dimensional, as we saw in Chapter 2, and could thus be laid flat for inspection in a triangular coordinate system. In most applications, however, the table of interest has many more rows and columns and the profiles lie in a space of much higher dimensionality. Since we cannot easily observe or even imagine points in a space with more than three dimensions, it becomes necessary to reduce the dimensionality of the points. This dimension-reducing step is the crucial analytical aspect of correspondence analysis (CA) and can be performed only with a certain loss of information, but the objective is to restrict this loss to a minimum so that a maximum amount of information is retained.

Contents

An example of a table of higher dimensionality is given in Exhibit 6.1, a cross-tabulation generated from the database of the Spanish National Health Survey (*Encuesta Nacional de la Salud*) in 1997. One of the questions in this survey concerns the opinions that respondents have of their own health, which they can judge to be "very good" (*muy bueno* in the original survey), "good" (*bueno*), "regular" (*regular*), "bad" (*malo*) or "very bad" (*muy malo*). The table cross-tabulates these responses with the age groups of the respondents. There are seven age groups (rows of Exhibit 6.1) and five health categories (columns). A total of 6371 respondents are cross-tabulated and give a representative snapshot of how the Spanish nation views its own health at this

Data set 3: Spanish National Health Survey

AGE GROUP	Very good	Good	Regular	Bad	Very bad	Sum
16–24	243	789	167	18	6	*1223*
25–34	220	809	164	35	6	*1234*
35–44	147	658	181	41	8	*1035*
45–54	90	469	236	50	16	*861*
55–64	53	414	306	106	30	*909*
65–74	44	267	284	98	20	*713*
75+	20	136	157	66	17	*396*
Sum	*817*	*3542*	*1495*	*414*	*103*	*6371*

AGE GROUP	Very good	Good	Regular	Bad	Very bad	Sum
16–24	19.9	64.5	13.7	1.5	0.5	*100.0*
25–34	17.8	65.6	13.3	2.8	0.5	*100.0*
35–44	14.2	63.6	17.5	4.0	0.8	*100.0*
45–54	10.5	54.5	27.4	5.8	1.9	*100.0*
55–64	5.8	45.5	33.7	11.7	3.3	*100.0*
65–74	6.2	37.4	39.8	13.7	2.8	*100.0*
75+	5.1	34.3	39.6	16.7	4.3	*100.0*
Average	*12.8*	*55.6*	*23.5*	*6.5*	*1.6*	*100.0*

point in time. But what is that view, and how does it change with age? Using CA we will be able to understand very quickly the relationship between age and self-assessment of health.

Comparison of age group (row) profiles

Let us suppose for the moment that we are interested in the profiles of the age groups across the health categories, i.e. the row profiles. The row profiles are given in percentage form in Exhibit 6.2. The last row is the average row profile, or the profile across the health categories for the sample as a whole, without distinguishing between age groups. Thus we can see, for example, that of the total of 6371 Spaniards sampled in this study, 12.8% regarded themselves as in *very good* health, 55.6% in *good* health, and so on. Looking at specific age groups we see that there are the differences that one would expect; for example, the youngest age group has higher percentages of these categories (19.9% *very good* and 64.5% *good*) whereas the oldest group has lower percentages (5.1% and 34.3% respectively). Perusing this table we quickly come to the conclusion that self-assessed health becomes worse with age, which is no surprise at all. It is not so easy, however, to see in the numbers how fast or slow this change is occurring; for example, where the changes in self-assessed health from one age group to the next are bigger or smaller.

It is possible to compute χ^2-distances between the row (age group) profiles, but the problem is that one cannot visualize these profiles exactly, since they are points situated in a five-dimensional space. As we saw in the previous three-dimensional examples, where profiles lay in a planar triangle, the age group profiles lie in a space of one less dimension because the elements of each profile add to 1, but even direct visualization in four-dimensional space is impossible. We might be able to visualize the profiles approximately, however, hoping that they do not "fill" the whole four-dimensional space but rather lie approximately in some low-dimensional subspace of one, two or three dimensions. This is the essence of CA, the identification of a low-dimensional subspace which approximately contains the profiles. Putting this the opposite way, CA identifies dimensions along which there is very little dispersion of the profile points and eliminates these low-variation directions of spread, thereby reducing the dimensionality of the cloud of points so that we can more easily visualize their relative positions.

Identifying lower-dimensional subspaces

In this example it turns out that the profiles actually lie very close to a line, so that the points can be imagined as forming an elongated cigar-shaped cloud of points in the four-dimensional profile space. If we now identify the line which comes "closest" to the points (we define the measure of closeness soon), we can drop (or *project*) the points perpendicularly onto this line, take the line out of the multidimensional space and lay it from left to right on a display which is now much easier to interpret. In Exhibit 6.3 we see this one-dimensional representation of the age group profiles, with the age groups lying in their inherent order from oldest on the left to youngest on the right, even though the method has no knowledge of the ordering of the categories. In this display we can see immediately that there are smaller differences amongst the younger age groups, and bigger differences in the middle-age groups.

Projecting profiles onto subspaces

Exhibit 6.3:
Optimal one-dimensional projection of the age group profiles, using CA.

Since the lower-dimensional projections of the profiles are no longer at their true positions, we need to know how large a discrepancy there is between their exact positions and their approximate ones. To do this we use the total inertia of the profiles as a measure of the total variation, or geometric dispersion, of the points in their true four-dimensional positions. Both quality of display and its counterpart, the loss, or error of display, are measured in the form of percentages of the total inertia, and they add up to 100%: the lower the loss,

Measuring quality of display

the higher the quality, and the higher the loss the lower the quality. In the present example the loss incurred by projecting the points onto the straight line of Exhibit 6.3 turns out to be only 2.7% of the total inertia; in other words the quality of the unidimensional approximation of the profiles is equal to 97.3%. This is a very favourable result — we started with a 7×5 table of numbers with a total dimensionality of 4 and, by sacrificing only 2.7% of the dispersion of the points in three dimensions of the space, the remaining 97.3% is represented by a scatter of points along a single dimension! This percentage can be interpreted exactly as in regression as a "percentage of explained variance": the single dimension showing the seven projected profile points in Exhibit 6.3 explains 97.3% of the inertia of the true profiles (or 97.3% of the total inertia of the table in Exhibit 6.1).

Exhibit 6.4:
Observed interpoint distances measured in Exhibit 6.3 between all pairs of points, plotted against the true χ^2-distances between the row profiles of Exhibit 6.3.

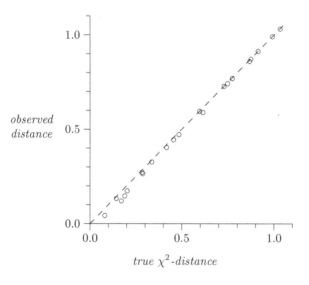

Approximation of interprofile distances The distances between the projected profiles in Exhibit 6.3 are approximations of the true χ^2-distances between the row profiles in their full four-dimensional spatial positions. Exact χ^2-distances, computed directly from Exhibit 6.2, can be compared with those displayed in Exhibit 6.3, and this comparison is made graphically in Exhibit 6.4. Because there are 7 points there are $\frac{1}{2} \times 7 \times 6 = 21$ pairs of interpoint distances. Clearly the agreement is excellent, which was expected because of the relatively small loss in accuracy of 2.7% incurred in reducing the profiles to a one-dimensional display. Notice in Exhibit 6.4 that the observed distances are always less than or equal to the true distances — we say that the distances are approximated "from below". This is because the square of the true distance is the sum of a set of squared components, one for each dimension of the profile space, whereas the square of the projected distance is the sum of a reduced number of squared components, which in this unidimensional projection is just a single component. The "unexplained" part of the distances is shown by the deviations of the points from the 45° line in Exhibit 6.4, mostly for the small distances.

In the space of the seven age group profiles, there are five vertex points representing the health categories. Recall once more that each of these extreme profile points represents a fictitious profile totally concentrated into one health category; for example, the vertex point [1 0 0 0 0] represents a group which has only *very good* self-assessed health. These vertex points can also be projected onto the optimal dimension in Exhibit 6.3 — see Exhibit 6.5. Notice the change in scale compared to Exhibit 6.3 — the age group profiles are in exactly the same positions in both these maps. The vertices are much more spread out than the profiles because they are the most extreme profiles obtainable.

Display of the projected vertex points

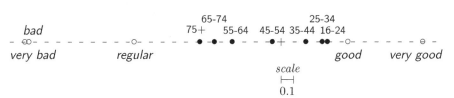

Exhibit 6.5:
Optimal map of Exhibit 6.3, showing the projected vertices of the health categories.

Notice in the joint display of Exhibit 6.5 how the health categories are also spread out in their intrinsic order, with the *very bad* health category on the extreme left and the *very good* on the extreme right. The positions of these reference points along the dimension gives us the key to the interpretation of the association between the rows (age groups) and columns (health categories), with the youngest age group farthest towards good health and the oldest group farthest towards bad health. The origin (or zero point, indicated by a + on the dashed line in Exhibits 6.3 and 6.5) represents the average profile; thus we can deduce that the age groups up to 44 years are on the "good" side of average, and groups 45 years and older on the "bad" side. The fact that *very bad* is so far away from the age group profiles shows that no age group is close to this extreme — indeed in Exhibit 6.2 we can see percentage values of 0.5–4.3% and an average of 1.6% (the average value is at the origin). The category *bad* is almost at the same position, but with a range of 1.5–16.7% and an average of 6.5% at the origin (more details about the joint interpretation will be given in Chapters 8 and 13). The relationship between the row profiles and column vertices in this one-dimensional projection is the same as we described for the triangular space in Chapters 2 and 3 — each age group profile is at the weighted average of the health category vertices, using the profile elements as weights. Hence the youngest age group 16–24 is at the rightmost position of the age groups because it has the highest profile values on the health categories *very good* and *good* on the right.

Joint interpretation of profiles and vertices

Exhibit 6.6:
*Profile points in a
multidimensional
space and a plane
cutting through the
space; the
best-fitting plane in
the sense of least
squares must pass
through the centroid
of the points.*

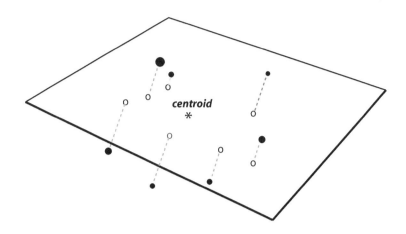

*Definition of
closeness of points
to a subspace*

The present example is simpler than usual because a single dimension adequately summarizes the data. In most cases we shall look for at least a two-dimensional plane that comes "closest to", or "best fits", the high-dimensional cloud of profiles. The profiles are then projected onto this plane, and the extreme vertices of the profile space as well. Exhibit 6.6 shows several profile points in an imaginary high-dimensional space, and their projections onto a plane cutting through the space. Whether we project the profiles onto a best-fitting line (a one-dimensional subspace), a plane (two-dimensional subspace) or even a subspace of higher dimensionality, we need to define what we mean by "closeness" of the points to that subspace.

Imagine any straight line in the multidimensional space of the profiles. The shortest distance from each profile to the line can be computed (by distance in this context we implicitly mean the χ^2-distance). Then to arrive at a single measure of closeness of all points to the line, an obvious choice would be to sum up the distances from all profiles to the imaginary line. Then our task would be to find the line for which this sum-of-distances is the smallest. In principle there is nothing stopping us from optimizing this criterion, but the mathematics involved in minimizing such a sum-of-distances is quite complicated. As in many other areas of statistics, the problem simplifies greatly if one defines a criterion in terms of sum of squared distances, rather than the distances alone, leading to what is called a *least-squares* optimization problem. In the present case, we also have a mass associated with each profile which quantifies the importance of the profile in the analysis. The criterion used in correspondence analysis is thus a weighted sum of squared distances, in other words the inertia, and the lines and planes of best fit are found by minimizing the inertia that is lost when projecting the points onto spaces of lower dimension.

Suppose that we have I profile points in a multidimensional space and that a candidate low-dimensional subspace is denoted by S. For the i-th profile point, with mass m_i, we compute the χ^2-distance between the point and S, denoted by $d_i(S)$. The closeness of this profile to the subspace is then $m_i[d_i(S)]^2$; i.e. the squared distance weighted by the mass. The closeness of all the profiles to S is the sum of these quantities:

$$\text{closeness to } S = \sum_{i=1}^{I} m_i[d_i(S)]^2 \qquad (6.1)$$

Formal definition of criterion optimized in CA

The objective of CA is to find the subspace S which minimizes this criterion. It can be shown that the subspace S being sought necessarily passes through the centroid of the points, as depicted in Exhibit 6.6, so we need to consider only subspaces that contain the centroid.

It is not necessary here to enter into the mathematical operations involved in this minimization. It suffices to say here that the most elegant way to define the theory of CA as well as to compute the solution to the above minimization problem is to use what is known in mathematics as the *singular value decomposition*, or SVD for short. The SVD is one of the most useful results in matrix theory, and has special relevance to all the methods of dimension reduction in statistics. It is to rectangular matrices what the eigenvalue–eigenvector decomposition is to square matrices, namely a way to break down a matrix into components, from the most to least important. The algebraic notion of *rank* of a matrix is equivalent to our geometric notion of dimension (or dimensionality), and the SVD provides a straightforward mechanism of approximating a rectangular matrix with another matrix of lower rank (i.e. lower dimension) by least squares. These results transfer directly into the theory of CA, and all the entities we need, the inertia, the definition of the optimal subspace, the coordinates, etc., are obtained directly from the SVD. Since the SVD is available in many computing languages, the analytical part of CA is easily executed. In the Computational Appendix we shall show how compactly CA can be programmed using the SVD function in the computing language R — see pages 259–260.

Singular value decomposition (SVD)

We have been describing the search for low-dimensional subspaces, for example, lines and planes, by least squares, and this sounds just like the objective of regression analysis, which also fits lines and planes to data points which can be situated in multidimensional space. But there is a major difference between regression and what we are doing here. In regression one of the variables is regarded as a response variable and the distances that are minimized are parallel to the response variable axis. In the present situation, by contrast, there is no response variable, and fitting is done by minimizing the distances perpendicular to the subspace being fitted (see Exhibit 6.6 where the projections are perpendicular onto the plane, giving the shortest distances between the

Finding the optimal subspace is not regression

points and the plane). Sometimes this way of fitting a low-dimensional sub-space to points is referred to as "orthogonal regression", where the dimensions of the subspace are regarded as new variables explaining the data points.

SUMMARY:
Reduction of
Dimensionality

1. Dimensionality refers to the number of inherent dimensions in a table of data. Usually, multivariate observations on m variables are m-dimensional, i.e. require m coordinate axes for their exact visualization. Reduction of dimensionality (also called dimension reduction) is the process of approximating multivariate data of high dimensionality by lower-dimensional versions, conserving as much as possible the properties of the original data.

2. Profiles consisting of m elements, which have the particular property that these elements sum to 1, are situated exactly in a space of dimensionality $m-1$. Hence, profiles with more than four elements are situated in spaces of dimensionality greater than three, which we cannot observe directly.

3. If we can identify a subspace of lower dimensionality, preferably not more than two or three dimensions, which lies close to all the profile points, then we can project the profiles onto such a subspace and look at the profiles' projected positions in this subspace as an approximation to their true higher-dimensional positions.

4. What is lost in this process of dimension reduction is the knowledge of how far and in what direction the profiles lie "off" this subspace. What is gained is a view of the profiles that would not be possible otherwise.

5. The accuracy of display is measured by a quantity called the *percentage of inertia*. For example, if 85% of the inertia of the profiles is represented in the subspace, then the residual inertia, or error, which lies external to the subspace, is 15%.

6. The vertices, or unit profiles, can also be projected onto the optimal subspace. The object is not to represent the vertices accurately but to use them as reference points for interpreting the displayed profiles.

7. The actual computation of the low-dimensional subspace relies on measuring the closeness between a set of points and a subspace as the weighted-sum-of-squared χ^2-distances between the points and the subspace, where the points are weighted by their respective masses.

Optimal Scaling

## 7

So far correspondence analysis (CA) has been presented as a geometric method of data analysis, stressing the three basic concepts of profile, mass and χ^2-distance, and the four derived concepts of centroid (weighted average), inertia, subspace and projection. Profiles are multidimensional points, weighted by masses, and distances between profiles are measured using the χ^2-distance. The profiles are visualized by projecting them onto a subspace of low dimensionality which best fits the profiles, and then projecting the vertex profiles onto the subspace as reference points for the interpretation. There are, however, many other ways to define and interpret CA and this is why the same underlying methodology has been rediscovered many times in different contexts. One of these alternative interpretations is called *optimal scaling* and a discussion of this approach at this point will provide additional insight into the properties of CA.

Contents

We refer again to the example in Exhibit 6.1, the cross-tabulation of age groups by self-assessed health categories. Both the row and column variables are categorical variables and are stored in a computer data file using codes 1 to 7 for age, and 1 to 5 for health. If we wanted to calculate statistics on the health variable such as mean and variance, or to use self-assessed health as a variable in a statistical analysis such as regression, it would be necessary to have acceptable values for each health category. It may not be true that each of the health categories is exactly one unit apart on such a scale, as is implicitly assumed if we use the values 1 to 5. The health categories are ordered (i.e. self-assessed health is an ordinal categorical variable), which indeed gives

Quantifying a set of categories

some minimal justification for using the values 1 to 5, but what if the variable were nominal, such as the country variable in Chapter 1 (see Exhibit 1.3) or a variable such as marital status?* The age group variable is also ordinal, established by defining intervals on the original age scale, so we could use the midpoints of each age interval as reasonable scale values, but it is not obvious what value to assign to age group 7, which is open-ended (75+ years). Failing any alternative, when categories are ordered as in this case, the integer values (1 to 7 and 1 to 5 here) are often used as default values in calculations. Optimal scaling provides a way of obtaining quantitative scale values for a categorical variable, subject to a specific criterion of optimality.

Computation of overall mean using integer scale

Initially we will use the default integer values in some simple calculations, but let us first reverse the coding of the health categories so that the higher value corresponds to better health — hence 5 indicates *very good* health, down to 1 indicating *very bad* health. In the data set as a whole, there are 817 respondents with *very good* health (code 5), 3542 with *good* health (code 4), and so on, out of a total sample of 6371 respondents. Using these integer codes as scale values for the health categories, an *average health* in this sample can be calculated as follows:

$$[(817 \times 5) + (3542 \times 4) + ... + (103 \times 1)]/6371 = 3.72$$

i.e.
$$(0.128 \times 5) + (0.556 \times 4) + ... + (0.016 \times 1) = 3.72 \qquad (7.1)$$

where $817/6371 = 0.128$, $3542/6371 = 0.556$, etc. are the elements of the average row profile (see the last row of Exhibit 6.2). Therefore, this average across all the respondents is simply the weighted average of the scale values where the weights are the elements of the average profile.

Computation of group means using integer scale

Considering a particular age group now, say 16–24 years, we see from the first row of data in Exhibit 6.1 that there are 243 respondents with *very good* health, 789 with *good*, and so on, out of a total of 1223 in this young age group. Again, using the integer scale values from 5 down to 1 for the health categories, the *average health* for the 16–24 group is:

$$[(243 \times 5) + (789 \times 4) + ... + (6 \times 1)]/1223 = 4.02$$

i.e.
$$(0.199 \times 5) + (0.645 \times 4) + ... + (0.005 \times 7) = 4.02 \qquad (7.2)$$

where the second line again shows the profile values (for age group 16–24) being used as weights: $243/1223 = 0.199$, $789/1223 = 0.645$, etc. Thus we could say that the youngest age group has an average self-assessed health higher than the average: 4.02 compared to the average of 3.72. We could repeat the above calculation for the other six age groups and obtain averages

---

* In my experience as a statistical consultant I once did see a survey with a variable "Religious Affiliation: 0=none, 1=Catholic, 2=Protestant, etc." and the researcher seriously calculated an "average religion" for the sample!

as follows:

16–24	25–34	35–44	45–54	55–64	65–74	75+	*Overall*
4.02	3.97	3.86	3.66	3.39	3.30	3.19	3.72

Now that we have calculated health category means for the age groups using the integer scale values, we can compute the health category variance across the age groups. This is similar to the inertia calculation of Chapter 4 because each age group will be weighted proportional to its sample size. Alternatively you can think of all 6371 respondents being assigned the values corresponding to their respective age group, followed by the usual calculation of between-group variance. The variance is calculated as (see Exhibit 6.1 for row totals):

Computation of variance using integer scale

$$\frac{1223}{6371}(4.02 - 3.72)^2 + \frac{1234}{6371}(3.97 - 3.72)^2 + \cdots + \frac{396}{6371}(3.19 - 3.72)^2 = 0.0857$$

with standard deviation $\sqrt{0.0857} = 0.293$.

Given any set of scale values for the health categories, we can assign to each respondent a *score*, i.e. the scale value corresponding to the respondent's chosen category. This leads to an overall average score for the whole sample as well as age group averages and the variance of the scores across the age groups. All the previous calculations depend on the initial use of the 1-to-5 integer scale for the health categories, from **very bad** to **very good**, an arbitrary choice which is, admittedly, difficult to justify, especially after seeing the results of Chapter 6. The question is whether there are more justifiable scale values that lead to more informative group average scores. Answering this question depends on what is meant by "more informative", so we now consider one possible criterion which leads to scale values that turn out to be directly related to CA. Let us suppose that the scale values for the health categories are denoted by unknown quantities v_1, v_2, v_3, v_4 and v_5, which are to be determined. Then the average score for all respondents would be, in terms of these unknowns, as in (7.1):

Calculating scores with unknown scale values

$$\text{average health overall} = (0.128 \times v_1) + (0.556 \times v_2) + \cdots + (0.016 \times v_5) \quad (7.3)$$

while the average score for age group 16–24, for example, would be, as in (7.2),

$$\text{average health 16–24 years} = (0.199 \times v_1) + (0.645 \times v_2) + \cdots + (0.005 \times v_5) \quad (7.4)$$

Let s_1 denote the average score (7.4) for the first age group. For each age group, the score can be formulated in the same way, leading to seven scores $s_1, s_2, \ldots, s_7$, each defined in terms of the unknown scale values. The between-group variance is then computed as before: for example, the 1223 respondents in the 16–24 years age group are imagined notionally as piling up at the score s_1 in (7.4), and similarly for the other age groups at their respective group average scores. The variance is then computed between the group means, but still depends on the original unknown scale values.

Maximizing variance gives optimal scale

Now in order to determine a set of "informative" scale values, we can propose a property which we would like these age group average scores to have, namely that the age groups should be as distinct from one another as possible. Putting this the opposite way, what we do not want is to obtain age group scores so close to one another that it is difficult to distinguish between them. One way of phrasing this requirement more precisely is that we want scale values that lead to the maximum possible variance between age groups. Scale values v_1, v_2, ..., v_5 that lead to scores s_1, s_2, ..., s_7 and thus age group averages with maximum variance will define what we call an *optimal scale*.

Optimal scale values from the best-fitting dimension of CA

Fortunately, it turns out that the positions of the health categories along the best-fitting CA dimension solve this optimal scaling problem exactly: first, the coordinate values of the vertices in Exhibit 6.5 provide the optimal scale values, v_1 to v_5; second, the coordinate values of the profiles provide the corresponding group mean scores, s_1 to s_7; and third, the maximum between-group variance is equal to the inertia on this optimal CA dimension. The actual coordinate values are given in Exhibit 7.1. We already know from Chapter 3 that an age group lies at the centroid of the five health category vertices, and this property carries over to any projection of the points onto a subspace. For example, if the profile of age group 16–24 (Exhibit 6.2) is used to weight the positions of the vertices of the five health categories (Exhibit 7.1), the following score is obtained:

$$(0.199 \times 1.144) + (0.645 \times 0.537) + \ldots + (0.005 \times -2.076) = 0.371$$

which agrees with the coordinate of the profile 16–24 in Exhibit 7.1.

Exhibit 7.1:
Coordinate values of the points in Exhibit 6.5, i.e. the coordinates of the column vertices and the row profiles on the dimension that best fits the row profiles.

HEALTH CATEGORY	*Vertex coordinate*	AGE GROUP	*Profile coordinate*
Very good	1.144	16–24	0.371
Good	0.537	25–34	0.330
Regular	−1.188	35–44	0.199
Bad	−2.043	45–54	−0.071
Very bad	−2.076	55–64	−0.396
		65–74	−0.541
		75+	−0.658

The optimal scaling problem can be turned around by making a similar search for scale values for the age groups which maximize the variance of the health categories. The solution is given by the vertex coordinates for the age groups, and the scores for the health categories are their profile coordinates. The symmetry in the row and column problems is discussed further in the next chapter. This symmetry, or *duality*, of the scaling problems has led to calling the method *dual scaling*, a synonym for this form of correspondence analysis.

The optimal scale does not position the five health categories at equal distances from one another, like the original integer scale. Exhibit 6.5 showed that there is a big difference between *good* and *regular* and a very small difference between *bad* and *very bad*. These scale values lead to average health scores for the age groups that are the most separated in terms of the variance criterion, in other words we have the maximum discrimination between the age groups using the optimal scale for the health categories. In Exhibit 6.3, which displays only the age group scores, we can see that there are small changes in self-assessed health up to the age group 34-45 years, followed by large changes in the middle age categories, especially from 45–54 to 55–64 years, and then slower changes in the older groups. Checking back to the profile data in Exhibit 6.2, we can verify that from the 45–54 to 55–64 age group there is an approximate 50% drop in the *very good* category and a more than doubling of the *bad* category, which accounts for this large change in the scores.

Interpretation of optimal scale

The optimal health category scale values obtained are 1.144, 0.537, −1.188, −2.043 and −2.076 respectively (Exhibit 7.1). These numbers are calculated under certain restrictions which are required in order that a unique solution can be found. These restrictions are that, for all 6371 respondents, the average on the health scale is 0 and the variance is 1:

Identification conditions for an optimal scale

$$(0.128 \times 1.144) + (0.556 \times 0.537) + \ldots + (0.016 \times -2.076) = 0 \qquad \text{(mean 0)}$$

and

$$(0.128 \times 1.144^2) + (0.556 \times 0.537^2) + \ldots + (0.016 \times (-2.076)^2) = 1 \qquad \text{(variance 1)}$$

These prerequisites for the scale values are known as *identification conditions* or *constraints* in the jargon of mathematical optimization theory. The first condition is necessary since it is possible for two different sets of scale values to have different means but the same variance, so that it would be impossible to fix (or identify) a solution without specifying the mean. The second condition is required because if we arbitrarily multiplied the scale values by any large number, the variance of the eventual scores would be greatly increased as well — this would make no sense at all since we are trying to maximize the variance. Hence, it is necessary to look for a solution amongst scale values which have a fixed mean and fixed range of variation. The "mean 0, variance 1" condition is a conventional choice in such a situation, and conveniently leads to the vertex coordinates in CA, which satisfy the same conditions.

To determine the optimal scale, the two identification conditions described above are simply technical devices to ensure a unique mathematical solution to our problem. Having obtained the scale values, however, we are at liberty to transform them to a more convenient scale, as long as we remember that the mean and variance of the transformed scale are chosen for convenience and have no substantive or statistical relevance. The redefinition of this scale is usually performed by fixing the endpoints at some substantively meaningful

Any linear transformation of the scale is still optimal

values. For example, in the present case we could fix the *very bad* health category at 0 and the *very good* one at 100. So we need to make a transformation that takes the value of −2.076 to 0 and the 1.144 to 100. We can first add 2.076 to all five scale values, so that the lowest value is zero. The scale now ranges from 0 to 1.144 + 2.076 = 3.220. In order to have the highest value equal to 100, we then multiply all values by 100/3.220. So the formula in this particular case for computing a new scale value from the old one is simply

$$ new = (old + 2.076) \times \frac{100}{3.220} $$

or, in the general case,

$$ new = \left[(old - \text{old lower limit}) \times \frac{\text{new range}}{\text{old range}} \right] + \text{new lower limit} \qquad (7.5) $$

(in our example the new lower limit is zero). Applying this formula to all five optimal scale values results in the following transformed values (Exhibit 7.2):

Exhibit 7.2:
Optimal scale values from CA and the values transformed to lie between 0 and 100.

HEALTH CATEGORY	*Optimal scale value*	*Transformed scale value*
Very good	1.144	100.0
Good	0.537	81.1
Regular	−1.188	27.6
Bad	−2.043	1.0
Very bad	−2.076	0.0

The previous five-point integer scale with four equal intervals between the scale points would imply values 0 (*very bad*), 25, 50, 75, 100 (*very good*) on the scale with range 100. The optimal transformed values show that *regular* is not at the midpoint (50) of the scale, but much closer to the "bad" end of the scale.

Optimal scale is not unique

We should stress that the optimal scale depends on the criterion laid down for its determination as well as the chosen identification conditions. Apart from these purely technical issues, it clearly also depends on the particular cross-tabulation on which it is based. If we had a table which cross-tabulates health with another demographic variable, say education group, we would obtain a difference set of optimal scale values for the health categories, since they would now be optimally discriminating between the education groups.

A criterion based on row-to-column distances

Finally, in contrast to the maximization criterion described above for optimal scaling, we present a minimization criterion for finding scale values which also leads to the CA solution, based on the distances from each row to each column — in the present example these will be distances between the health categories and age groups. Firstly, imagine the health categories on any scale,

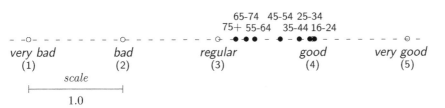

Exhibit 7.3:
The 1-to-5 scale of the health categories, and the corresponding weighted averages of the age groups.

for example the 1-to-5 integer scale from *very bad* to *very good* health, shown in Exhibit 7.3. Then the objective is to find the positions of the age groups on the same scale so that they come as "close" as possible to the health categories in the sense that an age group that has higher frequency of a particular health category tends to be closer on the scale to that category. Suppose the health category values are h_1, h_2,...,h_5 (in this initial example, the values 1 to 5) and the age group values a_1, a_2,...,a_7. The distance between an age group and a health category is the absolute difference $|a_i - h_j|$, but we will prefer to use the squared distance $(a_i - h_j)^2$ as a measure of closeness[†]. To make distances count more depending on the frequency of occurrence in the cross-tabulation each squared distance is weighted by p_{ij}, the relative frequency as defined in Chapter 4, page 31, i.e. the counts in Exhibit 6.1 divided by the grand total 6371 (hence all the p_{ij}'s sum to 1). Our objective would then be to minimize the following function:

$$\sum_i \sum_j p_{ij} d_{ij}^2 = \sum_i p_{ij}(a_i - h_j)^2 \qquad (7.6)$$

showing that distances should be shorter when p_{ij} is higher and longer when p_{ij} is lower. It is straightforward to show that, for any fixed set of health category points h_j, the minimum of (7.6) is achieved by the weighted averages for each age group. For the 1-to-5 scale values these weighted averages are just the set of age group scores calculated before (see formula (7.2) and the set of scores shortly afterwards), which are also shown in Exhibit 7.3. But the positions of the two sets of points in Exhibit 7.3 minimize (7.6) conditional on the fixed set of health categories, so the question is what the minimum would be over all possible configurations of scale values for the health categories. Again we need identification conditions for this question to make sense, otherwise the solution would simply put all health categories at the same point. If we add the same identification conditions that we had before, namely mean 0 and variance 1 for the scale values of the health categories, then the minimum is achieved by the optimal CA dimension once again.

Comparing the positions of the age groups in Exhibit 7.3 with the optimal positions in Exhibit 6.5, it is clear that the spread is higher in Exhibit 6.5,

---

[†] Again, as before, it is always easier to work with squared distances than distances — the square root in the Euclidean distance function causes many difficulties in optimization, and these disappear when we consider least-squares optimization.

which means that all age group points are closest to the health category points in terms of criterion (7.6). The value of the minimum achieved in Exhibit 6.5 is equal to 1 minus the (maximized) variance on the optimal CA dimension, and is sometimes referred to as the *loss of homogeneity* — we will return to this concept in Chapter 20 when discussing homogeneity analysis. Notice that the criterion (7.6) is easily generalized to two dimensions or more, say K dimensions, simply by replacing a_i and h_j by vectors of K elements and the squared differences $(a_i - h_j)^2$ by squared Euclidean distances in K-dimensional space.

SUMMARY:
Optimal Scaling

1. *Optimal scaling* is concerned with assigning scale values to the categories (or attributes) of a categorical variable to optimize some criterion which separates, or discriminates between, groups of cases, where these groups have been cross-tabulated with that variable.

2. The positions of the categories as vertex points on the optimal dimension of a CA provide optimal scale values in terms of a criterion that maximizes the variance between groups. The maximization is performed under the identification conditions that the mean and variance of the scale values are 0 and 1 respectively, which are conveniently satisfied by the vertex points.

3. The average scores for the groups are the projections of their profiles on this dimension and their maximized variance is equal to the inertia of these projected profiles.

4. The coordinate positions of the scale categories on the optimal dimension are standardized to mean 0 and variance 1, a standardization that is particular to the geometry of the vertex points in CA. For purposes of optimal scaling the mean and variance of the scale can be redefined; hence the scale values may be recentred and rescaled to conform to any scale convenient to the user, for example 0-to-1, or 0-to-100.

5. The optimal scale also satisfies a criterion based on the distances from each row point to each column point: that is, where the objective is to place the row and column points in a map such that the row-to-column distances, weighted by the frequencies in the contingency table, are minimized. This minimum, also called the loss of homogeneity, is equal to 1 minus the maximum variance achieved in optimal scaling.

Symmetry of Row and Column Analyses

## 8

In all the examples and analyses presented so far, we have dealt with the analysis of the rows of a table, visualizing the row profiles and using the columns as reference points for the interpretation: let's call this the "row analysis". All this can be applied in a completely symmetric way to the columns of the same table. This can be thought of as transposing the table, making the columns the rows and vice versa, and then repeating all the procedures described in Chapters 2 to 7. In this chapter we shall show that the row analysis and column analysis are intimately connected. In fact, if a row analysis is performed, then the column analysis is actually being performed as well, and vice versa. Correspondence analysis (CA) can thus be regarded as the simultaneous analysis of the rows and columns.

Contents

Let us again consider the data in Exhibit 6.1 on self-assessment of health. In Chapter 6 we performed the row analysis of these data since the object was to display the profiles of the age groups across the health categories. These seven profiles are contained in a four-dimensional space bounded by the five vertices that represent the extreme unit profiles corresponding to each health category. Then we diagnosed that most of the spatial variation of the profiles was along a straight line (Exhibit 6.3). The profiles were projected onto that line and the relative positions of these projections were interpreted as well as the projections of the five vertices (Exhibit 6.5).

Summary of row analysis

Exhibit 8.1:
Column profiles of
health categories
across the age
groups, expressed as
percentages.

AGE GROUP	Very good	Good	Regular	Bad	Very bad	Average
16–24	29.7	22.3	11.2	4.3	5.8	*19.2*
25–34	26.9	22.8	11.0	8.5	5.8	*19.4*
35–44	18.0	18.6	12.1	9.9	7.8	*16.2*
45–54	11.0	13.2	15.8	12.1	15.5	*13.5*
55–64	6.5	11.7	20.5	25.6	29.1	*14.3*
65–74	5.4	7.5	19.0	23.7	19.4	*11.2*
75+	2.4	3.8	10.5	15.9	16.5	*6.2*
Sum	*100*	*100*	*100*	*100*	*100*	*100*

Column analysis — profile values have symmetric interpretation

We now consider the alternative problem of displaying the column profiles of Exhibit 6.1, i.e. the profiles of the health categories across the age groups, shown as percentages in Exhibit 8.1. The column profiles give, for each health category, the percentages of respondents across the age groups: for example, in the *bad* health category, 4.3% are 16–24 years, 8.5% 25–34 years, and so on. Although this table of column profiles looks completely different to the row profiles in Exhibit 6.2, when we look at specific values and compare them to their averages we can see that they contain the same information (we already noticed this in Chapter 2, page 11, for the travel data set). For example, consider the value in the *bad* column for the age group 65–74 years: 23.7%. Compare this value with the average percentage of 65–74 year olds in the whole sample, given in the last column: 11.2%. Thus we can conclude that in the 65–74 age group there are just over twice as many respondents saying their health is *bad* compared to the overall percentage in this age group — in fact, the ratio is $23.7/11.2 = 2.1$. If we look at the same cell of Exhibit 6.2, we see that, of the 65–74 year olds, 13.7% assess their health as *bad*, while the percentage of *bad* responses in the sample is 6.5% (last row of Exhibit 6.2). Again, compared to the marginal percentage, just over twice as many say their health is *bad*, and in fact the ratio is identical: $13.7/6.5 = 2.1$. We can show this result theoretically, using the notation at the end of Chapter 4: the first ratio we computed was $(p_{ij}/c_j)/r_i$, whereas the second was $(p_{ij}/c_j)/r_i$, both equal to $p_{ij}/(r_i c_j)$ — this ratio, called the *contingency ratio*, is an important concept in the theory of correspondence analysis (see the last section of Chapter 22 and the Theoretical Appendix, pages 244 and 250).

Column analysis — same total inertia

In Chapter 4 it was shown that the total inertia of the column profiles is equal to the total inertia of the row profiles — the two calculations are just alternative ways of writing the same formula, the χ^2 statistic divided by the sample size. For the health assessment data, the total inertia is equal to 0.1404.

Column analysis — same dimensionality

The column profiles define a cloud of five points, each with seven components, which should then lie in a space of dimension six, using the same argument as before because the components add up to 1. It turns out, however, that the five points do not fill all six dimensions of this space, but only four of the dimensions. One way to grasp this fact intuitively is to realize that two points lie exactly on a one-dimensional line, three points lie in a two-dimensional plane, four points lie in a three-dimensional space, and so five points lie in a four-dimensional space. Hence, although the row profiles and column profiles lie in different spaces, the dimensionality of these two clouds of points is identical — in this case it is equal to four. This is the first geometric way in which the analyses of the row and of the column profiles are the same. Many more similarities will soon become apparent.

Still considering the five health category profiles in four-dimensional space, we now ask the same question as before: Can these points be approximately displayed in a lower-dimensional subspace and what is the quality of this approximation? By performing an analogous set of computations as was required in Chapter 6, it turns out that the column profiles are well approximated by a one-dimensional line, and the quality of the approximation is 97.3%, exactly the same percentage that was obtained in the case of the row profiles. This is the second geometric property that is common to the two analyses.

Column analysis — same low-dimensional approximation

Exhibit 8.2: *Optimal one-dimensional map of the health category profiles.*

The projections of the column profiles onto the best-fitting line are shown in Exhibit 8.2. Here we see that the health categories lie in an order which concurs exactly with the positions of the vertices in Exhibit 6.5. The actual values of their coordinate positions are not the same but the relative positions are identical. According to the scale of Exhibit 8.2 and comparing the positions of the health categories in Exhibit 6.5, it appears that the coordinates of the profiles are a contracted, or shrunken, version of the vertices. A specific interpretation of this "contraction factor" will be given soon. Furthermore, in Exhibit 8.3 the projections onto this line of the seven outer vertices, representing the age groups, are displayed. Comparing the positions of the vertices here with those of the age group profiles in Exhibit 6.5 (or Exhibit 6.3 where the scale is larger) reveals exactly the same result for the rows — the positions

Column analysis — same coordinate values, rescaled

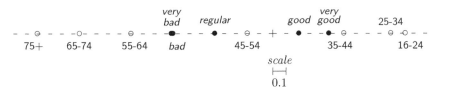

Exhibit 8.3: *Same map as Exhibit 8.2, showing the projected positions of the age group vertices.*

of the row profiles with respect to their best-fitting line in Exhibit 6.5 are a
contracted version of the positions of the age group vertices projected onto
the best-fitting line of the health category profiles in Exhibit 8.2. Putting this
the opposite way, the positions of the row vertices in the column analysis are a
simple expansion of the positions of the row profiles in the row analysis. This
is the third and most important way in which the two analyses are related
geometrically.

Principal axes The best-fitting line in each analysis is called a *principal axis*. More specifically
and principal it is referred to as the *first* principal axis, since there are other principal axes,
inertias as we shall see in the following chapters. We have seen that in both row and
column analyses the total inertia is equal to 0.1404 and that the percentage of
inertia accounted for by the first axis is 97.3%. The specific part of inertia that
is accounted for by the first axis is equal to 0.1366 in both cases, which gives
the percentage explained as $100 \times 0.1366/0.1404 = 97.3\%$. The inertia amount
(0.1366) accounted for by a principal axis is called a *principal inertia*, in this
case the first principal inertia because it refers to the first principal axis. It is
also often called an *eigenvalue* because of the way it can be calculated, as an
eigenvalue of a specific square symmetric matrix (see Theoretical Appendix,
pages 243–244).

Scaling factor It seems, then, that we have to do only one analysis — either the row analysis
is the square root or the column analysis. The results of the one can be obtained from those of
of the principal the other. But what is the exact connection between the two; in other words
inertia what is the scaling factor which can be used to pass from vertex positions in
one analysis to profile positions in the other? This scaling factor turns out to be
equal to the square root of the principal inertia itself (i.e. the square root of an
eigenvalue, which is also called a *singular value*, as explained in the Theoretical
Appendix A, page 244); e.g. in this example it is $\sqrt{0.1366} = 0.3696$. Thus to
pass from the row vertices in Exhibit 8.3 to the row profiles in Exhibits 6.3
or 6.5, we simply multiply the coordinate values by 0.3696, which is just
over one-third. Conversely, to pass from the column profiles in Exhibit 8.3 to
the column vertices in Exhibit 6.5, we multiply the coordinate values by the
inverse, namely $1/0.3696 = 2.706$. The numerical values of all the profile and
vertex coordinates are given in Exhibits 7.1 and 8.4, and the following simple
relationship for both rows and columns can easily be verified comparing these
two exhibits:

$$\text{profile coordinate} = \text{vertex coordinate} \times \sqrt{\text{principal inertia}}$$

Notice in Exhibits 6.5 and 8.3 that the profile points are more bunched up than
the vertex points. The scaling factor is a direct measure of how bunched up the
"inner" profiles are compared to the "outer" vertices. In this case, the scaling
factor of 0.3696 implies that the spread of the profiles is about one-third that
of the vertices. At the end of Chapter 4 the total inertia was interpreted as

HEALTH CATEGORY	Profile coordinate	AGE GROUP	Vertex coordinate
Very good	0.423	16–24	1.004
Good	0.198	25–34	0.893
Regular	−0.439	35–44	0.538
Bad	−0.755	45–54	−0.192
Very bad	−0.767	55–64	−1.070
		65–74	−1.463
		75+	−1.782

Exhibit 8.4:
Coordinate values of the points in Exhibit 8.2, i.e. the coordinates of the column profiles and the row vertices on the first principal axis of the column profiles (cf. Exhibit 7.1).

the amount of dispersion in a set of profiles relative to the outer vertices (see Exhibit 4.2). The principal inertias (or their square roots considered here) are also measures of dispersion but refer to individual principal axes rather than to the whole profile space. The higher the principal inertia is, and thus the higher the scaling factor is, the more spread out the profiles are relative to the vertices, along the respective principal axis. It should now be obvious that a principal inertia cannot be greater than 1 — the profiles must be in positions "interior" to their corresponding vertices.

The square root of the principal inertia, which as we already pointed out is always less than 1, has an alternative interpretation as a correlation coefficient. A correlation coefficient is usually calculated between pairs of measurements, for example the correlation between income and age. In the present example there are two observations on each respondent — age group and health category — but these are categorical observations, not measurements. A correlation coefficient between these two variables can be computed using the default integer codes of 1 to 7 for the age groups and 1 to 5 for the health categories. The correlation is then computed to be 0.3456. Using any other set of scale values would give a different correlation, so the following question arises: Which scale values can be assigned to the age groups and health categories such that the correlation is the highest? The maximum correlation found in this way is sometimes called a *canonical correlation*. In the present example, the canonical correlation turns out to be 0.3696, exactly the square root of the principal inertia, i.e. the scaling factor linking the row and column analyses. The scale values for the age groups and the health categories that yield this maximum correlation are just the coordinate values of the age groups and health categories on the first CA principal axis, given in Exhibits 7.1 and 8.4 and displayed in Exhibits, 6.3, 6.5, 8.2 and 8.3. We can use profile or vertex coordinates, since correlation is unaffected by recentring or rescaling the scale values. It is conventional to use standardized scale values, with mean 0 and variance 1, to identify the solution uniquely, i.e. the coordinate values of the vertices.

Correlation interpretation of the principal inertia

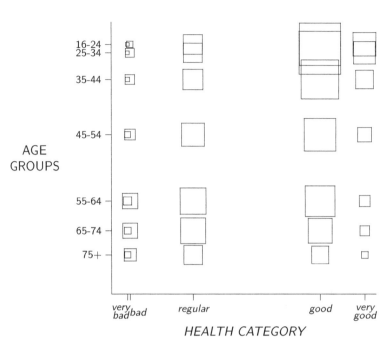

Exhibit 8.5:

Scatterplot according to the scale values that maximize the correlation between health category and age group; squares are shown at each combination of values, with area proportional to the number of respondents. The correlation is equal to 0.3696, the square root of the first principal inertia.

Graph of the correlation

A correlation between two variables is usually illustrated graphically by a scatterplot of the cases, e.g. age group (y-axis) by health category (x-axis). Although we have 6371 cases in the scatterplot, there are only 7 values along the y axis and 5 values along the x axis, thus only $7 \times 5 = 35$ possible points in this scatterplot (Exhibit 8.5). At a specific point corresponding to a health category and age group lie all the cases in the respective cell of the original cross-table (Exhibit 6.1), displayed here in the form of a square with an area proportional to the cell frequency. The canonical correlation is then the usual Pearson correlation of all 6371 cases in this scatterplot. The optimal property of the canonical correlation means that there is no other way of scaling the row and column categories which would yield a higher correlation coefficient in such a scatterplot. A canonical correlation of 1 would be attained if all points were lying on a straight line, which means that each age group is associated with only one health category (i.e. the profiles are all unit profiles, or vertex points).

Principal coordinates and standard coordinates

It is convenient to introduce some terminology at this stage to avoid constant repetition of the phrases "coordinate positions of the vertices" and "coordinate positions of the profiles". Since the former coordinates are standardized to have mean 0 and variance 1, we call them *standard coordinates*. Since the latter coordinates refer to the profiles with respect to principal axes, we call them *principal coordinates*. For example, the first column of numerical results in Exhibit 8.4 contains the principal coordinates of the health categories (columns), while the second column contains the standard coordinates of the age groups. In both cases these are coordinates on the first principal axis of the CA; in future chapters we shall usually have more than one principal axis.

Another way of thinking about the correlation definition of CA is that each of the 6371 individuals in the health survey example can be assigned a pair of scale values, one (a_i, say) for age group and one (h_j) for health category. As before, these scale values are unknown but we can define a criterion to optimize in order to determine them. Each individual has a score equal to the sum of these two scale values, $a_i + h_j$; for example, someone in the 25–34 age group with **very good** health (second age group and first health category) would have a score of $a_2 + h_1$. Suppose that the correlation of the pairs of values $\{a_i, a_i + h_j\}$ is denoted by $\mathrm{cor}(a, a + h)$, where a and h denote the 6371 scale values for the whole sample; similarly, the correlation for the pairs $\{h_j, a_i + h_j\}$ is denoted by $\mathrm{cor}(h, a + h)$. A criterion to optimize would be to find the scale values that optimize these two correlations in some way. It can be shown that the first dimension of the CA solution gives scale values that are optimal in the sense that they maximize the average of the squares of these correlations:

$$\text{average squared correlation} = \frac{1}{2}[\mathrm{cor}^2(a, a + h) + \mathrm{cor}^2(h, a + h)] \qquad (8.1)$$

Since $\mathrm{cor}(X, X + Y) = \sqrt{[1 + \mathrm{cor}(X, Y)]/2}$ for any two standardized variables X and Y, the average squared correlation in (8.1) is equal to:

$$\text{average squared correlation} = \frac{1 + \mathrm{cor}(a, h)}{2} \qquad (8.2)$$

Therefore, when the CA solution maximizes $\mathrm{cor}(a, h)$, i.e. the canonical correlation, it also maximizes (8.2), equivalently (8.1). This result will be useful later because it can easily be generalized to more than two variables — see Chapter 20.

Yet another equivalent way of expressing the optimality of the CA solution is as follows, using the notation of the previous section. Instead of sums of scale values, calculate an average for each person: $\frac{1}{2}(a_i + h_j)$. Then calculate the respective differences between each person's age value and health value with respect to the average: $a_i - \frac{1}{2}(a_i + h_j)$ and $h_j - \frac{1}{2}(a_i + h_j)$. A measure of how similar the age values are to the health values is the sum of squares of these two differences, again averaged, leading to a measure of variance of the two values a_i and h_j:

$$\text{variance (for one case)} = \frac{1}{2}\left([a_i - \frac{1}{2}(a_i + h_j)]^2 + [h_j - \frac{1}{2}(a_i + h_j)]^2\right) \qquad (8.3)$$

The term *homogeneity* is used in this context because if the scale values a_i and h_j were the same, their variance would be zero; hence an individual with such a pair of values is called *homogeneous*. An alternative term for homogeneity is *internal consistency*. Averaging the values (8.3) for the whole sample, we obtain an amount which is called the *loss of homogeneity* (see page 56, where this term was used in the same sense). If all the age values coincided with the health values, the loss of homogeneity would be zero, that is the sample would be completely homogeneous (or internally consistent). The aim is to find scale values which minimize this loss, and once more the solution coincides with the coordinates of the age and health points on the first CA dimension. Again it is clear that this definition is easily extended to more than two variables, as we shall do in Chapter 20.

SUMMARY:
Symmetry of Row
and Column
Analyses

1. The series of operations and displays in the row analysis can be performed in a completely symmetric fashion to the columns, as if the table were transposed and everything repeated.

2. The column analysis thus leads to the visualization of the column profiles in their optimal subspace of display, along with the display of the vertices representing the rows.

3. In either analysis the best-fitting line, or dimension, is called the first *principal axis* of the profiles. The amount of inertia this dimension accounts for is called the first *principal inertia*.

4. These two "dual" analyses are equivalent in the sense that each has the same total inertia, the same dimensionality and the same part of inertia along the first principal axis in each analysis (in the following chapter, this last property extends to additional principal axes).

5. The coordinate positions of profiles with respect to a principal axis are called *principal coordinates* and the coordinate positions of vertices with respect to a principal axis are called *standard coordinates*.

6. Furthermore, the profiles and vertices in the two analyses are intimately related as follows: along the first principal axis, for example, profile positions (in principal coordinates) have exactly the same relative positions as the corresponding vertices (in standard coordinates) in the dual analysis, but are reduced in scale. The scaling factor involved is the square root of the principal inertia along that axis, which is always less than 1.

7. The scaling factor (or its square, the principal inertia itself) quantifies how spread out the row profiles are along a principal axis compared to the outer column vertices, and equivalently how spread out the column profiles are compared to the outer row vertices.

8. This scaling factor can also be interpreted as a *canonical correlation*. It is the maximum correlation that can be attained between the row and column variables as a result of assigning numerical quantifications to the categories of these variables.

Two-Dimensional Displays

## 9

We have discussed at some length the projections of a cloud of profiles onto a single principal axis, the best-fitting straight line. In practice you will find that most of the reported correspondence analysis (CA) displays are two-dimensional, usually with the first principal axis displayed horizontally (the x-axis) and the second principal axis vertically (the y-axis). In general, the projections may take place onto any low-dimensional subspace, but the two-dimensional case is, of course, rather special because of our two-dimensional style of displaying graphics on computer screens or on paper. In the Computational Appendix there are also some examples using the R programming language to do CA graphics in three dimensions (e.g. Exhibit B.4 on page 268; see also pages 257–258).

Contents

The next example, which appeared originally in my 1984 book *Theory and Applications of Correspondence Analysis*, has been adopted as a test example in many implementations of CA in major commercial statistical packages. This example still serves as an excellent introduction to two-dimensional displays and has also been referred to in several journal articles, even though it is an artificial data set. It concerns a survey of all 193 staff members of a fictitious company, in order to formulate a smoking policy. The staff members are cross-tabulated according to their rank (five levels) and a categorization of their smoking habits (four groups) — the contingency table is reproduced in Exhibit 9.1. Because it is a 5×4 table, its row profiles and column profiles lie exactly in three-dimensional spaces.

Data set 4: Smoking habits of staff groups

Exhibit 9.1:
*Cross-tabulation of
staff group by
smoking category,
showing row profiles
and average row
profile in
parentheses, and the
row masses.*

STAFF GROUPS	SMOKING CATEGORIES				Row Totals	Masses
	None	*Light*	*Medium*	*Heavy*		
Senior Managers SM	4 (0.364)	2 (0.182)	3 (0.273)	2 (0.182)	*11*	*0.057*
Junior Managers JM	4 (0.222)	3 (0.167)	7 (0.389)	4 (0.222)	*18*	*0.093*
Senior Employees SE	25 (0.490)	10 (0.196)	12 (0.235)	4 (0.078)	*51*	*0.279*
Junior Employees JE	18 (0.205)	24 (0.273)	33 (0.375)	13 (0.148)	*88*	*0.456*
Secretaries SC	10 (0.400)	6 (0.240)	7 (0.280)	2 (0.080)	*25*	*0.130*
Total	*61*	*45*	*62*	*25*	*193*	
Average Profile	*(0.316)*	*(0.233)*	*(0.321)*	*(0.130)*		

Row analysis   As before, this table may be thought of as a set of rows or a set of columns. We assume that the row analysis is more relevant; that is, we are interested in displaying for each staff group what percentage are non-smokers, what percentage are light smokers, and so on. The row profile space is a four-pointed simplex, a tetrahedron, in three dimensions, which is the three-dimensional equivalent of the triangular space in two dimensions (this can be seen using the three-dimensional graphics described in the Computational Appendix). To reduce the dimensionality of the profiles, they should be projected onto the best-fitting plane (see Exhibit 6.6 on page 46). The map, shown in Exhibit 9.2, also shows the projections of the four vertex points representing the smoking groups. Notice that the first principal axis customarily defines the horizontal axis of the map, and the second principal axis the vertical axis. On the axes the respective principal inertias are given (0.07476 and 0.01002 respectively), as well as the corresponding percentages of inertia. These values can be accumulated to give the amount and percentage of inertia accounted for by the plane of the two axes. Thus the inertia in the plane is 0.08478, which is 99.5% of the total inertia of 0.08519. This means that by sacrificing one dimension we have lost only 0.5% of the inertia of the profile points. Putting this another way, the five row profiles lie very close to this plane of representation, so close that we can effectively ignore their distance from the plane when interpreting their relative positions.

*Interpretation
of row profiles and
column vertices*   Looking only at the profiles' positions for a moment, we can see that the groups farthest apart are Junior Employees (JE) and Junior Managers (JM) on the left-hand side, opposed to Senior Employees (SE) on the right-hand side — hence the greatest differences in smoking habits are between these extremes. Senior Managers (SM) appear to lie between Junior Managers and

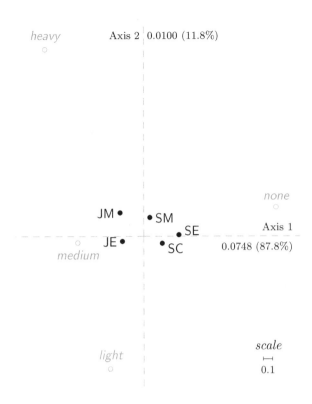

Exhibit 9.2:
Optimal two-dimensional CA map of the smoking data of Exhibit 9.1, with rows in principal coordinates (projections of profiles) and columns in standard coordinates (projections of vertices).

Senior Employees, while Secretaries (SC) are quite close to Senior Employees. In order to explain the similarities and differences between the staff groups, it is necessary to inspect the positions of the profiles relative to the vertices. Since the three smoking categories are on the left and the non-smoking category is on the right, the left-to-right distinction is tantamount to smokers versus non-smokers. The groups JE and JM are different to SE because the former groups have relatively more smokers, and SE has relatively more non-smokers. The centre of such a display is always the average profile, so that we can also consider the deviations of the staff groups outwards from the average profile in different directions, the main deviations being from left to right.

Nesting of principal axes

The two-dimensional display is such that it actually contains the best one-dimensional display in it as well. If all the points in Exhibit 9.2 were projected vertically onto the horizontal axis, then this unidimensional display would be the one obtained by looking for the best one-dimensional display right from the start. The principal axes are said to be *nested*, in other words an optimal display of a certain dimensionality contains all the optimal displays of lower dimensionality. Notice that the three smoking groups on the left will project very close together on the first axis, a long way from the non-smoking point on the right. This is the greatest single feature in the data. Putting this in the optimal scaling terminology of Chapter 7, a "smoking scale" which best differentiates the five staff groups is not one which assumes equal intervals between the four smoking categories, but rather one which places all three smoking categories quite close to one another but far from the non-smoking category, effectively a smoking/non-smoking dichotomy.

Interpretation of second dimension

Continuing with the two-dimensional interpretation, we see that the second (vertical) principal axis pulls apart the three smoking levels. The profiles do not differ as much vertically as horizontally, as indicated by the much lower percentage of inertia on the second axis. Nevertheless, we can conclude that the profile of JE has relatively more light smokers than heavy smokers compared to that of JM, even though both these groups have similar percentages of smokers as seen by their similar positions on the horizontal axis.

Verifying the profile–vertex interpretation

Each row profile point (staff group) is at a weighted average position of the column vertex points (smoking categories), where the weights are the elements of the respective row profile. As a general rule, assuming that the display is of good quality, which is true in this case, the closer a profile is to that vertex, the higher its profile value is for that category. One way of verifying the interpretation of the positions of the profiles relative to the vertices is to measure the profile-to-vertex distances in Exhibit 9.2 and then compare these to the profile values. This verification should be performed one vertex at a time, for example the five distances from the staff groups to the vertex *light*. Therefore, the interpretation that we made in the previous paragraph can be confirmed in another way: because JE lies more towards *light* than JM, JE should have relatively more light smokers than JM. The actual data are that 24/88 or 27% of JEs are light smokers, whereas 3/18 or 17% of JMs are light smokers, so this agrees with our interpretation. Exhibit 9.3 graphically compares all profile-to-vertex distances to their corresponding profile values. The abbreviation 42, for example, is used for JE-to-*light* (row 4, column 2) and 22 for JM-to-*light* (row 2, column 2). Clearly, the higher profile element of 0.27 for 42 corresponds to a smaller distance than the profile of 0.17 for 22. For each vertex, we say that the profile elements are *monotonically inversely* related to the profile-to-vertex distances, which in graphical terms means that each set of five points in Exhibit 9.3 corresponding to a particular vertex forms a descending pattern from top left to bottom right. For example, the set of five points corresponding to the fourth vertex point (*heavy*), with labels 34, 54, 44, 14 and 24, are arranged in such a descending sequence.

Asymmetric maps

We say that the Exhibit 9.2 is an *asymmetric map*, or a map which is *asymmetrically scaled*, because it is the joint display of profile and vertex points. In an asymmetric map, therefore, one of the sets of points, in this case the rows, is scaled in principal coordinates, while the other is scaled in standard coordinates. If we were more interested in the column analysis, then the column

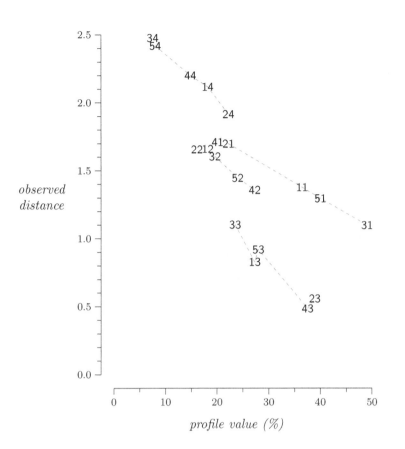

observed
distance

profile value (%)

Exhibit 9.3:
The measured profile-to-vertex distances in Exhibit 9.2 plotted against the corresponding values of the row profiles of Exhibit 9.1. Each row-column pair is labelled with their respective category numbers: for example, row profile 3 (senior employees) and column vertex 4 (heavy smoking) are denoted by **34**. *Notice the descending pattern with increasing profile value for each set of distances corresponding to a particular vertex, with some small exceptions.*

points would be in principal coordinates and the row points in standard coordinates. What we said in Chapter 8 about the scaling factor between the row and column problems holds for each principal axis. Thus the two-dimensional display of the column profiles would be a shrunken version of the positions of the column vertices given in Exhibit 9.2, but the "shrinking factors" (i.e., the canonical correlations, equal to the square roots of the principal inertias) along the two axes are not the same: $\sqrt{0.07476} = 0.273$ and $\sqrt{0.01002} = 0.100$ respectively. Thus along the first axis the shrinking is by a factor of 0.273 (i.e. just over a quarter) and along the second axis by a factor of 0.1 (i.e. a tenth). By the same argument, to pass from the row profiles in Exhibit 9.2 to their vertex positions in the column problem we would simply expand them nearly fourfold along the first axis and tenfold along the second axis. Apart from these scaling factors the relative positions of the profiles and the vertices are the same. Exhibit 9.4 shows the other possible asymmetric map, where the columns are represented as profiles in principal coordinates and the rows as vertices in standard coordinates. In this map the column points are at weighted averages of the row points using the elements of the column profiles as weights. The asymmetric map of Exhibit 9.2 is often called the *row prin-*

Exhibit 9.4:
*Asymmetric CA
map of the smoking
data of Exhibit 9.1,
with columns in
principal
coordinates and
rows in standard
coordinates.*

cipal map (because row points are in principal coordinates) and Exhibit 9.4 the *column principal* map.

Symmetric map   Having gone to great lengths to explain the geometry of asymmetric displays, we now introduce an alternative way of mapping the results, called the *symmetric map*. This option is by far the most popular in the CA literature, especially amongst French researchers. In a symmetric map the separate configurations of row profiles and column profiles are overlaid in a joint display, even though they emanate, strictly speaking, from different spaces. In a symmetric map, therefore, both row and column points are displayed in principal coordinates. Exhibit 9.5 shows the symmetric map of the smoking data, and is thus an overlay of the two sets of "inner" points in black in Exhibits 9.2 and 9.4. This simultaneous display of rows and columns finds some justification in the intimate relationship between the row and column analyses, involving a simple scaling factor between profiles and corresponding vertices. The convenience of such a display is that, whatever the absolute level of association might be, we always have both clouds of points equally spread out across the plotting area, hence there is less possibility of overlapping labels in the display. In asymmetric maps, by contrast, the profile points (which are usually the points of primary interest) are often bunched up in the middle of the display, far from the outer vertices, and the visualization is generally less aesthetic.

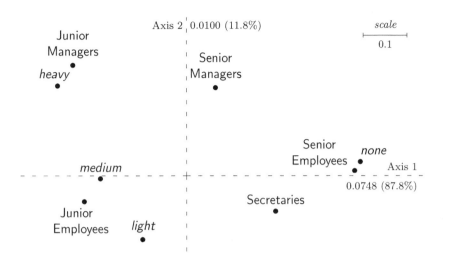

Exhibit 9.5:
Symmetric map of smoking data; both rows and columns are in principal coordinates.

Since both clouds of profiles are displayed simultaneously in Exhibit 9.5, the plotted row-to-row distances approximate the inter-row χ^2-distances and the plotted column-to-column distances approximate the inter-column χ^2-distances. Of course, the inter-row distance interpretation applies to the points in Exhibit 9.2 as well, since this is the identical display of the rows which is used in Exhibit 9.5 (note the difference in scales between these two maps) — a similar remark applies to the column points in Exhibit 9.4. The interpoint χ^2-distances can be verified by plotting the observed distances versus the true ones (Exhibit 9.6). For the five row points there are $10 = 5 \times 4/2$ interpoint distances, and for the four column points there are $6 = 4 \times 3/2$ interpoint distances. There is an excellent agreement, which was to be expected since the quality of display of the profiles is 99.5% in both cases.

Verification of interpoint chi-squared distances in symmetric map

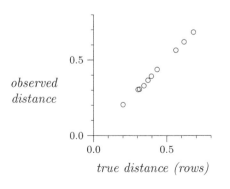

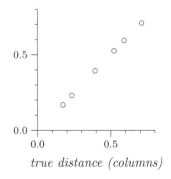

Exhibit 9.6:
Observed interpoint row distances and interpoint column distances measured in Exhibit 9.5, plotted against the true χ^2-distances between the row profiles and between the column profiles, respectively, of Exhibit 9.1.

Danger in interpreting row-to-column distances in a symmetric map

There is a price to pay for the convenience of the symmetric map which comes in the form of a danger in interpreting row-to-column distances directly. No such distance is defined or intended in this map. This is the aspect of CA that is often misunderstood and has caused some confusion amongst users who would like to make clusters of row and column points in a symmetric map (see the Epilogue, page 296). Strictly speaking, it is not possible to deduce from the closeness of a row and column point the fact that the corresponding row and column necessarily have a high association. Such an interpretation is justified to a certain extent in the case of the asymmetric map, as illustrated in Exhibit 9.3. A golden rule in interpreting maps of this type is that interpoint distances can be interpreted whenever the points concerned are situated in the same space, for example row profiles along with the vertex points representing the columns in the row profile space. When interpreting a symmetric map, the fact that this is the overlay of two separate maps should always be borne in mind. In Chapter 13 the row–column interpretation called the "biplot" will be described — this is the more accurate way of thinking about the joint display of rows and columns.

SUMMARY: Two-Dimensional Displays

1. As the dimensionality of the subspace of display is increased, so the capacity of the display to represent the profile points accurately is improved. There is, however, a trade-off in the sense that the visualization of the points becomes more and more complex beyond two dimensions. Two-dimensional displays are usually the displays of choice.

2. The principal axes are *nested*; i.e. the first principal axis found in the one-dimensional solution is identical to the first principal axis in the two-dimensional solution, and so on. Increasing the dimensionality of the display simply implies adding new principal axes to those already found.

3. An *asymmetric map* is one in which the row and column points are scaled differently, e.g. the row points in principal coordinates (representing the row profiles) and the column points in standard coordinates (representing the column vertices). There are thus two asymmetric plots possible, depending on whether the row or column analysis is of chief interest.

4. In an asymmetric map where the rows, for example, are in principal coordinates (i.e. the row analysis), distances between displayed row points are approximate χ^2-distances between row profiles; and distances from the row profile points to a column vertex point are, as a general rule, inversely related to the row profile elements for that column.

5. A more common type of display, however, is the *symmetric map* where both rows and columns are scaled in principal coordinates.

6. In a symmetric map, the row-to-row and column-to-column distances are approximate χ^2-distances between the respective profiles. There is no specific row-to-column distance interpretation in a symmetric map.

Three More Examples

## 10

To conclude these first 10 introductory chapters, three additional applications of correspondence analysis (CA) are now given: (i) a table which summarizes the classification of scientists from 10 research areas into different categories of research funding; (ii) a table of counts of 92 marine species at a number of sampling points on the ocean floor; (iii) a linguistic example, where the letters of the alphabet have been counted in samples of texts by six English authors. In the course of these examples we shall discuss some further issues concerning two-dimensional displays, such as the interpretation of dimensions, the difference between asymmetric and symmetric maps, and the importance of the aspect ratio of the map.

Contents

The data come from a scientific research and development organization which has classified 796 scientific researchers into five categories for purposes of allocating research funds (Exhibit 10.1). The researchers are cross-classified according to their scientific discipline (the 10 rows of the table) and funding category (the 5 columns of the table). The categories are labelled *A*, *B*, *C*, *D* and *E*, and are in order from highest to lowest categories of funding. Categories *A* to *D* are for researchers who are receiving research grants, from *A* (most funded) to *D* (least funded), while *E* is a category assigned to researchers whose applications have not been successful (i.e. funding application rejected).

Data set 5: Evaluation of scientific researchers

Exhibit 10.1:
Frequencies of
funding categories
for 796 researchers
who applied to a
research agency: A is
the most funded, D
is the least funded
and E is not funded.
The last row shows
the average row
profile, i.e. relative
frequencies of the
column sums, in
percentage form.

SCIENTIFIC AREAS	FUNDING CATEGORIES					
	A	*B*	*C*	*D*	*E*	*Sum*
Geology	3	19	39	14	10	85
Biochemistry	1	2	13	1	12	29
Chemistry	6	25	49	21	29	130
Zoology	3	15	41	35	26	120
Physics	10	22	47	9	26	114
Engineering	3	11	25	15	34	88
Microbiology	1	6	14	5	11	37
Botany	0	12	34	17	23	86
Statistics	2	5	11	4	7	29
Mathematics	2	11	37	8	20	78
Sum	*31*	*128*	*310*	*129*	*198*	*796*
Average row profile	*3.9%*	*16.1%*	*38.9%*	*16.2%*	*24.9%*	

This 10×5 table lies exactly in four-dimensional space and the decomposition of inertia along the four principal axes is as follows:

Dimension	Principal inertia	Percentage of inertia
1	0.03912	47.2%
2	0.03038	36.7%
3	0.01087	13.1%
4	0.00251	3.0%

Each axis accounts for a part of the inertia, expressed as a percentage. Thus the first two dimensions account for almost 84% of the inertia. The sum of the principal inertias is 0.08288, so the χ^2 statistic is $0.08288 \times 796 = 65.97$. If one wants to perform the statistical test using the χ^2 distribution with $9 \times 4 = 36$ degrees of freedom, this value is highly significant ($p = 0.002$).

Exhibit 10.2 shows the asymmetric map of the row profiles and the column vertices. In this display we can see that the magnitude of the association between the disciplines and the research categories is fairly low; in other words the profiles do not deviate too much from the average (cf. Exhibit 4.2). This situation is fairly typical of social science data, so the asymmetric map is not so successful because all the profile points are bunched up in the middle of the display — in fact, they are so close to one another that we cannot write the full labels and have just put the first two letters of each discipline. Nevertheless, we can interpret the space easily looking at the positions of the vertices. The horizontal dimension lines up the four categories of funding in their inherent ordering, from *D* (least funded) to *A* (most funded), with *B* and *C* close together in the middle. The vertical dimension opposes category *E* (not funded) against the others, so the interpretation is fairly straightforward. The more a discipline is high up in this display the less its researchers are granted funding. The more a discipline lies to the right of this display, the more funding its funded researchers receive. Using marketing research terminology,

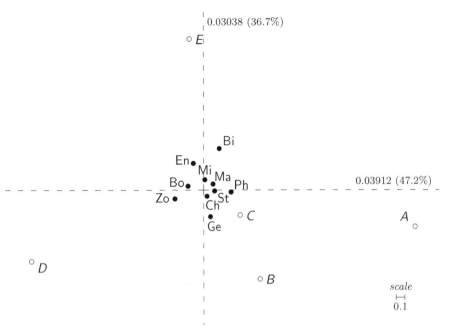

0.03038 (36.7%)

0.03912 (47.2%)

scale
⊢⊣
0.1

Exhibit 10.2:
*Asymmetric CA
map of the row
profiles of Exhibit
10.1 (scientific
funding data).*

the "ideal point" is in the lower right of the map: more grant applications accepted (low down), and those accepted receiving good classifications (to the right). Hence, if we were doing a trend study over time, disciplines would need to move towards the lower right-hand side to show an improvement in their funding status. At the moment there are no disciplines in this direction, although Physics is the most to the right (highest percentage — 10 out of 114, or 8.8% — of type *A* researchers). But Physics is at the middle vertically since it has a percentage of non-funded researchers close to average (26 out of 114 not funded, or 22.8%, compared to the average of 198 out of 796, or 26.5%).

Exhibit 10.3 shows the symmetric map of the same data, so that the only difference between this display and that of Exhibit 10.2 is that the column profiles are now displayed rather than the column vertices, leading to a change in scale which magnifies the display of the row profiles. This zooming in on the configuration of disciplines facilitates the interpretation of their relative positions and also gives space for fuller labels. The relative positions of the disciplines can now be seen more easily: for example, Geology, Statistics, Mathematics and Biochemistry are all at a similar position on the first axis, but widely different on the second. This means that the researchers in these fields whose grants have been accepted have similar positions with respect to the funded categories *A* to *D*, but Geology has many fewer rejections (11.8% of category *E*) than Biochemistry (41.4%). In this symmetric display we cannot assess graphically the overall level of association (inertia) between the rows and the columns. This can be assessed only from the numerical value of the principal

Symmetric map

Exhibit 10.3:
*Symmetric CA map
of Exhibit 10.1
(scientific funding
data).*

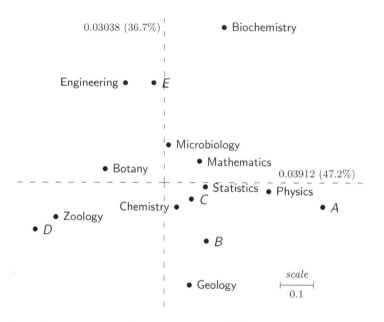

inertias along the axes, or their square roots which are the canonical correlations along each axis, namely $\sqrt{0.039117} = 0.198$ and $\sqrt{0.030381} = 0.174$ respectively. The level of row–column association can be judged graphically only in an asymmetric map such as Exhibit 10.2 (compare again the different levels of association illustrated in Exhibit 4.2).

*Dimensional
interpretation of
maps*

Whether the joint map is produced using asymmetric or symmetric scaling, the *dimensional* style of interpretation remains universally valid. This involves interpreting one axis at a time, as we did above and as is customary in factor analysis. The relative positions of one set of points — the "variables" of the table, in this case the funding categories — are used to give a descriptive name to the axis. Then the positions of the scientific discipline points are interpreted with respect to each axis. All statements in such an interpretation are relative and it is not possible to judge the absolute difference in funding profiles between the disciplines unless we refer to the original data. Putting this another way, symmetric maps similar to Exhibit 10.3 could be obtained for other data sets where there are much larger (or smaller) levels of association between the funding categories and the disciplines.

*Data set 6:
Abundances of
marine species in
seabed samples*

CA is used extensively to analyse ecological data, and the second example represents a typical data set in marine biology. The data, given partially in Exhibit 10.4, are the counts of 92 marine species identified in 13 samples from the seabed in the North Sea. Most of the samples are taken close to an oil-drilling platform where there is some pollution of the seabed, while two samples, regarded as reference samples and assumed unpolluted, are taken far from the drilling activities. These data, and biological data of this kind in

	STATIONS (SAMPLES)												
SPECIES	S4	S8	S9	S12	S13	S14	S15	S18	S19	S23	S24	R40	R42
Gala.ocul.	193	79	150	72	141	302	114	136	267	271	992	5	12
Chae.seto.	34	4	247	19	52	250	331	12	125	37	12	8	3
Amph.falc.	49	58	66	47	78	92	113	38	96	76	37	0	5
Myse.bide.	30	11	36	65	35	37	21	3	20	156	12	58	43
Goni.macu.	35	39	41	37	32	45	41	41	31	29	64	32	23
⋮	⋮	⋮	⋮	⋮	⋮	⋮	⋮	⋮	⋮	⋮	⋮	⋮	⋮
Ophi.flex.	0	1	1	0	0	0	0	1	0	0	0	0	0
Eucl.sp.	0	0	0	0	1	0	0	1	1	0	0	0	0
Scal.infl.	0	1	0	0	0	1	0	0	0	0	0	0	1
Eumi.ocke.	0	0	1	0	0	1	1	0	0	0	0	0	0
Modi.modi.	0	0	0	1	1	0	0	1	0	0	0	0	0

Exhibit 10.4:
Frequencies of 92 marine species in 13 samples (the last two are reference samples); the species (rows) have been ordered in descending order of total abundance; hence, the five most abundant and five least abundant species are shown here.

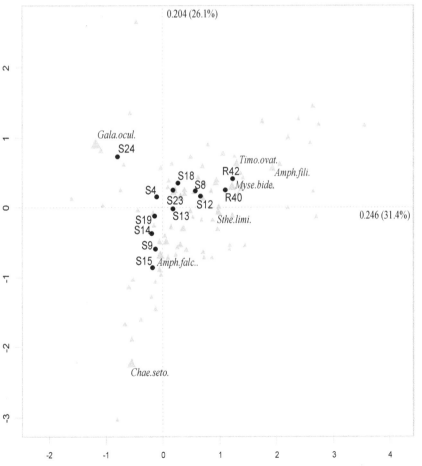

Exhibit 10.5:
Asymmetric CA map of abundance data of Exhibit 10.4, with stations in principal coordinates and the species in standard coordinates. The species symbols have sizes proportional to the total species abundance (mass) — some important species for the interpretation are labelled, with the first letter of the label being close to its corresponding triangular symbol (see text why these species are singled out). Inertia explained in map: 57.5%.

general, are characterized by high variability, which can already be seen by simple inspection of the small part of the data given here. The total inertia of this table is 0.7826, much higher than in the previous examples, so we can expect the profiles to be more spread out relative to the vertices. Notice that in this example the χ^2-test is not applicable, since the data do not constitute a true contingency table — each individual count is not independent of the others, since the marine organisms often occur in groups at a sampling point.

Asymmetric CA map of species abundance data

Exhibit 10.5 shows the asymmetric map of the sample (column) profiles and species (row) vertices. In ecology this is frequently referred to as an *ordination* of the samples and species. Since there are 92 species points, it is impossible to label each point so we have labelled only the points which have a high contribution to the map; these are generally the most abundant ones. (The topic of how to measure this contribution is described in Chapter 11; for the moment let us simply report that 10 out of the 92 species contribute over 85% to the construction of this map, the other 82 could effectively be removed without the map changing very much.) The stations form a curve from bottom left (actually, the most polluted stations) to top right (the least polluted), with the reference stations far from the drilling area at upper right. An exception is station 24, which separates out notably from the others, mainly because of the very high abundance of species *Gala.ocul.* (*Galathowenia oculata*) which can be seen in the first row of Exhibit 10.4. Notice that the asymmetric map functions well in this example because the inertia is so high, which is typical of ecological data where there is high variability between the samples. The next example is the complete opposite!

Data set 7: Frequencies of letters in books by six authors

This surprising example is a data set provided in the **ca** package of the R program (see Computational Appendix, pages 260–265). The data form a 12×26 matrix with the rows representing 12 texts which form six pairs, each pair by the same author (Exhibit 10.6 shows a part of the matrix). The columns are the 26 letters of the alphabet, *a* to *z*. The data are the counts of these letters in a sample of text from each of the books. There are approximately 6500–7500 letter counts for each book or chapter.

One of the lowest inertias one can get, but with a significant structure

This data set has one of the lowest total inertias I have seen in my experience with CA: the total inertia is 0.01873, which means that the data are very close to the expected values calculated from the marginal frequencies; i.e. the profiles are almost identical. The asymmetric map of these data is shown in Exhibit 10.7, showing the letters in their vertex positions and the 12 texts as a tiny blob of points around the origin, showing how little variation there is between the texts in terms of letter distributions, as expected. If one expands the tiny blob of points, it is surprising to see how much structure there is within such tiny variation. Each pair of texts by the same author lies in the same vicinity, and the result is highly significant from a statistical viewpoint (this permutation test is described in Chapter 30, pages 235–236).

BOOKS	a	b	c	d	e	$\cdots$	w	x	y	z	Sum
TD-Buck	550	116	147	374	1015	$\cdots$	155	5	150	3	*7144*
EW-Buck	557	129	128	343	996	$\cdots$	187	10	184	4	*7479*
Dr-Mich	515	109	172	311	827	$\cdots$	156	14	137	5	*6669*
As-Mich	554	108	206	243	797	$\cdots$	149	2	80	6	*6510*
LW-Clar	590	112	181	265	940	$\cdots$	146	13	162	10	*7100*
PF-Clar	592	151	251	238	985	$\cdots$	106	15	142	20	*7505*
FA-Hemi	589	72	129	339	866	$\cdots$	225	1	155	2	*6877*
Is-Hemi	576	120	136	404	873	$\cdots$	250	3	104	5	*6924*
SF7-Faul	541	109	136	228	763	$\cdots$	160	11	280	1	*6885*
SF6-Faul	517	96	127	356	771	$\cdots$	216	12	171	5	*6971*
Pe3-Holt	557	97	145	354	909	$\cdots$	194	9	140	4	*6650*
Pe2-Holt	541	93	149	390	887	$\cdots$	218	2	127	2	*6933*

Exhibit 10.6:
Letter counts in 12 samples of texts from books by 6 different authors, showing data for 9 out of 26 letters.

Abbreviations:
TD (Three Daughters), EW (East Wind) -Buck (Pearl S. Buck)
Dr (Drifters), As (Asia) -Mich (James Michener)
LW (Lost World), PF (Profiles of Future) -Clar (Arthur C. Clarke)
FA (Farewell to Arms), Is (Islands) -Hemi (Ernest Hemingway)
SF7 and SF6 (Sound and Fury, chapters 7 and 6) -Faul (William Faulkner)
Pen3 and Pen2 (Bride of Pendorric, chapters 3 and 2) -Holt (Victoria Holt)

An important final remark concerns the physical plotting of two-dimensional CA maps. Since distances in the map are of central interest, it is clear that a unit on the horizontal axis of a plot should be equal to a unit on the vertical axis. Even though this requirement seems obvious, it is commonly overlooked in many software packages and spreadsheet programs that produce scatterplots of points with different scales on the axes. For example, the points might in reality have little variation on the vertical second axis, but the map is printed in a pre-defined rectangle which then exaggerates the second axis. We say that the *aspect ratio* of the map, that is the ratio of one unit length horizontally to one unit vertically, should be equal to 1. A few options for producing maps are described at the end of the Computational Appendix, pages 283–284.

Importance of preserving a unit aspect ratio in maps

1. When applicable, it is useful to test a contingency table for significant association, using the χ^2 test. However, statistical significance is not a crucial requirement for justifying an inspection of the maps. CA should be regarded as a way of re-expressing the data in pictorial form for ease of interpretation — with this objective any table of data is worth looking at.

2. In both asymmetric and symmetric maps the dimensional style of interpretation is valid. This applies to one axis at a time and consists of using the relative positions of one set of points on a principal axis to give the dimension a conceptual name, and then separately interpreting the relative positions of the other set of points along this named dimension.

SUMMARY: Three More Examples

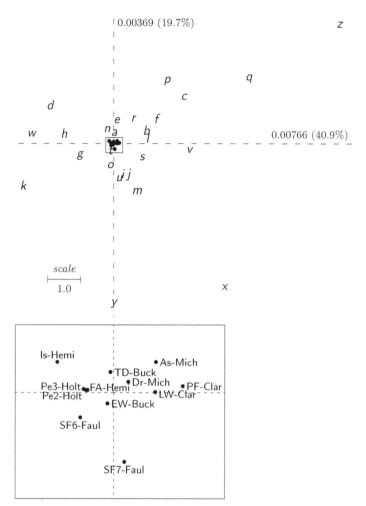

3. The asymmetric map functions well when total inertia is high, but it is problematic when total inertia is small because the profile points in principal coordinates occupy a small space around the origin.

4. It is important to have plotting facilities which preserve the *aspect ratio* of the display. A unit on the horizontal axis must be as close as possible to a unit on the vertical axis of the map; otherwise distances will be distorted if the unit lengths are different.

Contributions to Inertia

## 11

The total inertia of a cross-tabulation is a measure of how much variation there is in the table. We have seen how this inertia is decomposed along principal axes and also how it is decomposed amongst the rows or amongst the columns. The inertia can be further broken down into row and column components along individual principal axes. The investigation of these parts of inertia, analogous to an analysis of variance, plays an important supporting role in the interpretation of correspondence analysis (CA). They provide diagnostics which allow the user to identify which points are the major contributors to a principal axis and to assess how well individual points are displayed.

Contents

In Chapter 4, Equation (4.7), we saw that the total inertia can be interpreted geometrically as the weighted average of squared χ^2-distances between the profiles and their average profile, and is identical for row profiles and for column profiles. If there are only small differences between the profiles and their average, then the inertia is close to zero; i.e. there is low dispersion (see Exhibit 4.2, top left display). At the other extreme, if each profile is highly concentrated in a few categories, and in different categories from profile to profile, then the inertia is high (Exhibit 4.2, lower right display). The inertia is a measure of how spread out the profiles are in the profile space, which is a simplex delimited by the vertex points. If all the profiles are at the vertices of the space, i.e. each in one category only, the inertia is equal to the dimensionality of the space: 2 for a 3-vertex triangle, 3 for a tetrahedron, etc.

Total inertia measures overall variation of the profiles

Row and column inertias

There are various ways that the inertia can be decomposed into the sum of positive components, and this provides a numerical "analysis of inertia" which is helpful in interpreting the results of CA. According to Equation (4.7), each row makes a positive contribution to the inertia in the form of its mass times squared distance to the row centroid — we call this the *row inertia*. The same applies to the columns, leading to *column inertias*. The actual values of these parts of inertia are numbers that are inconvenient to interpret and it is easier to judge them relative to the total inertia, expressed as proportions, percentages or more conveniently as *permills* (i.e. thousandths, denoted by $^0/_{00}$). Exhibit 11.1 gives the row and column inertias for the scientific funding data of Exhibit 10.1, first in their "raw" form, and then in their relative form expressed as permills ($^0/_{00}$). Permills are used extensively throughout the numerical results of the "ca" package in R (see Computational Appendix, pages 262–263), because they provide three significant digits without using a decimal point, thus improving legibility of the results.

Exhibit 11.1:
Row and column contributions to inertia, in raw amounts which sum up to the total inertia, or expressed relatively as permills ($^0/_{00}$) which add up to 1000.

ROWS	Inertia	$^0/_{00}$ inertia	COLUMNS	Inertia	$^0/_{00}$ inertia
Geology	0.01135	137	A	0.01551	187
Biochemistry	0.00990	119	B	0.00911	110
Chemistry	0.00172	21	C	0.00778	94
Zoology	0.01909	230	D	0.02877	347
Physics	0.01621	196	E	0.02171	262
Engineering	0.01256	152			
Microbiology	0.00083	10			
Botany	0.00552	67			
Statistics	0.00102	12			
Mathematics	0.00466	56			
Total	*0.08288*	*1000*	*Total*	*0.08288*	*1000*

Large and small contributions

From the "$^0/_{00}$ *inertia*" columns in Exhibit 11.1 we can see at a glance that the major contributors to inertia are the rows Zoology, Physics, Engineering, Geology and Biochemistry, in that order, while for the columns the major contributors are categories D and E. As a general guideline for deciding which contributions are large and which are small, we use the average as a threshold. For example, there are 10 rows, so on a permill scale this would be 100 on average per row; hence we regard rows with contributions higher than 100$^0/_{00}$ as major contributors. On the other hand, there are five columns, which give an average of 200$^0/_{00}$, so the two columns D and E are the major contributors.

Cell contributions to inertia

A finer look at the inertia contributions can be made by looking at each individual cell's contribution. As described in Chapter 4, each cell of the table contributes a positive amount to the total inertia, which can again be expressed on a permill scale — see Exhibit 11.2. Here we can see specific cells such as [Zoology,D] and [Physics,A], that are contributing highly to the inertia

SCIENTIFIC AREAS	FUNDING CATEGORIES					
	A	*B*	*C*	*D*	*E*	*Sum*
Geology	0	32	16	0	89	137
Biochemistry	0	23	4	44	48	119
Chemistry	3	12	1	0	5	21
Zoology	9	15	11	189	8	230
Physics	106	11	2	74	3	196
Engineering	1	11	38	1	102	152
Microbiology	2	0	0	3	5	10
Botany	51	4	0	10	2	67
Statistics	10	0	0	2	0	12
Mathematics	5	3	22	26	0	56
Sum	*187*	*110*	*94*	*347*	*262*	*1000*

Exhibit 11.2: *Cell contributions to inertia, expressed as permills; the row and column sums of this table are identical to the row and column inertias in permills given n Exhibit 11.1.*

— just these two cells together account for almost 30% of the table's total inertia ($189 + 106 = 295‰$, i.e. 29.5%). The cell contributions to inertia are sometimes called chi-square contributions because they are identical to the relative contributions of each cell to the χ^2 statistic. The row and column sums of this table give the same permill contributions of Exhibit 11.1.

The other major decomposition of inertia is with respect to, or "along", principal axes. On page 74 we gave the first two principal inertias for this 10×5 table, which has four dimensions. Exhibit 11.3 gives all the principal inertias, their precentages and a bar chart (this type of bar chart is often called a *scree plot*). We have seen that the principal inertias have an interpretation in their own right, for example as squared canonical correlations (see Chapter 8, page 61), but we mainly interpret their values relative to the total, usually expressed as percentages rather than permills in this particular case.

Decomposition along principal axes

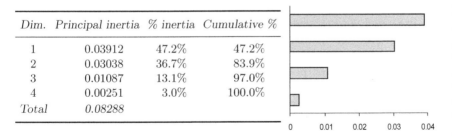

Dim.	Principal inertia	% inertia	Cumulative %
1	0.03912	47.2%	47.2%
2	0.03038	36.7%	83.9%
3	0.01087	13.1%	97.0%
4	0.00251	3.0%	100.0%
Total	*0.08288*		

Exhibit 11.3: *Principal inertias, percentages and cumulative percentages for all dimensions of the scientific-funding data, and a scree plot.*

Each principal inertia is itself an inertia, calculated for the projections of the row profiles (or column profiles) onto a principal axis. For example, the 10 row profiles of the scientific funding data lie in a *full space* of dimensionality 4 — as we have seen before, this is one less than the number of columns, since there are less columns than rows. Their weighted sum of squared distances to the row centroid is equal to the total inertia, with value 0.08288. The first principal axis is the straight line that comes closest to the profile points in

Components of each principal inertia

ROWS	Inertia	‰ inertia	COLUMNS	Inertia	‰ inertia
Geology	0.00062	16	A	0.00890	228
Biochemistry	0.00118	30	B	0.00260	67
Chemistry	0.00023	6	C	0.00265	68
Zoology	0.01616	413	D	0.02471	632
Physics	0.01426	365	E	0.00025	6
Engineering	0.00153	39			
Microbiology	0.00001	0			
Botany	0.00345	88			
Statistics	0.00057	14			
Mathematics	0.00112	29			
Total	*0.03912*	*1000*	*Total*	*0.03912*	*1000*

SCIENTIFIC AREAS	PRINCIPAL AXES				
	Axis 1	Axis 2	Axis 3	Axis 4	Total
Geology	0.00062	0.00978	0.00082	0.00013	*0.01135*
Biochemistry	0.00118	0.00754	0.00084	0.00034	*0.00990*
Chemistry	0.00023	0.00088	0.00029	0.00032	*0.00172*
Zoology	0.01616	0.00158	0.00063	0.00073	*0.01909*
Physics	0.01426	0.00010	0.00169	0.00016	*0.01621*
Engineering	0.00153	0.00941	0.00127	0.00036	*0.01256*
Microbiology	0.00001	0.00056	0.00008	0.00019	*0.00083*
Botany	0.00345	0.00016	0.00180	0.00011	*0.00552*
Statistics	0.00057	0.00001	0.00042	0.00003	*0.00102*
Mathematics	0.00112	0.00037	0.00302	0.00015	*0.00466*
Total	*0.03912*	*0.03038*	*0.01087*	*0.00251*	*0.08288*

the sense of least squares. This axis passes through the row centroid, which is at the *origin*, or zero point, of the display. Suppose that all the row profiles are projected onto this axis, as was done for the health categories in Exhibit 6.3. The first principal inertia, equal to 0.03912, is then the weighted sum of squared distances from these projections to the centroid, i.e. the inertia of the set of projected points on the first principal axis. Using the principal coordinates on the axis we obtain the row and column components of the first principal inertia, shown in Exhibit 11.4. This shows that category D is the dominant contributor to the first axis, followed by A, while the other categories contribute very little. As for the rows, Zoology (highly associated with D) and Physics (highly associated with A) contribute almost 78% of the inertia on the first axis.

*Complete
decomposition of
inertia over profiles
and principal axes*

We can repeat the above for all the principal axes, and Exhibit 11.5 shows the raw components of inertia of the rows for all four axes, (a similar table can be constructed for the columns). Just as the raw inertias in Exhibit 11.4 have been expressed in permills relative to the first principal inertia, we could do the same for axes 2 to 4 as well. For example, the major row contributions to the second axis are Geology, Engineering and Biochemistry. Inspecting these contributions of each row point (and, similarly, each column point) to the principal axes gives numerical support to our interpretation of the map, showing which rows and which columns are important in constructing the axes.

Whereas the column sums of Exhibit 11.5 give the principal inertias on respective axes, the row sums give the inertias of the profiles (hence these row sums are the same as the first column of Exhibit 11.1). We can also express these components relative to the row inertias, again either as proportions, percentages or permills. These will tell us how well each row is explained by each principal axis. This is a pointwise version of the way we interpreted the principal inertias, which quantified the percentage of the total inertia that was contained on each axis — here we do the same for each point separately. Exhibit 11.6 gives these relative amounts in permills, so that each row now adds up to 1000. For example, Geology is mostly explained by axis 2, whereas Physics mostly by axis 1. Mathematics, on the other hand, is not well explained by axis 1 or axis 2; in fact, its inertia is mostly in the third dimension.

Components of each profile's inertia

SCIENTIFIC AREAS	*PRINCIPAL AXES*				
	Axis 1	*Axis 2*	*Axis 3*	*Axis 4*	*Total*
Geology	55	861	72	11	*1000*
Biochemistry	119	762	85	35	*1000*
Chemistry	134	510	170	186	*1000*
Zoology	846	83	33	38	*1000*
Physics	880	6	104	10	*1000*
Engineering	121	749	101	28	*1000*
Microbiology	9	671	96	224	*1000*
Botany	625	29	326	20	*1000*
Statistics	554	7	410	30	*1000*
Mathematics	240	79	649	33	*1000*
Average	*472*	*367*	*131*	*30*	*1000*

Exhibit 11.6: *Relative contributions (in permills, i.e. ‰) of each principal axis to the inertia of individual points; the last row shows the same calculation for the principal inertias (cf. Exhibit 11.3), which can be regarded as average relative contributions.*

Exhibit 11.7 illustrates the decomposition of inertia geometrically and introduces some notation at the same time. The point $\mathbf{a}_i$ is a general profile point in multidimensional space, i.e. the i-th row profile, with mass r_i, at a distance of d_i from the average row profile $\mathbf{c}$. Hence, using formula (4.7), the total inertia is equal to $\sum_i r_i d_i^2$. A general principal axis k is shown and the point's principal coordinate on this axis is denoted by f_{ik}. Thus the inertia along this axis (i.e. the k-th principal inertia) is $\sum_i r_i f_{ik}^2$, usually denoted by λ_k. The contribution of each point i to the principal inertia of axis k is $r_i f_{ik}^2$ relative to λ_k (these proportions are given in permills for axis 1 in Exhibit 11.4). Exhibit 11.5 is actually the table of values $r_i f_{ik}^2$ for the 10 rows and 4 principal axes of the scientific funding data, with column sums equal to the λ_k's and the

Algebra of inertia decomposition

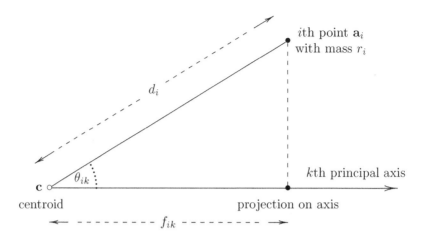

row sums equal to the row inertias $r_i d_i^2$. Thanks to Pythagoras' theorem, we have $d_i^2 = \sum_k f_{ik}^2$, which is why the rows of Exhibit 11.5 sum up to the row inertias:

$$\sum_k r_i f_{ik}^2 = r_i d_i^2$$

Hence, the contribution of axis k to the inertia of point i is $r_i f_{ik}^2$ relative to $r_i d_i^2$ (these proportions are given in Exhibit 11.6).

*Relative
contributions as
squared angle
cosines*

There is an alternative geometric interpretation of the relative contributions in Exhibit 11.6. Since the proportion of inertia of point i explained by axis k is $r_i f_{ik}^2 / r_i d_i^2 = (f_{ik}/d_i)^2$, it is clear from Exhibit 11.7 that this is the square of the angle cosine between the point and the axis. Suppose this angle is denoted by θ_{ik}, then $\cos(\theta_{ik}) = f_{ik}/d_i$ and the relative contribution is $\cos^2(\theta_{ik})$. For example, axis 1 has a relative contribution of 0.880 to the point Physics; hence $\cos^2(\theta_{51}) = 0.880$, from which we can evaluate $\cos(\theta_{51}) = 0.938$ and the angle $\theta_{51} = 20°$. This shows that the point Physics, which is mostly explained by axis 1, is close to axis 1, subtending a small angle of $20°$ with it. A point such as Geology, with a relative contribution of 0.055, subtends a large angle of $\theta_{11} = 76°$ with axis 1, so it is not at all aligned with this axis but lying closer to different dimensions of the space (in fact, mostly along axis 2, as can be seen by the high relative contribution to Geology of 0.861 by the second axis).

*Relative
contributions as
squared
correlations*

There is a further interpretation of the relative contributions: angle cosines between vectors can be interpreted as correlation coefficients; hence the relative contributions are also squared correlations. We can thus say that Physics has a high correlation of $\sqrt{0.880} = 0.938$ with axis 1, whereas Geology has a low correlation of $\sqrt{0.055} = 0.234$. If the correlation is 1, the profile point lies on the principal axis, and if the correlation is 0 the profile is perpendicular to the principal axis (angle of $90°$).

SCIENTIFIC AREAS	Quality	FUNDING CATEGORIES	Quality
Geology	916	A	587
Biochemistry	881	B	816
Chemistry	644	C	465
Zoology	929	D	968
Physics	886	E	990
Engineering	870		
Microbiology	680		
Botany	654		
Statistics	561		
Mathematics	319		
Overall	*839*	*Overall*	*839*

Exhibit 11.8:
Quality of display (in permills) of individual row profile points in two dimensions; only Mathematics has less than 50% (i.e. 500‰) of its inertia explained.

Thanks to Pythagoras' theorem, the squared cosines of the angles between a point and each of a set of axes can be added together to give squared cosines between the point and the subspace generated by those axes. For example, the angle between a row profile and the principal plane can be computed from the sum of the relative contributions along the first two principal axes. Exhibit 11.8 gives the sum of the first two columns of Exhibit 11.6, and these are interpreted as measures of *quality* of individual points in the two-dimensional CA maps, just as the sum of the first two percentages of inertia (83.9%) is interpreted as a measure of overall (or average) quality of display. Here we can see which points are well represented in the two-dimensional display and which are not. Some profiles will not be accurately represented because they lie more along the third and fourth axes than along the first two. Thus Mathematics is poorly displayed, with over two-thirds of its inertia lying off the plane of Exhibits 10.2 and 10.3. The position of Mathematics looks quite similar to that of Statistics, but this projected position of Mathematics is not an accurate reflection of its true position.

Quality of display in a subspace

This section is mainly aimed at readers with a knowledge of factor analysis — several entities in CA have direct analogues with those in factor analysis.

Analogy with factor analysis

- The analogue of a *factor loading* is the angle cosine between a point and an axis, i.e. the square root of the squared correlation along with the sign of the point's coordinate. For example, from Exhibits 11.1 and 11.4, the squared correlations of the categories A to E are:

$$A: \frac{0.00890}{0.01551} = 0.574 \quad B: \frac{0.00260}{0.00911} = 0.286 \quad C: \frac{0.00265}{0.00778} = 0.341$$

$$D: \frac{0.02471}{0.02877} = 0.859 \quad E: \frac{0.00025}{0.02171} = 0.012$$

Using the signs of the column coordinates in Exhibit 10.3, the CA "factor loadings" would be the signed square roots:

$$A: 0.758 \quad B: 0.535 \quad C: 0.584 \quad D: -0.927 \quad E: -0.108$$

- The analogue of a *communality* is the quality measure on a scale of 0 to 1. For example, in the two-dimensional solution, the CA "communalities" of the five column categories are given by the last column of Exhibit 11.8 on the original scale: 0.587, 0.816, 0.465, 0.968 and 0.990 respectively.

- The analogue of a *specificity* is 1 minus the quality measure on a scale of 0 to 1. For example, in the two-dimensional solution, the CA "specificities" of the five column categories are 0.413, 0.184, 0.535, 0.032 and 0.010 respectively.

SUMMARY:
Contributions to
Inertia

1. The (total) inertia of a table quantifies how much variation is present in the set of row profiles or in the set of column profiles.

2. Each row and each column makes a contribution to the total inertia, called a *row inertia* and a *column inertia* respectively.

3. CA is performed with the objective of accounting for a maximum amount of inertia along the first axis. The second axis accounts for a maximum of the remaining inertia, and so on. Thus, the total inertia is also decomposed into components along principal axes, called *principal inertias.*

4. The principal inertias are themselves decomposed over the rows and the columns. These *inertia contributions* are more readily expressed in relative amounts, and there are two possibilities:

   (a) express each contribution to the k-th axis relative to the corresponding principal inertia;

   (b) express each contribution to the k-th axis relative to the corresponding point's inertia.

5. Possibility (a) allows diagnosing which points have played a major part in determining the orientation of the principal axes. These contributions facilitate the interpretation of each principal axis.

6. Possibility (b) allows diagnosing the position of each point and whether a point is well represented in the map, in which case the point is interpreted with confidence, or poorly represented, in which case its position is interpreted with more caution. These quantities are squared cosines between the points and the principal axes, also interpreted as squared correlations.

7. The sum of squared correlations for a point in a low-dimensional solution space gives a measure of *quality* of representation of the point in that space.

8. The correlations of the points with the axes are the analogues of factor loadings, the qualities are analogues of communalities and 1 minus the qualities the analogues of specificities.

Supplementary Points

## 12

It frequently happens that there are additional rows and columns of data that are not the primary data of interest, but which are useful in interpreting features discovered in the primary data. Any additional row (or column) of a data matrix can be positioned on an existing map, as long as the profile of this row (or column) is meaningfully comparable to the existing row profiles (or column profiles) that have determined the map. These extra rows or columns that are added to the map afterwards are called *supplementary points*.

Contents

Up to now all rows and all columns of a particular table of data have been used to determine the principal axes and hence the map — we say that all rows and columns are *active* in the analysis. One can think of each active point having a different force of attraction for the principal axes, where this force depends on the position of the point as well as its mass. Profiles farther from the average have more "leverage" in orienting the map towards them, and higher mass profiles have a greater "pull" on the map.

Active points

There are situations, however, when we wish to suppress some points from the actual computation of the solution while still being able to inspect their projections onto the map which best fits the active points. The simplest way to think of such points is that they have a position but no mass at all, so that their contribution to the inertia is zero and they have no influence on the principal axes. Such zero mass points are called *supplementary points*, sometimes also called *passive points* to distinguish them from the active points that have positive mass. There are three common situations when supplementary rows or columns can be useful, and we now illustrate each of these in the context of the scientific funding data set of the previous chapters. Exhibit 12.1 shows

Definition of a supplementary point

Exhibit 12.1:
Frequencies of funding categories for 796 researchers (Exhibit 10.1), with additional column Y for a new category of "promising young researchers", an additional row for researchers at museums, and a row of cumulated frequencies for Statistics *and* Mathematics, *labelled* Math Sciences

| SCIENTIFIC | FUNDING CATEGORIES | | | | | |
AREAS	*A*	*B*	*C*	*D*	*E*	*Y*
Geology	3	19	39	14	10	0
Biochemistry	1	2	13	1	12	1
Chemistry	6	25	49	21	29	0
Zoology	3	15	41	35	26	0
Physics	10	22	47	9	26	1
Engineering	3	11	25	15	34	1
Microbiology	1	6	14	5	11	1
Botany	0	12	34	17	23	1
Statistics	2	5	11	4	7	0
Mathematics	2	11	37	8	20	1
Museums	*4*	*12*	*11*	*19*	*7*	
Math Sciences	*4*	*16*	*48*	*12*	*27*	

an expanded version of that data set where we have added:

1. an additional column, labelled Y, which is a special category of funding for young researchers, a category which had just been introduced into the funding system;

2. an additional row, labelled *Museums*, containing the frequencies of researchers working at museums (as opposed to universities, in the rest of the table);

3. another row, labelled *Math Sciences*, which is the sum of the rows Statistics and Mathematics.

First case — a point inherently different from the rest

The study from which these data are derived was primarily aimed at university researchers. Researchers from museums, however, were similarly graded and sponsored by the same funding organization, hence the frequencies of 53 museum researchers in the five funding categories. While it is necessary to consider the museum researchers separately from those at universities, it is still of interest to visualize the profile of museum researchers in the "space" of the university researchers, which can be done by declaring the row *Museums* to be a supplementary point. Its profile does not participate in the determination of the principal axes, but its profile can be projected onto the map. Exhibit 12.2 shows the symmetric map, as in Exhibit 10.2, with the additional point *Museums* in the lower left-hand side of the map. This point has no contribution to the principal inertia, but we can still look at the contributions of the axes to the point (i.e. the relative contributions or squared cosines or squared correlations). It turns out that this point is quite well displayed in the map, with over 50% of its inertia explained by it. Its position indicates that relatively few of the museum researchers have their applications rejected,

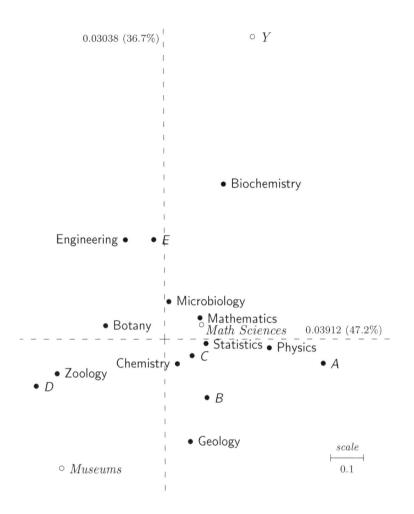

Exhibit 12.2:
Symmetric map of the data of Exhibit 12.1 (cf. Exhibit 10.2), showing in addition the profile positions of the supplementary column Y and supplementary rows Museums and Math Sciences.

while those that do receive funding tend towards the lower categories. Various types of supplementary information may be added to an active data set. Such information may be part of the same study, as in the case of *Museums* above, or it may come from separate but similar studies. For example, a similar table of frequencies may be available from a previous, or subsequent, classification of scientific researchers, and may be added as a set of supplementary rows in order to trace the evolution of each discipline's funding position over time. Another example is when some target profiles for the disciplines are specified, e.g. in the favourable bottom-right quadrant, and we want to judge how far away their actual positions are from these targets. This concept of an "ideal point" is frequently used in product positioning studies in marketing research.

Because the additional *Y* category had only just been introduced into the funding system, very few researchers were allocated to that category, in fact

Second case — an outlier of low mass

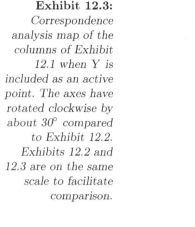

Exhibit 12.3:
Correspondence analysis map of the columns of Exhibit 12.1 when Y is included as an active point. The axes have rotated clockwise by about 30° compared to Exhibit 12.2. Exhibits 12.2 and 12.3 are on the same scale to facilitate comparison.

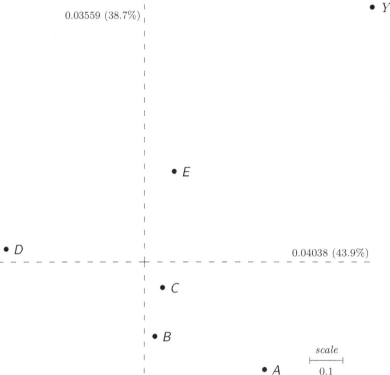

only six researchers and each one in a different discipline. This means that the profile of this column is quite unusual: six of the profile values have the value $\frac{1}{6} = 0.167$ and the others are 0. No other column profile has the slightest resemblance to this one, so it is to be expected that it has a very unusual position in the multidimensional space. As we see in Exhibit 12.2, this point is an *outlier* and if it were to be included as an active point in the analysis, it may contribute greatly to the map. This would not be a satisfactory situation since only six people are contained in the column Y — hence, apart from the substantive reason for making it supplementary, there is also a technical one. In this particular case, if we do include Y as an active point, its mass is less than 1% of the columns, but the total inertia of the table increases from 0.0829 to 0.0920, an increase of 11%. In addition, the map changes substantially, as can be seen in Exhibit 12.3 — there appears to be an approximate 30° clockwise rotation in the solution compared to the previous solution; hence the inclusion of Y has swung the axes around. We should be on the lookout for such outlying points with low mass that contribute highly to the inertia of the solution. In some extreme cases outliers can start to dominate a map so much that the more interesting contrasts between the more frequently occurring categories are completely masked. By declaring outliers supplementary, their positions can still be approximately visualized without influencing the solution space. Another way of dealing with rows or columns of low mass is to combine them with other rows or columns in a way which conforms to the data context: if we had an additional discipline, for example "Computer Science", with very few researchers in this category and a possibly strange profile as a result, we could combine them with an allied field, say Statistics. Having said this, it is a fact that outliers of low mass are often not such a serious problem in CA, since influence is measured by mass times squared distance and the low mass decreases the influence. The real problem is the fact that they lie so far from the other points — we return to this subject in Chapter 13 when we discuss alternative scalings for the map.

Supplementary points can be used to display a group of categories or to display subdivisions of a category. For example, the additional row *Math Sciences* in Exhibit 12.1 is the sum of the frequencies for Mathematics and Statistics, two disciplines which are frequently grouped together. The profile of this new row is the centroid of the two component rows, which are weighted by their respective masses. Since there are 78 and 29 researchers in Mathematics and Statistics respectively, the profile of Math Sciences would thus be:

Third case — displaying groups or partitions of points

$$\textit{Math Sciences} \text{ profile} = \frac{78}{107} \times \text{Mathematics profile} + \frac{29}{107} \times \text{Statistics profile}$$

so that the *Math Sciences* profile would be more like the Mathematics profile than the Statistics one. Geometrically, this means that the point representing the profile of *Math Sciences* is on a line between the Mathematics and Statistics points, but closer to Mathematics (cf. Exhibit 3.5 on page 23). In order to display the point *Math Sciences*, as in Exhibit 12.2, we do not have to actually compute the weighted average; the new row is simply declared to be a supplementary point. We would not make this point active along with its two component rows, since this would mean that the 107 researchers in these two disciplines would be counted twice in the analysis. In the same way, subdivisions of categories may be displayed on existing CA maps. Suppose that data were available for a breakdown of Engineering into its different branches, for example mechanical, civil, electrical, etc. Then, to investigate whether the profiles of these subgroups lie in the same general region, these additional rows of frequencies can simply be declared supplementary. The result described above still applies: in the map the active Engineering point would be at the centroid of all the points representing its different branches.

In the above we have described supplementary points as additional profile points that are projected onto a previously computed map. An alternative way of obtaining their positions is to position them relative to the set of vertex points in an asymmetric map. For example, in Chapter 3 it was shown that the position of a row profile, say, is a weighted average of the column vertices, where the weights are the profile elements. A supplementary point can be positioned in exactly the same way. Once the principal axes of the row profiles have been determined, the standard coordinates of the columns

Positioning a supplementary point relative to the vertices

are the projections of the column vertex points onto the principal axes. An extra row profile can now be placed on any map by evaluating the appropriate centroid of the vertices on each principal axis of the map, using the elements of the new profile as weights. For example, to calculate the position of the supplementary point *Museums*,

$$\text{position of } Museums = \frac{4}{53} \times \text{vertex } \textbf{\textit{A}} + \frac{12}{53} \times \text{vertex } \textbf{\textit{B}} + \cdots \text{etc.}$$

i.e. calculate the weighted average of the standard coordinates of the columns along each principal axis.

Contributions of supplementary points

Since supplementary points have zero mass, they also have zero inertia and make no contribution to the principal inertias. Their relative contributions, which relate to the angles between profiles and axes and do not involve masses, can still be interpreted to diagnose how well they are represented. The relative contributions of the three supplementary points described above, and their qualities in the two-dimensional space are as follows (Exhibit 12.4):

Exhibit 12.4:
Contributions of supplementary row and column points to the first two principal axes of Exhibit 12.2.

SUPPLEMENTARY POINTS	*Relative contributions*		*Quality in 2 dimensions*
	Axis 1	*Axis 2*	
Museums	225	331	556
Math Sciences	493	66	559
Y	54	587	641

These quantities describe how well these additional points are being displayed. For example, the supplementary point Y subtends an angle whose squared cosine is 0.054 with the first axis and 0.587 with the second axis. Its quality of display in the plane is thus $0.054 + 0.587 = 0.641$, so that 64.1% of its position is contained in the plane, and 35.9% in the remaining dimensions. Or we can say that Y is correlated $\sqrt{0.641} = 0.801$ with the plane.

Vertices are supplementary points

We have already encountered supplementary points in the form of the vertex points that were projected onto maps for purposes of interpretation, but whose positions were not taken into account in computing the map itself. This suggests an alternative way of determining the positions of the vertices: firstly, increase the data set by a number of rows, shown in Exhibit 12.5, as many rows as there are columns of data, each of which consists of zeros except

Exhibit 12.5:
Supplementary rows which could be added to the table in Exhibit 12.1 — their positions are identical to the column vertex points.

FUNDING CATEGORIES	*A*	*B*	*C*	*D*	*E*
A	1	0	0	0	0
B	0	1	0	0	0
C	0	0	1	0	0
D	0	0	0	1	0
E	0	0	0	0	1

for a single 1 (this is called an *identity matrix*); and secondly, declare these additional rows to be supplementary points. The positions of these supplementary rows are identical to those of the column vertices; in other words their coordinates will be the standard coordinates of the columns.

The example of column Y and the vertex points in Exhibits 12.2 and 12.4 should not be confused with what is called "dummy variable" coding, a subject which we shall treat in detail when we come to multiple correspondence analysis in later chapters. For example, suppose that we had a classification of the scientific areas into "Natural Sciences" (*NS*) and "Biological Sciences" (*BS*), the latter group including Biochemistry, Zoology, Microbiology and Botany and the former group containing the rest. A standard way of coding this in CA is as a pair of dummy variables, *NS* and *BS* say, zero-one variables with the values $NS = 1$ and $BS = 0$ for Geology (a natural science), for example, and $NS = 0$ and $BS = 1$ for Biochemistry (a biological science), and so on. One might be tempted to add these dummy variables as columns of the table and display them as supplementary points, but this would not be correct. This is not a count variable like the Y variable, which happened to have had 0's and 1's as well; in that case the data were real counts and could have been other integer values. The correct way to display this *NS/BS* information is as a pair of rows, similar to the way we displayed *Math Sciences* above. That is, sum up the frequencies for the *NS* rows and add an extra row called *NS* to the table, and do the same for the *BS* rows. In this way the *NS* and *BS* points will be weighted averages of the points representing the two sets of scientific areas (we shall return to this subject in Chapter 18).

Categorical supplementary variables and dummy variables

Additional information in the form of continuous variables also needs special consideration. Suppose we had some external information about each scientific area, for example, the average impact factor of all papers published in these areas in international journals. This would also be stored as a column of data, and because these are all positive numbers one might be tempted to represent the profile of this column in the standard supplementary point fashion. But remember that it is the profile of the column that is represented, not the original numbers, so the values should be nonnegative and expressing them as proportions of the total should make sense in the context of the study. But what if the data were changes in the average impact factors over a period of time, so that some changes were positive and some negative? Clearly, expressing these changes relative to their sum makes no sense. In this situation the continuous variable can be depicted in the map in a completely different way, using regression analysis (for an example, see Computational Appendix, page 264). This subject will be treated in more detail in later chapters, especially in Chapter 27 on canonical correspondence analysis, which is a combination of CA and regression — for the moment, we merely alert the reader to this possibility.

Continuous supplementary variables

SUMMARY:
Supplementary
Points

1. The rows and columns of a table analysed by CA are called *active* points. These are the points that determine the orientation of the principal axes and thus the construction of low-dimensional maps. The active rows and columns are projected onto the map.

2. *Supplementary* (or *passive*) points are additional rows or columns of a table which have meaningful profiles and which exist in the full spaces of the row and column profiles respectively. They can also be projected onto a low-dimensional map in order to interpret their positions relative to the active points.

3. Since supplementary points have zero mass, all quantities involving the mass, the point inertia and the contribution of the point to an axis are also zero.

4. The contribution of each principal axis to a supplementary point (i.e. squared cosine or squared correlation) can be computed and allows an assessment of whether a supplementary point lies to a larger or lesser extent in the subspace of the map. For example, the map might explain the supplementary point quite well even though the supplementary point has not determined the solution.

5. Be on the lookout for outliers with low mass — their presence in the analysis might have high influence on the solution. If they do, they should be made supplementary or combined with other rows or columns in a substantively sensible way.

6. An additional column categorical variable should not be coded in the form of dummy variables and added as supplementary columns, but rather rows of frequencies should be aggregated according to its categories and then these aggregated rows treated as supplementary rows.

7. Care is needed when adding an external continuous variable as a supplementary point: its values have to be nonnegative and its profile must make sense in the context of the example. An alternative approach to displaying such variables, using regression analysis, will be described in the following chapters.

Correspondence Analysis Biplots

## 13

Up to now we have drawn and interpreted correspondence analysis (CA) maps in two possible ways. In the asymmetric map, for example in the row analysis, the χ^2-distances between row profiles are displayed as accurately as possible, taking into account the masses of each profile, while the column vertices serve as references for the interpretation. In the symmetric map, the rows and the columns are both represented as profiles, thus the χ^2-distances between row profiles and between column profiles are approximated. The *biplot* is an alternative way of interpreting a joint map of row and column points. This approach is based on the scalar products between row vectors and column vectors, which depend on the lengths of the vectors and the angles between them rather than their interpoint distances. In the biplot only one of the profile sets, either the rows or the columns, are represented in principal coordinates. In fact, asymmetric CA maps, with one set in principal coordinates and the other in standard coordinates, are biplots. But there are alternative choices of coordinates for the other set of points serving as the references for the interpretation.

Contents

Definition of a scalar product

In Euclidean geometry a *scalar product* between two vectors $\mathbf{x}$ and $\mathbf{y}$ with coordinates x_1, x_2, ... and y_1, y_2, ... is the sum of products of respective elements $x_k y_k$, denoted by $\mathbf{x}^\mathsf{T}\mathbf{y} = \sum_k x_k y_k$ (T is the notation for the transpose of a vector or a matrix). Geometrically the scalar product is equal to the product of the lengths of the two vectors, multiplied by the cosine of the angle between them:

$$\mathbf{x}^\mathsf{T}\mathbf{y} = \sum_k x_k y_k = \|\mathbf{x}\| \cdot \|\mathbf{y}\| \cdot \cos\theta \qquad (13.1)$$

where $\|\mathbf{x}\|$ denotes the length of the vector $\mathbf{x}$, i.e. the distance between the point $\mathbf{x}$ and the zero point, and similarly for $\|\mathbf{y}\|$. This result is illustrated in Exhibit 13.1 (notice that two vectors in multidimensional space can always be represented exactly in a plane).

Exhibit 13.1:
Example of two points $\mathbf{x}$ and $\mathbf{y}$ whose vectors subtend an angle of θ with respect to an origin (either the zero point or the centroid of a cloud of points). The scalar product between the points is the length of the projection of $\mathbf{x}$, say, onto $\mathbf{y}$, multiplied by the length of $\mathbf{y}$.

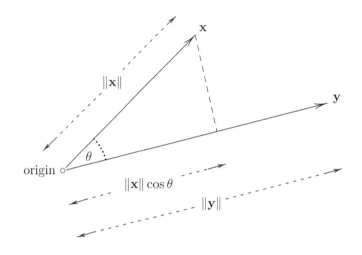

Relationship between scalar product and projection — Another standard geometric result is that the perpendicular projection of a vector $\mathbf{x}$ onto a direction defined by a vector $\mathbf{y}$ has a length equal to the length of $\mathbf{x}$ multiplied by the cosine of the angle between $\mathbf{x}$ and $\mathbf{y}$, i.e. the $\|\mathbf{x}\| \cdot \cos\theta$ part of (13.1). Thus, the scalar product between $\mathbf{x}$ and $\mathbf{y}$ can be thought of as the projected length of $\mathbf{x}$ onto $\mathbf{y}$ multiplied by the length of $\mathbf{y}$ (illustrated in Exhibit 13.1). Equivalently, it is the projected length of $\mathbf{y}$ onto $\mathbf{x}$, i.e. $\|\mathbf{y}\| \cdot \cos\theta$ multiplied by the length of $\mathbf{x}$. If the length of one of the vectors, say $\mathbf{y}$, is one (i.e. $\mathbf{y}$ is a unit vector), then the scalar product is simply the length of the projection of the other vector $\mathbf{x}$ onto $\mathbf{y}$.

For fixed reference vector, scalar products are proportional to projections — If we think of $\mathbf{y}$ as a fixed reference vector, and then imagine several vectors $\mathbf{x}_1, \mathbf{x}_2, \ldots$ projecting onto $\mathbf{y}$, then it is clear that

- the scalar products $\mathbf{x}_1^\mathsf{T}\mathbf{y}, \mathbf{x}_2^\mathsf{T}\mathbf{y}, \ldots$ have magnitudes proportional to the projections, since they are the projections multiplied by the fixed length of $\mathbf{y}$;

- the sign of a scalar product is positive if the vector $\mathbf{x}$ makes an acute angle ($< 90°$) with $\mathbf{y}$ and it is negative if the angle is obtuse ($> 90°$).

These properties are the basis for the biplot interpretation of CA.

The *biplot* is a low-dimensional display of a rectangular data matrix where the rows and columns are represented by points, with a specific interpretation in terms of scalar products. The idea is to recover the individual elements of the data matrix approximately in these scalar products. As an initial example of a biplot that recovers the data exactly, consider the following 5×4 table, denoted by $\mathbf{T}$:

$$\mathbf{T} = \begin{bmatrix} 8 & 2 & 2 & -6 \\ 5 & 0 & 3 & -4 \\ -2 & -3 & 3 & 1 \\ 2 & 3 & -3 & -1 \\ 4 & 6 & -6 & -2 \end{bmatrix} \qquad (13.2)$$

and then compare it to the map in Exhibit 13.2, which also gives the coordinates of each point. (Notice the convention in matrix algebra to denote vectors as columns, so that a vector is transposed if it is written as a row.)

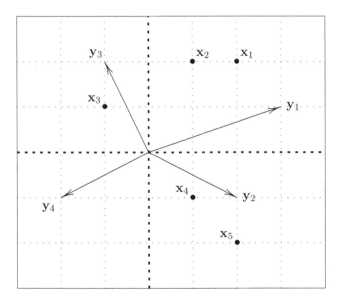

$$\mathbf{x}_1 = [\; 2 \quad 2 \;]^\mathsf{T}$$
$$\mathbf{x}_2 = [\; 1 \quad 2 \;]^\mathsf{T}$$
$$\mathbf{x}_3 = [\; -1 \quad 1 \;]^\mathsf{T}$$
$$\mathbf{x}_4 = [\; 1 \quad -1 \;]^\mathsf{T}$$
$$\mathbf{x}_5 = [\; 2 \quad -2 \;]^\mathsf{T}$$

$$\mathbf{y}_1 = [\; 3 \quad 1 \;]^\mathsf{T}$$
$$\mathbf{y}_2 = [\; 2 \quad -1 \;]^\mathsf{T}$$
$$\mathbf{y}_3 = [\; -1 \quad 2 \;]^\mathsf{T}$$
$$\mathbf{y}_4 = [\; -2 \quad -1 \;]^\mathsf{T}$$

Exhibit 13.2:
Map of five row points $\mathbf{x}_i$ and four column points $\mathbf{y}_j$. The scalar product between the i-th row point and the j-th column point gives the (i,j)-th value t_{ij} of the table in (13.2). The column points are drawn as vectors to encourage the interpretation of the scalar products as projections of the points onto the vectors, multiplied by the respective lengths of the vectors.

For example, the scalar product between $\mathbf{x}_1$ and $\mathbf{y}_1$ is equal to $2 \times 3 + 2 \times 1 = 8$, the first element of $\mathbf{T}$. Just to show that (13.1) can also be used, although with much more trouble, first calculate the respective angles that $\mathbf{x}_1$ and $\mathbf{y}_1$ make with the horizontal axis, using basic trigonometry: $\arctan(2/2) = 45°$ and $\arctan(1/3) = 18.43°$ respectively; hence the angle between $\mathbf{x}_1$ and $\mathbf{y}_1$ is $45 - 18.43 = 26.57°$. Equation (13.1) thus gives the scalar product as:

$$\mathbf{x}_1^\mathsf{T}\mathbf{y}_1 = \|\mathbf{x}_1\| \cdot \|\mathbf{y}_1\| \cdot \cos\theta = \sqrt{8} \cdot \sqrt{10} \cdot \cos(26.57°) = 8.00$$

so this checks. The projection of $\mathbf{x}_1$ onto $\mathbf{y}_1$ is equal to $\sqrt{8}\cos(26.57°) = 2.530$, and the length of $\mathbf{y}_1$ is $\sqrt{10} = 3.162$, the product of which is 8.00.

Some special   The "bi" in the name biplot comes from the fact that both rows and columns
patterns in biplots   are displayed in a map, not because of the bidimensionality of the map —
biplots could be of any dimensionality, but the most common case is the
planar one. The points in Exhibit 13.2 have been chosen to illustrate some
other properties of a biplot:

- $\mathbf{x}_2$ and $\mathbf{y}_2$ are at right angles, so $\mathbf{x}_2$ projects onto the origin; hence the
  value t_{22} in table $\mathbf{T}$ is 0.

- $\mathbf{x}_2$ and $\mathbf{x}_3$ project onto $\mathbf{y}_3$ at the same point; hence the values t_{23} and t_{33}
  are equal ($= 3$ in this case).

- $\mathbf{x}_4$ is a reflection of $\mathbf{x}_3$ with respect to the origin, which shows that the
  fourth row of $\mathbf{T}$ is the negative of the third row: $\mathbf{x}_4 = -\mathbf{x}_3$. Similarly, $\mathbf{x}_5$
  is opposite $\mathbf{x}_3$ with respect to the origin and twice as far away; hence the
  fifth row of table $\mathbf{T}$ is twice the third row, with a change of sign.

- $\mathbf{x}_3$, $\mathbf{x}_4$ and $\mathbf{x}_5$ are on a straight line (this could be any straight line, not
  necessarily through the origin), so they have a linear relationship, specifi-
  cally $\mathbf{x}_4 = \frac{1}{3}\mathbf{x}_3 + \frac{2}{3}\mathbf{x}_5$; this weighted average relationship carries over to the
  corresponding rows of $\mathbf{T}$, for example $t_{41} = \frac{1}{3}t_{31} + \frac{2}{3}t_{51} = \frac{1}{3}(-2) + \frac{2}{3}(4) = 2$.

Rank and   In mathematics we would say that the *rank* of the matrix $\mathbf{T}$ in (13.2) is equal to
dimensionality   2, and this is why the table can be perfectly reconstructed in a two-dimensional
biplot. In our geometric approach, rank is equivalent to dimensionality.

Biplots give   In real life, a data matrix has higher dimensionality and cannot be recon-
optimal   structed exactly in a low-dimensional biplot. The idea of the biplot is to find
approximations of   row points $\mathbf{x}_i$ and column points $\mathbf{y}_j$ such that the scalar products between the
real data   row and column vectors approximate the corresponding elements of the data
matrix as closely as possible. So we can say that the biplot models the data
t_{ij} as the sum of a scalar product in some low-dimensional subspace (say K^*
dimensions) and a residual "error" term:

$$t_{ij} = \mathbf{x}_i^\mathsf{T}\mathbf{y}_j + e_{ij}$$
$$= \sum_{k=1}^{K^*} x_{ik}y_{jk} + e_{ij} \qquad (13.3)$$

This biplot "model" is fitted by minimizing the errors, usually by least squares
where $\sum_i \sum_j e_{ij}^2$ is minimized. This looks just like a multiple linear regression
equation, except that there are two sets of unknown parameters, the row
coordinates $\{x_{ik}\}$ and the column coordinates $\{y_{jk}\}$ — we shall return to the
connection with regression analysis in Chapter 14.

The CA model   To understand the link between CA and the biplot, we need to introduce a
mathematical formula which expresses the original data n_{ij} in terms of the
row and column masses and coordinates. One version of this formula, known

as the *reconstitution formula* (see Theoretical Appendix, page 244), is:

$$p_{ij} = r_i c_j \left(1 + \sum_{k=1}^{K} \sqrt{\lambda_k} \phi_{ik} \gamma_{jk}\right) \qquad (13.4)$$

where

- p_{ij} are the relative proportions n_{ij}/n, where n is the grand total $\sum_i \sum_j n_{ij}$;
- r_i and c_j are the row and column masses;
- λ_k is the k-th principal inertia;
- ϕ_{ik} and γ_{jk} are row and column standard coordinates respectively.

In the summation in (13.4) there are as many terms K as there are dimensions in the data matrix, which we have seen to be equal to one less than the number of rows or columns, whichever is smaller. If we map the CA solution in K^* dimensions, where K^* is usually 2, then the fit is optimal since the terms in (13.4) from K^*+1 onwards are minimized — these latter terms thus constitute the "error", or residual.

Equation (13.4) can be slightly re-arranged so that the right-hand side is in the form of a scalar product in a space of dimensionality K^*, plus an error term, as in (13.3): *Biplot of contingency ratios*

$$\frac{p_{ij}}{r_i c_j} - 1 = \sum_{k=1}^{K^*} f_{ik} \gamma_{jk} + e_{ij} \qquad (13.5)$$

where $f_{ik} = \sqrt{\lambda_k} \phi_{ik}$, the principal coordinate of the i-th row on the k-th axis. This shows that the row asymmetric map, which displays row principal coordinates f_{ik} and column standard coordinates γ_{jk}, is an approximate biplot of the values on the left-hand side of (13.5). The ratios $p_{ij}/(r_i c_j)$ of observed proportions to expected ones are called *contingency ratios* — the closer these ratios are to 1, the closer the data are to the independence (or homogeneity) model.

Equation (13.5) can be rewritten in terms of the row profiles as: *Biplot from row profile point of view*

$$\left(\frac{p_{ij}}{r_i} - c_j\right)/c_j = \sum_{k=1}^{K^*} f_{ik} \gamma_{jk} + e_{ij} \qquad (13.6)$$

which shows that the row asymmetric map is also a biplot of the deviations of the row profiles from their average, relative to their average (see, for example, Exhibit 10.2). As we have seen, however, the asymmetric map can be quite unsatisfactory when inertia is small, because the row profiles (with coordinates f_{ik}) are concentrated into a small space at the centre of the map, while the column vertex points (with coordinates γ_{jk}) are very far out.

In the biplot it is the direction of each vertex point which is of interest, since this direction defines the line onto which the row profiles are projected. *The contribution biplot*

Exhibit 13.3:
*CA contribution
biplot of the
scientific
researchers' data of
Exhibit 10.1,
showing the rows in
principal
coordinates as
before, but the
columns in
contribution
coordinates, i.e. the
standard
coordinates
multiplied by the
square roots of the
respective column
masses. The vertex
position of A, for
example, in Exhibit
10.2 has been
multiplied by
$\sqrt{0.0389} = 0.197$ to
obtain the position
of A in this map.*

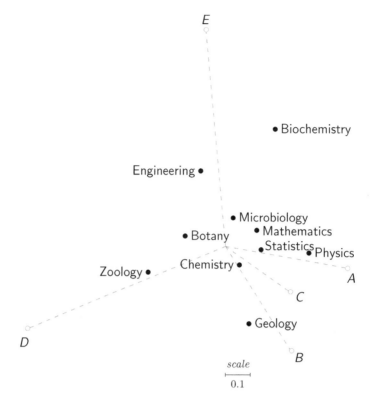

Different variations of this biplot have been proposed to redefine the lengths of these vectors. The most useful alternative is to rewrite (13.6) as:

$$(\frac{p_{ij}}{r_i} - c_j)/c_j^{1/2} = \sum_{k=1}^{K^*} f_{ik}(c_j^{1/2}\gamma_{jk}) + e_{ij} \qquad (13.7)$$

(notice that the residuals e_{ij} in (13.7) have a different definition and standardization compared to (13.6), although we use the same notation in each case). Thus we have expressed the left-hand side as a standardized deviation from the average, and then we absorb the remaining factor $c_j^{1/2}$ into the coordinate of the column point on the right-hand side. In this way, the vertex point gets pulled inwards by an amount equal to the square root of the mass of the corresponding category, so that the rarer categories are pulled in more, which is just what we want to improve the legibility of the asymmetric map. Moreover, with this alternative scaling of the column points, the squared coordinate values of the columns are exactly their contributions to the inertia of the respective axis, hence this is called a *contribution biplot*, with the columns in *contribution coordinates*. Exhibit 13.3 shows the CA contribution biplot for the research funding example; compare this map with Exhibits 10.2 and 10.3. In all these maps the positions of the row points are the same, it is the positions of the column points that change (compare the scales of each map). Exhibit 13.3 clearly shows that *D* and *E* are the largest contributors to the first and second axes respectively.

In Exhibit 13.3 the column points have no distance interpretation; they point in the directions of the biplot axes, and it is the projections of the row points onto the biplot axes which give estimates of the standardized values on the left-hand side of (13.7). Thus we can take a fixed reference direction, such as *D*, and then line up the projections of all the rows on this axis to estimate that Zoology has the highest profile element, then Botany, Geology, and so on, with Physics and Biochemistry having the lowest profile values on *D* (a few calculations on Exhibit 10.1 show this to be correct, with some small exceptions since this is an approximate biplot, representing 84% of the total inertia of the table).

Interpretation of the biplot

Since the projections of the rows onto the biplot axes are proportional to the values on the left-hand side of (13.7), each biplot axis can be calibrated in profile units. For example, to estimate the standardized profile values for category *A*, the projections of the row points have to be multiplied by the length of the *A* vector in Exhibit 13.3, equal to 0.484 (note that we are illustrating this calculation using the contribution coordinates). To unstandardize and return to the original profile scale, multiply this length by the square root of the mass ($\sqrt{0.0389} = 0.197$) to obtain the scale factor of 0.0955. The calibration of a biplot axis is computed by simply inverting this value: $1/0.0955 = 10.47$. This is the length of the full range of one unit on the profile scale. An interval of 1% (i.e. 0.01) on the biplot axis in Exhibit 13.4 is thus a hundredth of this length, i.e. 0.1047. So we know all three facts necessary for calibrating the biplot axis for *A*: (i) the origin of the map represents the average of 0.039 (or 3.9%) for *A*), which gives a fixed point on the axis; (ii) a length of 0.01 (1%) is equal to 0.1047, according to the scale shown in Exhibit 13.3; and (iii) the vector in Exhibit 13.3 indicates the positive direction of the axis. Exhibit 13.4 shows the calibrations on the *A* axis as well as the result of a similar calibration exercise on the *D* axis.

Calibration of biplots

Previously we thought of the overall quality of a two-dimensional CA map as the amount of inertia accounted for by the first two principal axes. The biplot provides another way of thinking about the map's quality, namely as the success of recovering the profile values in the map. The row profiles of the original data in Exhibit 10.1 can be approximately recovered by the two-dimensional biplot in Exhibit 13.3, for example by projecting all the row points onto the calibrated column axes. The closer the estimated profile values are to the true ones, the higher the quality of the map. Conversely, the differences between the true profile elements and the estimated ones can be accumulated to give an overall measure of error. When accumulated in a chi-squared fashion, i.e. by taking squared differences divided by the expected values, exactly the same measure of error will be obtained as before. In this particular example, the percentage of explained inertia in the two-dimensional map is 84%; hence the error is 16%.

Overall quality of display

Exhibit 13.4:
Contribution biplot of Eshibit 10.1 (scientific funding data), with calibrated axes for categories A and D. Notice that the calibrated axes pass through the column points in contribution coordinates (these are the same directions as the outlying vertex points in standard coordinates, since the contribution coordinates are a simple shrinking of the standard coordinates).

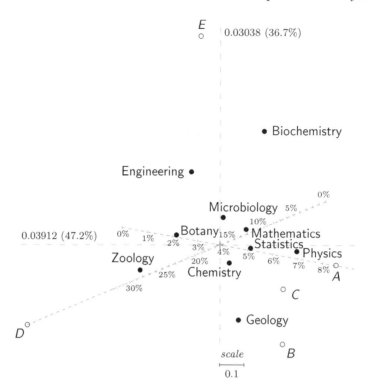

SUMMARY:
Correspondence Analysis Biplots

1. A *scalar product* between two vectors is the product of their lengths multiplied by the cosine of the angle between them.

2. The projected length of a vector $\mathbf{x}$ onto the direction defined by a second vector $\mathbf{y}$ is equal to the length of $\mathbf{x}$ multiplied by the cosine of the angle between $\mathbf{x}$ and $\mathbf{y}$. Thus the scalar product can be thought of as the product of the projected length of $\mathbf{x}$ onto $\mathbf{y}$ and the length of $\mathbf{y}$.

3. A *biplot* is a method of displaying a point for each row and column of a data matrix in a joint map such that the scalar products between the row vectors and the column vectors approximate the values in the corresponding cells of the matrix as closely as possible.

4. Asymmetric maps in CA are biplots. Strictly speaking, symmetric maps are not, although in practice the directions defined by the profile point and the corresponding vertex point are often not too different, in which case the biplot interpretation is still valid.

5. A variation of the asymmetric map which is a convenient biplot is the one where the position of each vertex point is pulled in towards the origin by an amount equal to the square root of the mass associated with the vertex category — this is called the *contribution biplot* for CA.

6. For a biplot the axes passing through the origin of the map may be calibrated in profile units (either proportions or percentages). This allows approximate profile values to be read directly off the map by projecting profile points onto the calibrated biplot axes.

Transition and Regression Relationships

## 14

Correspondence analysis (CA) produces a map where the rows and columns are depicted together as points, with an interpretation that depends on the choice amongst the many scaling options for the row and column points. Geometrically, we have seen how the positions of the row points depend on the positions of the column points, and vice versa. In this chapter we focus on the mathematical relationships between the row and column points, known as the transition equations. In addition, since regression analysis is a well-known method in statistics, we show how the row and column results and the original data can be connected through linear regression models. This chapter gives insight into the theory of CA and can be skipped without losing the thread of the geometric understanding of CA.

Contents

Coordinates on first axis of scientific funding example

In this chapter we are interested in the relationships between all the coordinates for the rows and columns — principal, standard or contribution — that emanate from a CA, as well as their relationships with the original data. Initially we look at the relationships that are valid for each principal axis separately. Using the scientific funding example again, Exhibit 14.1 reproduces all the results for the first principal axis (the contribution coordinates are given for the columns only). This axis has inertia $\lambda_1 = 0.03912$, with $\sqrt{\lambda_1} = 0.1978$. In Chapter 8 this latter value, which is the scaling factor which links the principal to the standard coordinates, was also seen to be the correlation coefficient between rows and columns in terms of their coordinates on the first dimension. Since correlation is related to regression, we first look at the regression of row coordinates on column coordinates and vice versa.

Exhibit 14.1: Principal and standard coordinates of the scientific disciplines and research funding categories on the first principal axis of the CA (original data in Exhibit 10.1), as well as the contribution coordinates of the funding categories.					

SCIENTIFIC DISCIPLINE	Princ. coord.	Stand. coord.	*FUNDING CATEGORY*	Princ. coord.	Stand. coord.	Contrib. coord.
Geology	0.076	0.386	*A*	0.478	2.417	0.477
Biochemistry	0.180	0.910	*B*	0.127	0.643	0.258
Chemistry	0.038	0.190	*C*	0.083	0.417	0.260
Zoology	−0.327	−1.655	*D*	−0.390	−1.974	−0.795
Physics	0.316	1.595	*E*	−0.032	−0.161	−0.080
Engineering	−0.117	−0.594				
Microbiology	0.013	0.065				
Botany	−0.179	−0.904				
Statistics	0.125	0.630				
Mathematics	0.107	0.540				

Regression between coordinates

In Exhibit 8.5, using the health survey data, we showed the scatterplot of values of the row and column coordinates on the first principal axis, for each individual constituting the contingency table. Exhibit 14.2 shows the same type of plot for the standard coordinates of the scientific funding data. There are 50 points in this plot, corresponding to the 50 cells in the contingency table of Exhibit 10.1. Each box is centred on the pair of values for the respective cell, and has area proportional to the number of individuals (scientists) in that cell. We know that the correlation, calculated for all 796 individuals that occur at the 50 points in this plot, is equal to 0.1978. Here we are interested in the regressions of scientific discipline on funding category, and funding category on scientific discipline. To compute a regression analysis we could string out all 796 scientists and assign their corresponding pair of values, for example a geologist in the *A* category would have the pair of standard coordinate values 0.386 (the y-variable, say) and 2.417 (the x-variable), according to Exhibit 14.1. Since there are only 50 unique pairs, an alternative is to list the 50 pairs of coordinate values along with their frequencies and then perform a weighted regression with the frequencies as weights. A standard result in simple linear regression is that the slope coefficient is equal to the correlation multiplied by the ratio of the standard deviations of the y-variable to the x-variable. The variances of the row and column standard coordinates are the same ($= 1$); hence the slope of the regression of y on x will be the same as the correlation, i.e. 0.1978 (Exhibit 14.2). In a symmetric way, the regression of x onto y will also have a slope of 0.1978, but this is with respect to the x-axis as vertical and the y-axis as horizontal — this is a slope of $1/0.1978 = 5.056$ in the plot of Exhibit 14.2 where y is the vertical axis.

The profile–vertex relationship

We saw as early as Chapter 3 that the row profiles are at weighted averages of the column vertices, where the weights are the values of the row profiles. The same relationship holds between column profiles and row vertices. These weighted average relationships hold in any projected space, in particular they

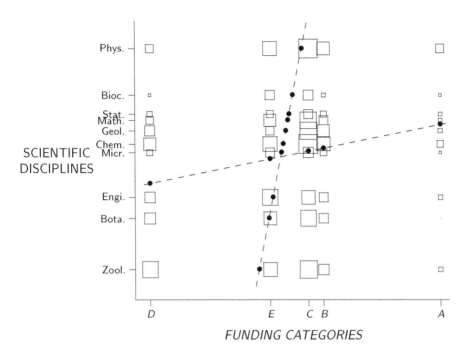

Exhibit 14.2:
Scatterplot based on standard coordinates of rows and columns on first CA dimension (Exhibit 14.1). Squares are shown at each combination of values, with area proportional to the number of respondents. The two regression lines, rows on columns and columns on rows, have slopes of 0.1978 and 5.056, inverses of each other. The dots • indicate conditional means (weighted averages), i.e. principal coordinates.

hold for the coordinates along the principal axes, as illustrated in Chapter 8. In other words, along an axis k, the principal coordinates of the row points are at weighted average positions of the standard coordinates of the column points, and vice versa. This relationship can be illustrated for the first principal axis by calculating weighted average positions for each row according to the standard coordinates of the columns and vice versa, shown by the dots on the two regression lines in Exhibit 14.2. This shows that the principal coordinates lie on the two regression lines.

Regression is a model for the conditional means of the response variable with respect to the predictor variable. The dots in Exhibit 14.2 are simply the discrete sets of conditional means of y on x (five means on the line with slope 0.1978) and x on y (ten means on the line with slope 5.056). These means are the principal coordinates, which thus define the two regression functions. For example, the row of squares for Physics depicts the corresponding row of frequencies in the data matrix of Exhibit 10.1, plotted horizontally according to the first standard coordinates of the five column categories, that is the vertex positions on principal axis 1. The conditional mean is just the weighted average, shown by the black dot at the top of the diagram, which is thus the first principal coordinate of Physics. Similarly, the column of squares for category A depicts the frequencies for the first column of the data matrix, plotted vertically at the first standard coordinate positions (i.e. vertices projected on the

Principal coordinates are conditional means in regression

first dimension) of the ten rows. The conditional mean, shown by the black dot at the right, is then the weighted average position, which is the principal coordinate of *A*. Thus Exhibit 14.2 shows both the principal and standard coordinates together; for example, to read the principal coordinates for the rows, read the values of the ten dots on the regression line on the horizontal axis according to the scale of the standard column coordinates, and vice versa.

Simultaneous linear regressions

The fact that the regressions of *y* on *x* (rows on columns) and *x* on *y* (columns on rows) turn out to be straight lines in the CA solution is one of the oldest definitions of CA, called *simultaneous linear regressions*. If the row–column correlation is high, then the two regression lines will subtend a smaller angle and the principal coordinates will be more spread out, i.e. the inertia will be higher (remember that the principal inertia is the square of the correlation). In other words, CA could be alternatively defined as trying to achieve simultaneous linear regressions (i.e. a scatterplot such as Exhibit 14.2) with the smallest possible angle between the two regression lines, which is equivalent to maximizing the row–column correlation.

Transition equations between rows and columns

Using notation defined on pages 31 and 101, we can write these weighted average (or conditional mean) relationships between rows and columns as follows, remembering that principal coordinates correspond to profiles and standard coordinates to vertices:

$$\text{row profile} \leftarrow \text{column vertices}: \quad f_{ik} = \sum_j \left(\frac{p_{ij}}{r_i} \right) \gamma_{jk} \quad (14.1)$$

$$\text{column profile} \leftarrow \text{row vertices}: \quad g_{jk} = \sum_i \left(\frac{p_{ij}}{c_j} \right) \phi_{ik} \quad (14.2)$$

(the ← stands for "is obtained from", for example "row profile ← column vertices" means that the principal coordinates of a row are obtained from the standard coordinates of all the columns using the relationship shown in the formula). Here we use the notation *f* and *g* for the row and column principal coordinates, *ϕ* and *γ* for the row and column standard coordinates, index *i* for rows, *j* for columns and *k* for dimensions. In the parentheses we have the elements of the row and column profiles in (14.1) and (14.2) respectively — these add up to 1 in each case and serve as weights. The weighted average relationships in (14.1) and (14.2) are called the *transition equations*. Recall the relationships between principal and standard coordinates:

$$\text{row profile} \leftarrow \text{row vertex}: \quad f_{ik} = \sqrt{\lambda_k} \phi_{ik} \quad (14.3)$$
$$\text{column profile} \leftarrow \text{column vertex}: \quad g_{jk} = \sqrt{\lambda_k} \gamma_{jk} \quad (14.4)$$

where λ_k is the principal inertia (eigenvalue) on the *k*-th axis. So we could write the transition equation between row and column principal coordinates as:

row profile ← column profiles :
$$f_{ik} = \frac{1}{\sqrt{\lambda_k}} \sum_j \left(\frac{p_{ij}}{r_i} \right) g_{jk} \quad (14.5)$$

column profile ← row profiles :
$$g_{jk} = \frac{1}{\sqrt{\lambda_k}} \sum_i \left(\frac{p_{ij}}{c_j} \right) f_{ik} \quad (14.6)$$

and similarly between row and column standard coordinates:

row vertex ← column vertices :
$$\phi_{ik} = \frac{1}{\sqrt{\lambda_k}} \sum_j \left(\frac{p_{ij}}{r_i} \right) \gamma_{jk} \quad (14.7)$$

column vertex ← row vertices :
$$\gamma_{jk} = \frac{1}{\sqrt{\lambda_k}} \sum_i \left(\frac{p_{ij}}{c_j} \right) \phi_{ik} \quad (14.8)$$

Any of the above transition equations can be used trivially in a standard lin- *Regression* ear regression analysis, with the profiles as predictors, in order to "estimate" *between* a set of coordinates. As an illustration, we recover the column standard co- *coordinates using* ordinates in (14.1), using the 10×5 matrix of row profiles of Exhibit 10.1 as *transition* five predictors and the first principal coordinates of the rows (first column of *equations* Exhibit 14.1) as response. The regression analysis gives the following results for the regression coefficient:

Source	Coefficient
Intercept	0.000
A	2.417
B	0.643
C	0.417
D	−1.974
E	−0.161

$$R^2 = 1.000$$

The variance explained is 100% and the regression coefficients are the column standard coordinates on the first axis (see last column of Exhibit 14.1).

The more interesting and relevant regression analysis is when the data are *Recall the CA* to be predicted from the coordinates, as summarized in the CA model given *bilinear model* in Chapter 13, Equations (13.4)–(13.7). We repeat this model here in three different versions, a "symmetric" version using only standard coordinates (see (13.4)), and the two asymmetric versions using row and column principal coordinates, respectively:

$$\frac{p_{ij}}{r_i c_j} = 1 + \sum_{k=1}^{K^*} \sqrt{\lambda_k} \phi_{ik} \gamma_{jk} + e_{ij} \quad (14.9)$$

$$\left(\frac{p_{ij}}{r_i} \right) / c_j = 1 + \sum_{k=1}^{K^*} f_{ik} \gamma_{jk} + e_{ij} \quad (14.10)$$

$$\left(\frac{p_{ij}}{c_j}\right)/r_i = 1 + \sum_{k=1}^{K^*} \phi_{ik}g_{jk} + e_{ij} \qquad (14.11)$$

This model is called *bilinear* because it is linear in the products of parameters. We shall, however, fix either the row or column standard coordinates, and show how to obtain the principal coordinates of the other set by multiple regression analysis.

Weighted On the left-hand sides of (14.9), (14.10) and (14.11) are the contingency ratios
regression defined in Chapter 13, written in three equivalent ways. Taking (14.10) as an example, and assuming that the standard coordinates γ_{jk} of the columns are known, we have on the right-hand side a regular regression model which is predicting the values of the rows on the left. Suppose that we are interested in the first row (Geology) and want to perform a regression for $K^* = 2$. To fit the CA model, we have to minimize a weighted sum of squared residuals, where the categories (columns) are weighted by their masses. Another way of understanding this is that in (14.10) the "predictors" γ_{jk} are normalized with respect to column masses as follows: $\sum_j c_j\gamma_{jk}^2 = 1$. Furthermore, the predictors are orthogonal as well when we weight by the column masses: $\sum_j c_j\gamma_{jk}\gamma_{j'k} = 0$ if $j \neq j'$. Hence, to perform the regression we set up the response vector as the 5×1 vector of contingency ratios for Geology, and the predictors as the 5×2 matrix of column standard coordinates on the first two principal axes. A weighted regression analysis is then performed, with regression weights equal to the column masses c_j. The data for the regression are as follows (contingency ratios $p_{1j}/(r_1c_j)$ for Geology, standard coordinates γ_1 and γ_2 for dimensions 1 and 2, and weights c_j):

Category	Geology	γ_1	γ_2	Weight
A	0.9063	2.4175	−0.4147	0.0389
B	1.3901	0.6434	−0.9948	0.1608
C	1.1781	0.4171	−0.2858	0.3894
D	1.0163	−1.9741	−0.7991	0.1621
E	0.4730	−0.1613	1.6762	0.2487

The results of the regression are:

Source	Coefficient	Standardized coefficient
Intercept	1.000	—
f_{11}	0.076	0.234
f_{12}	−0.303	−0.928

$$R^2 = 0.916$$

The coefficients are the principal coordinates f_{11} and f_{12} of Geology (see the first one, f_{11}, in Exhibit 14.1) while the variance explained (R^2) is the quality of Geology in the two-dimensional map (see Exhibit 11.8).

Since the predictors are standardized and orthogonal in the weighted regression, it is known that the standardized regression coefficients are also the partial correlations between the response and the predictors. The correlation matrix between all three variables is as follows (remember that the weights are included in the calculation):

Variables	Geology	γ_1	γ_2
Geology	1.000	0.234	-0.928
γ_1	0.234	1.000	0.000
γ_2	-0.928	0.000	1.000

The two predictors are uncorrelated, as expected, and the correlations between Geology and the two predictors are exactly the standardized regression coefficients. The squares of these correlations, $0.234^2 = 0.055$ and $(-0.928)^2 = 0.861$, are the squared cosines (relative contributions) given in Exhibit 11.6. The above series of results illustrates the property in regression that if the predictors are uncorrelated, then the variance explained R^2 is equal to the sum of squares of the partial correlations.

The transition equations (14.1) and (14.2) are the basis of a popular algorithm for finding the solution of a CA, called *reciprocal averaging*. The algorithm starts from any set of standardized values for the columns, say, where centring and normalizing are always with respect to weighted averages and weighted sum of squares. Then the averaging in (14.1) is applied to obtain a set of row values. The row values are then used in the averaging equation of (14.2) to obtain a new set of column values. The column values are restandardized (otherwise the sucessive averages would just collapse the values to zero). The above process is repeated until convergence, giving the coordinates on the first principal axis. Finding the second set of coordinates is more complicated because we have to ensure orthogonality with the first, but the idea is the same. We have shown in different ways that the passage from column to row coordinates and row to column coordinates can be described by a regression analysis in each case, so that this flip-flop process is also known as *alternating least-squares*, or alternating regressions. Numerically, it is better to perform the computations using the singular value decomposition (SVD) (see page 47 as well as the Theoretical and Computational Appendices), but for a fuller understanding of CA it is illuminating to be aware of these alternative algorithms.

Reciprocal averaging and alternating least squares

The alternative form of the CA biplot given in (13.7), repeated here,

Contribution coordinates as regression coefficients

$$(\frac{p_{ij}}{r_i} - c_j)/c_j^{1/2} = \sum_{k=1}^{K^*} f_{ik}(c_j^{1/2}\gamma_{jk}) + e_{ij}$$

expresses standardized row profile values as a bilinear function of row points in principal coordinates and column points in contribution coordinates. Thus, similar to the above arguments, this implies that if the standardized values for a particular column are regressed on the row principal coordinates (always using weighted least-squares regression), then the regression coefficients will be exactly the contribution coordinates. This gives an alternative intepretation for the contribution biplot as visualizing these regression coefficients in the map of the row profiles, just like standardized variables can be added as supplementary variables to an existing map.

SUMMARY:
Transition and
Regression
Relationships

1. For any values assigned to the row and column categories, the conditional means (i.e. regressions) can be computed for rows on columns or columns on rows.

2. The CA solution, using standard coordinates on a particular axis as the two sets of values, has the following properties:
   – the two regressions are linear (hence the name *simultaneous linear regressions*);
   – the angle between the two regression lines is minimized;
   – the conditional means that lie on the two regression lines are the principal coordinates.

3. The weighted average relationship between row and column coordinates, when the weights are the elements of profiles (row or column profiles as the case may be) are called *transition equations*. Successive applications of the pair of transition equations lead to an algorithm for finding the CA solution, called *reciprocal averaging*.

4. CA can be defined as a *bilinear regression model*, since the data can be recovered by a model that is linear in products of the row and column coordinates. This model becomes linear if either set of coordinates is regarded as fixed, leading to an algorithm for finding the CA solution called *alternating least-squares regressions* (which, in fact, is identical to the reciprocal averaging algorithm).

5. The CA contribution biplot, showing rows in principal coordinates, has coordinates for the columns with multiple interpretations: (i) the squared coordinates are column contributions to the inertias on corresponding principal axes, useful for deciding which columns are important for the construction of the axes and thus for their interpretation (see Chapter 13); (ii) the coordinates are regression coefficients in the weighted regression of the standardized row profile values for a particular column on the row principal coordinates. Both interpretations lead to the same conclusion: the higher the regression coefficient on a principal axis, the more important is the column's relationship with the axis.

Clustering Rows and Columns

## 15

Up to now we have been transforming data matrices to maps or biplots where the rows and columns are displayed as points in a continuous space, usually a two-dimensional plane. An alternative way of displaying structure consists in performing separate cluster analyses on the row and column profiles. This approach has close connections to correspondence analysis (CA) and decomposes the inertia according to the discrete groupings of the profiles rather than along continuous axes. In the case of a contingency table there is an interesting spin-off of this analysis in the form of a statistical test for significant clustering of the rows or columns.

Contents

Partitioning the rows or the columns

The idea of grouping objects is omnipresent in data analysis. The grouping might be a given classification, or it might be determined according to some criterion that clusters similar objects together. We first consider the former case, when the grouping is established according to a categorical variable which classifies the rows or columns of a table. Taking the scientific research funding example again, suppose that there is a predetermined grouping of the scientific disciplines into four groups, according to university faculties: {Geology, Physics, Statistics, Mathematics}, {Biochemistry, Chemistry}, {Zoology, Microbiology, Botany} and {Engineering}. As we pointed out in Chapter 12, when a categorical variable is defined on the rows, as in this example, each category defines a supplementary row of the table that aggregates the frequencies of the rows corresponding to that category. Thus the ten rows of Exhibit 10.1 are condensed into four rows corresponding to the four groups,

Exhibit 15.1:
*Frequencies of
funding categories
for 796 researchers
grouped into four
categories according
to scientific
discipline.*

SCIENTIFIC GROUPS	*FUNDING CATEGORIES*					
	A	*B*	*C*	*D*	*E*	*Sum*
Geology/Physics/Statistics/Mathematics	17	57	134	35	63	*306*
Biochemistry/Chemistry	7	27	62	22	41	*159*
Zoology/Microbiology/Botany	4	33	89	57	60	*243*
Engineering	3	11	25	15	34	*88*
Sum	*31*	*128*	*310*	*129*	*198*	*796*

shown in Exhibit 15.1. The CA of the original data of Exhibit 10.1 had a total inertia of 0.08288, whereas if we perform the CA of Exhibit 15.1, the total inertia turns out to be 0.04386. There is a loss of inertia when points are merged, or putting this the other way around, there is an increase in inertia if a row or column is split apart according to some subclassification.

*Between- and
within-groups
inertia*

The inertia of the merged table in Exhibit 15.1 is called the *between-groups inertia*, since it measures the variation in the table between the four groups of rows. The difference between the total inertia of 0.08288 and the between-groups inertia of 0.04386 is called the *within-groups inertia*, measuring the variation within the four groups which is lost when we merge the original ten rows into the four groups. This decomposition of inertia is a classic result of analysis of variance, usually applied to a single variable, but equally applicable to multivariate data. In the CA context each row profile, denoted by $\mathbf{a}_i$, has a mass r_i assigned to it, and the average row profile (centroid) is the vector $\mathbf{c}$ of column masses (see Chapter 4). Distances between row profiles are measured by the χ^2-distance: for example, if d_i denotes the χ^2-distance between $\mathbf{a}_i$ and $\mathbf{c}$, then the total inertia is $\sum_i r_i d_i^2$ (formula (4.7), page 29). Between-group inertia is a similar formula, but applied to the merged rows as follows. Suppose $\bar{\mathbf{a}}_g$ denotes the profiles of the merged rows, where $g = 1, \ldots, G$ is the index of the groups (here $G = 4$), and the mass of the g-th group, $\bar{r}_g$, is the sum of the masses of the members of the group. The group profiles $\bar{\mathbf{a}}_g$ still have centroid at $\mathbf{c}$ and, denoting their χ^2-distances to the centroid by $\bar{d}_g$, the between-group inertia is $\sum_g \bar{r}_g \bar{d}_g^2$. Finally, each group g has an inertia with respect to its own centroid $\bar{\mathbf{a}}_g$: if d_{ig} denotes the χ^2-distance from each profile i in group g to the centroid $\bar{\mathbf{a}}_g$, then the inertia within the g-th group is $\sum_{i \epsilon g} r_i d_{ig}^2$, where $i \epsilon g$ means the set of rows in group g. Summing this quantity over all the groups gives the within-groups inertia. The decomposition of inertia is thus:

total inertia = between-groups inertia + within-groups inertia

$$\sum_i r_i d_i^2 = \sum_g \bar{r}_g \bar{d}_g^2 + \sum_g \sum_{i \epsilon g} r_i d_{ig}^2 \qquad (15.1)$$

$$0.08288 = 0.04386 + 0.03902$$

The within-groups inertia is equal to 0.03902, according to the above, but what is the contribution from each of the four groups? One can calculate this directly, remembering to use the same values of **c** for the normalization in all χ^2-distance calculations, but a quicker way is to apply CA to a matrix where, one at a time, we merge the groups. For example, if we merge Geology, Physics, Statistics and Mathematics into the first group and then analyse this merged row along with the other (unmerged) rows (i.e. seven rows in total), the total inertia is 0.06446. Compared to the total inertia 0.08288 of the original data set, the reduction by 0.01842 represents the within-group inertia that was lost in the merging. Then we merge Biochemistry and Chemistry and analyse the matrix with six rows, and the inertia now drops to 0.05382, so the within-group inertia of that group is the difference, $0.06446 - 0.05382 = 0.01064$ and so on. Exhibit 15.2 gives the complete decomposition of inertia, in raw units and percentages. Notice that there is no within-group inertia for the group composed of one row, Engineering — it is 0.

Calculating the inertia within each group

Group	Definition	Component	% of part	% of total
Between-groups inertia:				
Geol/Phys/Stat/Math	$\bar{r}_1\bar{d}_1^2$	0.01482	33.8%	17.9%
Bioc/Chem	$\bar{r}_2\bar{d}_2^2$	0.00099	2.3%	1.2%
Zool/Micr/Bota	$\bar{r}_3\bar{d}_3^2$	0.01548	35.3%	18.7%
Engi	$\bar{r}_4\bar{d}_4^2$	0.01256	28.6%	15.2%
Total	$\sum_g \bar{r}_g\bar{d}_g^2$	0.04386	100.0%	52.9%
Within-groups inertia:				
Geol/Phys/Stat/Math	$\sum_{i\epsilon 1} r_i d_{i1}^2$	0.01842	47.2%	22.2%
Bioc/Chem	$\sum_{i\epsilon 2} r_i d_{i2}^2$	0.01064	27.3%	12.8%
Zool/Micr/Bota	$\sum_{i\epsilon 3} r_i d_{i3}^2$	0.00996	25.5%	12.0%
Engi	$\sum_{i\epsilon 4} r_i d_{i4}^2$	0	0%	0%
Total	$\sum_g \sum_{i\epsilon g} r_i d_{ig}^2$	0.03902	100.0%	47.1%

Exhibit 15.2: *Decomposition of inertia between and within groups, including components of each part and percentages with respect to the part and the total. The sum of the total of the between-groups inertia and the total of the within-groups inertia is the total inertia 0.08288 of the original table (Exhibit 10.1).*

In the above, the partition of the rows into groups was given by available information, but we now consider constructing groups using a particular type of cluster analysis. We use a small data matrix to illustrate the calculations involved. This example is taken from an actual sample of 700 shoppers at five different food stores. The sample has been tabulated according to store and age group, yielding the 5×4 table in Exhibit 15.3. The χ^2 statistic for this table is 25.06, which corresponds to a p-value of 0.015. Thus we would conclude that there exists a significant association between age and choice of store. Alongside the table we show the symmetric CA map. A market

Data set 8: Age distribution in food stores

| FOOD | AGE GROUP (years) | | | | |
STORE	16-24	25-34	35-49	50+	Sum
A	37	39	45	64	185
B	13	23	33	38	107
C	33	69	67	56	225
D	16	31	34	22	103
E	8	16	21	35	80
Sum	107	178	200	215	700

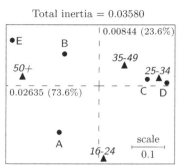

Total inertia = 0.03580

researcher would be interested to know where this significant association is concentrated; for example, which stores or group of stores have a significantly different age profile from the others. The major contrast in the data is between the oldest group on the left of the CA map and the second youngest group on the right. Store E is the most associated with the oldest group and stores C and D tend more towards the younger ages. The vertical axis contrasts the youngest age group with the others. Store A appears to separate from the others towards the youngest age group.

Clustering algorithm

We now construct a partition of the rows and columns using a clustering algorithm which tries to maximize the between-groups inertia and — simultaneously — minimize the within-groups inertia. The clustering algorithm is illustrated in Exhibit 15.4 for the rows. At the start of the process, each row is separate and the between-groups inertia is just the total inertia. Any merging will reduce the between-groups inertia, so the first step is to identify which pair of rows (stores) can be merged to result in the least reduction in the inertia. The two rows which are the most similar in this sense are stores C and D. When these rows are merged to form a new row, labelled (C,D), the inertia for the resultant 4 × 4 table is reduced by 0.00084 to 0.03496, or on the χ^2 scale by 0.59 to 25.06 (in Exhibit 15.4 we report the χ^2 values, which are always the inertia values multiplied by the sample size $N = 700$: $\chi^2 = 0.03496 \times 700 = 25.06$). In percentage terms this is a decrease of 2.3% in χ^2 or, equivalently, in inertia. The procedure is then repeated to find the rows in the new table that again decrease the between-groups inertia the least. These turn out to be stores B and E, leading to a further reduction in χ^2 of 1.53 (6.1%). The table now has three rows labelled A, (B,E) and (C,D). The procedure is repeated and the smallest reduction is found when store A joins the pair (B,E) to form a new row labelled (A,B,E), reducing χ^2 by a further 5.95 (23.7%). Finally, the two rows (A,B,E) and (C,D) merge to form a single row, consisting of the marginal column sums of the original table, for which the χ^2 is zero. The final reduction is thus 16.99 (67.8%), which was the inertia of the penultimate table in Exhibit 15.4. The whole procedure can be repeated on the columns of the table in an identical fashion.

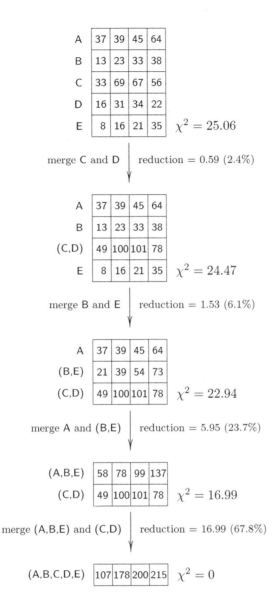

Exhibit 15.4:
Steps in the clustering of the rows of Exhibit 15.1: at each step two rows are merged, chosen to induce the minimum decrease in the χ^2 statistic, or equivalently in the between-group inertia (to convert the χ^2 values to inertias, divide by the sample size $N = 700$).

The successive merging of the rows, called *hierarchical clustering*, can be depicted graphically as a *binary tree* or *dendrogram* — this is shown in Exhibit 15.5 along with a similar hierarchical clustering of the columns. Notice that the ordering of the rows and columns of the original table usually requires modification to accommodate the tree displays, although in this particular example only the rows need to be reordered. The fact that stores C and D are the first to merge is apparent on the tree. The point at which this merging

Tree representations of the clusterings

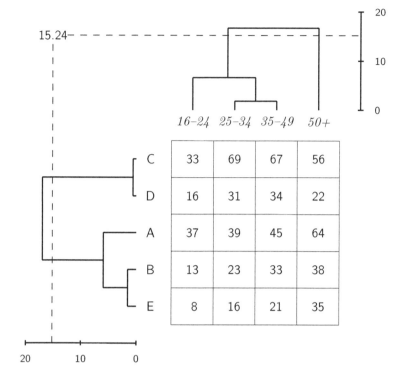

occurs is called a *node*, and in each case the level of the node corresponds to the associated reduction in χ^2.

*Decomposition
of inertia (or χ^2)*

Since the original χ^2 statistic is reduced to zero at the end of the clustering process, it is clear that the set of reductions forms a decomposition of χ^2: $25.06 = 16.99 + 5.95 + 1.53 + 0.59$. Dividing by the sample size 700 gives the corresponding decomposition of inertia: $0.03580 = 0.02427 + 0.00851 + 0.00218 + 0.00084$. The percentages remain the same, whatever decomposition is reported: 67.8%, 23.7%, 6.1% and 2.4%. The columns are merged in an identical fashion and the values of the nodes again constitute a decomposition of inertia (or χ^2): $0.03580 = 0.02383 + 0.00938 + 0.00259$, or in percentage form 66.6%, 26.2% and 7.2%.

*Deciding on the
partition*

In cluster analyses of this type, the trees are inspected to deduce the number of clusters of objects. For example, looking at the row clustering we see that there is a large difference between the two clusters of food stores (C,D) and (A,B,E), indicated by the high value at which these clusters merge. Thanks to the decomposition of inertia, we could say that 67.8% of the inertia is accounted for if we condense the rows into these two clusters. If we separate store A as a third cluster, then a further 23.7% of the inertia is accounted for, i.e. 91.5%. Percentages of inertia associated with the nodes of such a cluster analysis are thus interpreted in much the same way as percentages of inertia of principal axes in CA. The decision as to what percentage is great enough to halt the interpretation is usually an informal one, based on the sequence of percentages and the substantive interpretation of each node or principal axis.

The χ^2 statistic for the original contingency table was reported to be significant ($p = 0.015$); hence somewhere in the table there must be significant differences amongst the profiles. To pinpoint which profiles are significantly different in a statistical sense is not a simple question because there are many possible groupings of the stores that could be tested and the significance level has to be adjusted if many tests are performed on the same data set. Furthermore, particular groupings, for example of stores C with D and B with E, have been suggested by the data themselves and have not been set up as hypotheses before the data were collected.

Testing hypotheses on clusters of rows or columns

Here we are treading the fine line between exploratory and confirmatory data analysis, trying to draw statistical conclusions from data that were analysed in an exploratory fashion with no fixed *a priori* hypotheses. Fortunately, an area of statistical methodology has been developed to cope with this situation, called *multiple comparisons*. This approach is often used in the analysis of experiments where there are several "treatments" being tested against a "control". A multiple comparisons procedure allows any treatment (or group of treatments) to be tested against any other, and statistical decisions may be made at a prescribed significance level to protect all these tests from the so-called "Type I Error", i.e. finding a result which has arisen purely by chance. This is also known as controlling the *false discovery rate*.

Multiple comparisons

As in the case of different treatments in an experimental situation, we would like to test the differences between any two rows, say, of the table or any two groups of rows. If there were only one test to do, we would calculate the reduced table consisting of the two rows (or merged groups) and make a one-off χ^2 test in the usual way. The multiple comparisons procedure developed for this situation allows testing for differences between any two rows (or groups of rows). The usual χ^2 statistic for the reduced table is calculated but compared to different critical values for significance. In the Theoretical Appendix, Exhibit A.1 on page 254, a table of critical points is given for this test, at the 5% significance level, for contingency tables of different sizes. In the present 5×4 example the critical point can be read from the table as 15.24: so if the χ^2 statistic is superior to 15.24, then it can be deduced that the two rows (or groups of rows) are significantly different.

Multiple comparisons for contingency tables

The critical value for the multiple comparison test can be used for any subset or merging of the rows or columns of the table, in particular it can be used for the hierarchical clusterings, allowing us to separate out the statistically

Cut-off χ^2 value for significant clustering

significant groups, as shown in Exhibit 15.5. The interpretation of this cut-off point is that, amongst the age groups, it is really the contrast between the oldest age group and the rest that is statistically significant; and, concerning the food stores, the statistical differences lie between two groups, (A,B,E) and (C,D). Thus the contrast observed along the second axis of Exhibit 15.3 could be due to random variation in the observed data — the separation of the youngest age group from the others is not significant, and the distinction between age groups *16–24* and *35–49* years along the second axis is equally difficult to justify from a statistical point of view. This does not mean, of course, that we are prevented from inspecting the original information in the form of the two-dimensional map of Exhibit 15.3 — the data content is always worth considering irrespective of the statistically significant features. In Chapter 29, page 230, these same critical values will be used for a significance test on the principal inertias of a contingency table.

Ward
clustering

The clustering algorithm described in this chapter is a special case of *Ward clustering*. In this type of clustering, clusters are merged according to a minimum-distance criterion which takes into account the weights of each point being clustered. So, instead of thinking of this as a reduction in χ^2 (or inertia) at each step, the χ^2-distances between the profiles could be used along with the associated masses. For purposes of clustering the specific (squared) "distance" between two row clusters g and h, for example, is then:

$$\frac{\bar{r}_g \bar{r}_h}{\bar{r}_g + \bar{r}_h} \|\bar{\mathbf{a}}_g - \bar{\mathbf{a}}_h\|_c^2 \qquad (15.2)$$

where $\bar{r}_g$ and $\bar{r}_h$ are the masses of the respective clusters, and $\|\bar{\mathbf{a}}_g - \bar{\mathbf{a}}_h\|_c$ is the χ^2-distance between the profiles of the groups.

SUMMARY:
Clustering Rows
and Columns

1. Cluster analyses of the rows or columns provide an alternative way of looking for structure in the data, by collecting together similar rows (or columns) in discrete groups.

2. The results of the clusterings can be depicted graphically in a tree structure (*dendrogram* or *binary tree*), where nodes indicate the successive merging of the rows (or columns).

3. The total inertia (or equivalently the χ^2 statistic) of the table is reduced by a minimum at each successive level of merging of the rows (or columns). This *Ward clustering* procedure provides a decomposition of inertia with respect to the nodes of the tree, analogous to the decomposition of inertia with respect to principal axes in correspondence analysis.

4. Thanks to a *multiple comparisons* procedure, the inertia component accounted for by each node can be tested for significance, leading to statistical statements of difference between groups of rows (or groups of columns). This test applies to valid contingency tables only.

Multiway Tables

## 16

Up to now we have dealt exclusively with two-way tables in which the frequencies of co-occurrence of two variables have been mapped. We now start to consider situations where data are available on more than two variables and how we can explore such data graphically. One approach is to re-format the multiway frequency table as a two-way table and to use the usual simple correspondence analysis (CA) approach. The various ways of reformatting the table give different viewpoints of the association patterns in the data.

Contents

We return to the health assessment data (data set 3) that were discussed at length in Chapters 6 and 7, namely the representative sample of 6371 Spaniards cross-tabulated by age and self-assessment of their own health (Exhibit 6.1). Several other variables are available, for example gender, education, region of residence, and so on. We use the simplest of these, gender with two categories, as an example of how to introduce a third variable into the CA. Two further cross-tabulations can now be made: gender by age group and gender by health category. While the former table might be interesting from a demographic point of view, the latter table is more relevant to the substantive issue of health assessment — see Exhibit 16.1. There is no need to perform a CA of this table to see the pattern in the numbers — this 2×5 table is inherently one-dimensional and all the patterns are clear in the percentages. It is clear that males generally have a better opinion of their health; there are higher percentages of males in the *very good* and *good* categories, while the females are higher in the *regular*, *bad* and *very bad* categories.

Introducing a third variable in the health self-assessment data

We saw previously in Chapter 6 that self-assessed health deteriorated with age. Separately, Exhibit 16.1 shows a gender-related effect, with men on average more optimistic about their health than women. The question now is whether

Interaction between variables

	Very				Very	
GENDER	good	Good	Regular	Bad	bad	*Sum*
male	448	1789	636	177	39	*3089*
%	*14.5*	*57.9*	*20.6*	*5.7*	*1.3*	
female	369	1753	859	237	64	*3282*
%	*11.2*	*53.4*	*26.2*	*7.2*	*2.0*	
Sum	817	3542	1495	414	103	*6371*
%	*12.8*	*55.6*	*23.5*	*6.5*	*1.6*	

Exhibit 16.1:
Cross-tabulation of gender with self-assessed health, showing row profile values as percentages. Data source: Spanish National Health Survey, 1997.

the gender effect is the same across all age groups or whether it changes; for example, it could be that in a particular age group the gender effect is greater or is even reversed. This phenomenon is called an *interaction*, in this case an interaction between age and gender with respect to self-assessed health — notice that we think of the five-category profile of self-assessed health as a single response here. Absence of an interaction would mean that the same gender difference in the response exists across all age groups.

Interactive coding

To be able to visualize possible interactions between gender and age, we need to code the data in a more detailed manner. A new variable is created of all the combinations of the two genders and age groups, giving $2 \times 7 = 14$ combinations in total — this process is called *interactive coding*. The interactively coded variable is then cross-tabulated with the health categories to give the contingency table of Exhibit 16.2.

CA of the interactively coded cross-tabulation

Exhibit 16.3 shows the symmetric map of Exhibit 16.2. Here again the two-dimensional map is given although the result is still highly one-dimensional, as we saw in Chapter 6. In this map there are two points showing male–female differences for each age group. Comparing these pairs of gender points across the age groups, we see consistently that the female point is to the left of the male counterpart, illustrating the effect in Exhibit 16.1 that females are generally less optimistic about their health. There is no reversing of this phenomenon in any age group; however, there are some differences in the distances between the male and female points. At the younger ages the male–female distances are relatively small, up to the **35-44** age group. In the **45-54** age group, where large changes appear in self-assessed health, there is also a bigger difference between men and women. This change is maintained in the older age groups, and females in the **55-64** age group are seen to be more pessimistic than males in the higher **65-74** age group. Similarly, females in age group **65-74** are more pessimistic than males in the older group **75+**. This changing difference between men and women across the age groups is evidence of a gender–age interaction when it comes to self-assessment of health.

Data set 9: Attitudes to working women

As another illustration of interactive coding, we now introduce a data set that we shall be using several times in this and following chapters. These data are taken from the International Social Survey Programme (ISSP) survey of

GENDER–AGE	Very good	Good	Regular	bad	Very Bad	Sum
m16-24	145	402	84	5	3	639
m25-34	112	414	74	13	2	615
m35-44	80	331	82	24	4	521
m45-54	54	231	102	22	6	415
m55-64	30	219	119	53	12	433
m65-74	18	125	110	35	4	292
m75+	9	67	65	25	8	174
f 16-24	98	387	83	13	3	584
f 25-34	108	395	90	22	4	619
f 35-44	67	327	99	17	4	514
f 45-54	36	238	134	28	10	446
f 55-64	23	195	187	53	18	476
f 65-74	26	142	174	63	16	421
f 75+	11	69	92	41	9	222

Exhibit 16.2:
Cross-tabulation of interactively coded gender–age variable with self-assessed health (m=male, f=female, seven age groups as in Exhibit 6.1). Each row of Exhibit 6.1 has been subdivided into two rows according to gender.

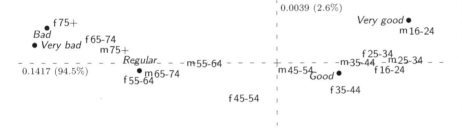

Exhibit 16.3:
Symmetric CA map of interactively coded gender–age variable cross-tabulated with health categories; the gender–age profiles are situated at the positions of the m and f labels.

Family and Changing Gender Roles in 1994, involving a total sample of 33590 respondents and conducted in 24 countries (former East and West Germany are still considered separately in the ISSP surveys, as are Great Britain and Northern Ireland). For our purposes here we consider the relationships between some demographic variables and the responses to the following question related to women's participation in the labour market: "Considering a woman who has a schoolchild at home, should she work full-time, work part-time, or stay at home?" As in all such questionnaire surveys, there is an additional response option "unsure/don't know" to which we have also added the few non-responses (dealing with non-responses will be discussed in more detail in Chapter 21). In addition to the responses to this question, we have data on several demographic variables for each respondent, of which the following three will be of interest here: gender (2 categories), age (6 categories) and country (24 categories). The response frequencies for each country are given in Exhibit 16.4.

COUNTRIES		*W*	*w*	*H*	*?*	*Sum (N)*
AUS	Australia	256	1156	176	191	*1779*
DW	West Germany	101	1394	581	248	*2324*
DE	East Germany	278	691	62	66	*1097*
GB	Great Britain	161	646	70	107	*984*
NIRL	Northern Ireland	126	394	75	52	*647*
USA	United States	482	686	107	172	*1447*
A	Austria	84	632	202	59	*977*
H	Hungary	285	736	447	32	*1500*
I	Italy	171	670	167	10	*1018*
IRL	Ireland	223	424	209	82	*938*
NL	Netherlands	539	1205	143	81	*1968*
N	Norway	487	1242	205	153	*2087*
S	Sweden	295	833	39	105	*1272*
CZ	Czechoslovakia	228	585	198	13	*1024*
SLO	Slovenia	341	428	222	41	*1032*
PL	Poland	431	425	589	152	*1597*
BG	Bulgaria	270	427	335	94	*1126*
RUS	Russia	175	1154	550	119	*1998*
NZ	New Zealand	120	754	72	101	*1047*
CDN	Canada	566	497	108	269	*1440*
RP	Philippines	243	448	484	25	*1200*
IL	Israel	468	664	92	63	*1287*
J	Japan	203	671	313	120	*1307*
E	Spain	738	1012	514	230	*2494*
Sum		*7271*	*17774*	*5960*	*2585*	*33590*
%		*21.7%*	*52.9%*	*17.7%*	*7.7%*	

Exhibit 16.4:
Frequencies of response to question on women working when they have a schoolchild at home, for 24 countries (source: ISSP survey on Family and Changing Gender Roles, 1994; West and East Germany are still kept separate), with the average profile in percentage form. The following abbreviations are used: W=work full-time, w=work part-time, H=stay at home, ?=don't know/unsure/missing.

Basic CA map of countries by responses — The CA map of this table is shown in Exhibit 16.5 (here the graphical style of the CA maps is changing slightly — different software options for producing the maps will be discussed at the end of the Computational Appendix). The interpretation of this map is quite clear; the contrast from left to right is between women working (on the left) versus women staying at home (on the right), while the vertical contrast is between women working full-time (at the top) versus women working part-time (at the bottom). Countries such as the Philippines and Poland are the most traditional on this issue, whereas countries such as Sweden, East Germany, Israel, New Zealand, Great Britain and Canada are the most liberal. On the left, the difference between the countries in the vertical direction separates out those such as Canada that are the most in favour of women working full-time versus New Zealand, for example, more in favour of part-time employment. Remember that the origin of the map represents the average profile in the last row of Exhibit 16.4, so that all countries on the left are more liberal than average, while if two countries

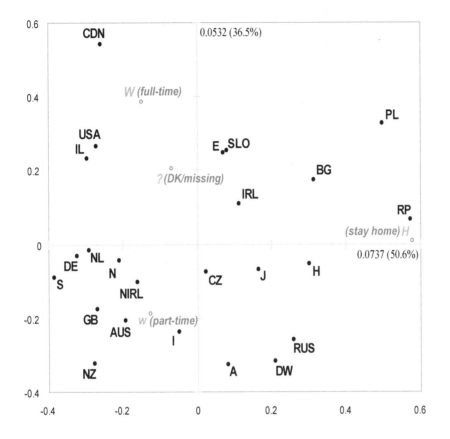

are at the same position on the horizontal axis (for example, USA and Great Britain) the country more positive on the vertical axis will be more in favour of women working full-time than part-time.

Gender can now be interactively coded with country in order to visualize male–female differences, as was done for the health assessment example. However, there is a weighting issue that needs attention, and which we avoided in the previous example: there are many more women respondents in these samples than men (in Exhibit 16.1, women formed 51.5% of the data set, but here they are 54.6% overall). We prefer to express each male and female sample relative to its respective total, which is in effect reweighting the samples. Male and female samples now have the same weight, as well as the countries, since there is no substantive reason why the countries should be weighted proportional to their sample sizes. Exhibit 16.6 shows the first and last rows of the 48 × 4 table of percentages. The map in Exhibit 16.7 has not changed much in terms of the positions of the reponse categories, but it is interesting to compare the pairs of points for each country. In almost all cases the female point is more to the left compared to the male counterpart (Bulgaria is the only

Introducing gender interactively

Exhibit 16.6:
Percentages of
response to question
on women working
when they have a
schoolchild at home,
for 24 countries,
i.e. Exhibit 16.4
split by gender and
then expressed as
row percentages
with respect to the
respective sample
size N. Small
differences in sample
sizes compared to
Exhibit 16.4 are due
to missing values for
gender.

COUNTRY	W	w	H	?	N
AUSm	12.9	65.6	12.5	9.0	*909*
AUSf	15.9	64.5	6.9	12.6	*866*
DWm	3.6	56.3	29.8	10.3	*1198*
DWf	5.2	63.9	19.9	11.1	*1126*
DEm	27.7	59.8	5.5	7.0	*528*
DEf	23.2	65.9	5.8	5.1	*569*
...	...	...	...	...	...
...	...	...	...	...	...
ILm	37.9	47.3	9.8	5.0	*581*
ILf	35.1	55.0	5.0	4.8	*703*
Jm	14.4	47.1	28.9	9.6	*592*
Jf	16.5	54.8	19.9	8.8	*715*
Em	29.0	37.2	24.6	9.3	*1197*
Ef	30.2	43.8	16.9	9.1	*1292*

exception). Gender attitudes within a country are surprisingly homogeneous compared to the large between-country differences. The countries where there is the biggest distance between male and female opinion are mostly on the conservative side of the map, for example the Philippines, Japan, Northern Ireland, West Germany and Spain, while on the left side of the map Australia shows one of the biggest male–female differences. In this analysis the inertia must be higher than the previous one since the splitting of the samples by gender must add inertia (see Chapter 15). We return to this subject of inertia components again in Chapter 23.

Introducing age
group and gender

Since the sample sizes in each country are so large, we can split the samples even further by age, that is each country–gender group is subdivided into six age groups: up to 25 years old, 26-35, 36-45, 46-55, 56-65, and 66+ years. Hence we code interactively three variables into one, with $24 \times 2 \times 6 = 288$ categories in total. The CA of the resultant 288×4 table is shown in Exhibit 16.8, and again remains remarkably stable as far as the response categories are concerned. The 288 row points are represented by dots since it is impossible to label each one. Some outlying points are labelled; for example, the most liberal group lying far out at top left is the youngest group of female Canadians up to 25 years old. Of this subsample of 168 women, 101 (60.1%) are in favour of women with a schoolchild at home working full-time, 32 (19.0%) respond part-time, 3 (1.8%) say women should stay at home, and 32 (19.0%) do not respond or are missing (as we shall see in Chapter 21, there are a lot of "don't knows" in the Canadian sample as a whole). The most liberal male group is the youngest East German male group. At the other extreme on the right we have the oldest group of Hungarian and Polish males; for example, of the 76 Polish men 66 years or older, 16 (21.1%) respond full-time, 13 (17.1%) part-time, 41 (53.9%) stay at home, with a non-response of 6 (7.9%). At the

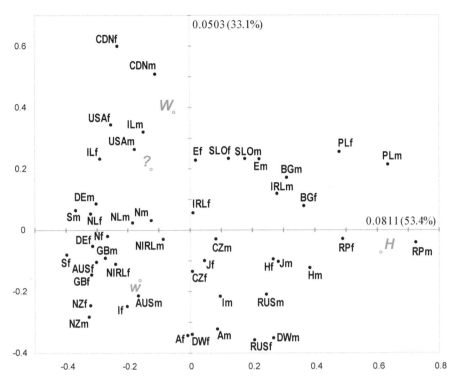

Exhibit 16.7:
Symmetric CA map of interactively coded data (Exhibit 16.6). The male points are consistently to the right of their female counterparts, with the exception of Bulgaria (BG), where females have a more conservative attitude than males (also New Zealand (NZ), but very slightly).

bottom we have the oldest group of New Zealand males — these will be the most in favour of part-time work.

Finally, notice the curve of the cloud of points in Exhibit 16.8, called the *arch effect* or the *horseshoe*, which follows the shape of the sequence of categories *W*–*?*–*w*–*H*. This commonly found phenomenon is a result of the profile space being a simplex, in the present case a tetrahedron in three dimensions since there are four columns (cf. the two-dimensional triangular geometry of profiles with three elements in Chapter 2). Any gradient of change from one extreme corner of the space (*W*: work full-time) to the other (*H*: stay at home), via the intermediate category (*w*: work part-time), will follow a curved path in this restricted space, rather than a straight line. Points that lie inside the arch, such as the labelled group of Polish males 26–35 years old, will tend to be polarized in the sense of being high on the two extreme responses. Of the 141 Polish respondents in this specific age group, 45 (31.9%) respond full-time, 31 (22.0%) part-time, 45 (31.9%) stay at home, and 20 did not respond (14.2%) — so this group does have above average responses on both extremes *W* and *H* — see the percentages in the last row of Exhibit 16.4, which show 21.7% as the overall average percentage for work full-time and 17.7% for stay at home.

Arch ("horseshoe") pattern in the map

Exhibit 16.8:
Symmetric CA map of three-way interactively coded data. The country–gender–age groups are represented by dots, and form a curved pattern that is encountered frequently in CA maps when the profiles fall on a gradient from one extreme (W) to the other (H).

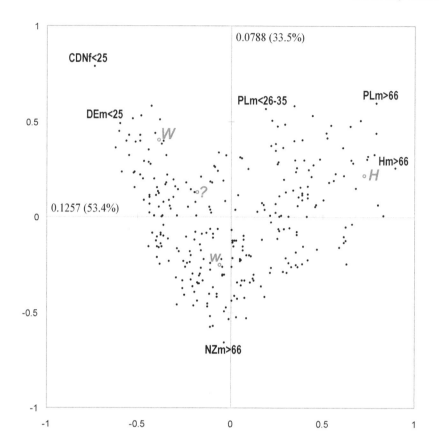

SUMMARY:
Multiway Tables

1. Two or more categorical variables can be *interactively coded* into a new variable that consists of all combinations of the categories. For example, two variables with J_1 and J_2 categories would be coded into a new variable with $J_1 J_2$ categories.

2. CA can be applied to a table that cross-tabulates an interactively coded variable with another variable. The resulting map shows the interaction pattern between the variables that have been interactively coded, in their relationship to the other variable.

3. Interactive coding of multiway tables would normally not proceed beyond three variables, since the number of categories increases rapidly, as well as the complexity of the map. The level of interaction that can be investigated depends on how much data are available, because interactive coding fragments the sample into subsamples with lower sample sizes, and these subsamples should not be too small, otherwise their positions in the map are imprecise.

Stacked Tables

## 17

Survey research in the social sciences usually involves a multitude of variables. For example, in a questionnaire survey there are many questions related to the survey objective as well as many demographic characteristics that are used to interpret and explain respondents' answers. The advantage of correspondence analysis (CA) is the ability to visualize many variables simultaneously, but there is a limit to the number of variables that can be interactively coded, as illustrated in the previous chapter, owing to the large number of category combinations. When there are many variables an alternative procedure is to code the data in the form of *stacked*, or *concatenated*, tables. The relationship between each demographic variable and each attitudinal variable can then be interpreted in a joint map. In this chapter we give examples of this approach, both when there are several demographic characteristics and when there are responses to several questions.

Contents

We now expand the data set from Chapter 16 on attitudes towards working women by including, in addition to country (24 categories, see Exhibit 16.4 for abbreviations), gender (2 categories, M and F) and age group (6 categories, A1 to A6), the two variables marital status (5 categories) and education level (7 categories), totalling five demographic variables. The definitions and abbreviations of the two additional variables are as follows:

Several demographic variables, one question

— *Marital status:* ma (married), wi (widowed), di (divorced), se (separated), si (single)

— *Education:* E1 (no formal education), E2 (incomplete primary), E3 (primary), E4 (incomplete secondary), E5 (secondary), E6 (incomplete tertiary), E7 (tertiary)

Exhibit 17.1:
Stacking of contingency tables which separately cross-tabulate five demographic variables with the responses to the question on women working (W=full-time, w=part-time, H=stay at home, ?=don't know/non-response).

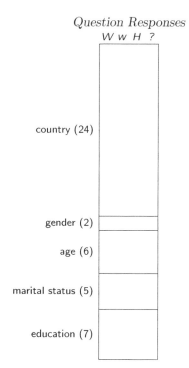

Stacking as an alternative to interactive coding

It is clearly not possible to code interactively all five variables: the number of combinations would be $24 \times 2 \times 6 \times 5 \times 7 = 10080$ combinations! As an alternative, each demographic variable can be cross-tabulated with the responses and the contingency tables can be stacked on top of one another, as depicted in Exhibit 17.1. The top table is the one in Exhibit 16.4, with countries as rows, then the table with two rows for gender, then six rows for age group and so on, constituting a table with $24 + 2 + 6 + 5 + 7 = 44$ rows, one for each demographic category. This type of coding will not reveal interactions and should be regarded as a type of average CA of the five individual tables.

CA of stacked tables

Applying CA to the 44×4 matrix of stacked tables results in the map of Exhibit 17.2. The relative positions of the four reponses, *W*, *w*, *H* and *?*, appear almost the same as in Exhibit 16.8. Compared to Exhibits 16.5 and 16.7 the positions are slightly rotated (rotations of CA solutions are discussed in the Epilogue). Each demographic category is defined by a profile of responses and finds its position in the map relative to the four response categories. The following features of the map are of special interest:

- The categories of an ordinal variable such as age can be connected, as shown in the map. Age follows the curved pattern of the responses *W–w–H* from liberal to traditional, as might be expected.

- Education has a similar pattern, but from right to left, except for category E1 (no formal education), which is near the average.

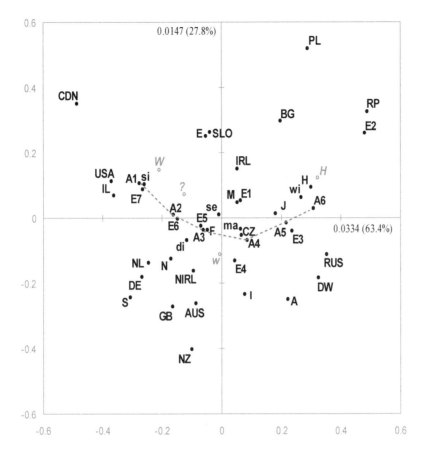

Exhibit 17.2:
*Symmetric CA map
of five stacked
contingency tables
shown schematically
in Exhibit 17.1;
total inertia =
0.05271, percentage
inertia in map:
91.2%.*

- Categories of marital status show si (single) on the liberal side, and wi (widowed) on the traditional side, probably correlated with age group.

- The male and female points M and F lie opposite each other with respect to the average, showing the overall differences between males and females across all countries (we saw the specific differences in Exhibit 16.7).

- Of all the demographical variables, the cross-national differences are still the most important on this issue.

- Countries such as Spain, Slovenia, Ireland and Bulgaria that lie within the arch are polarized countries with higher than average percentages of both *W* (work full-time) and *H* (stay at home) responses.

- The non-response point *?* lies more towards the liberal side of the map; i.e. its profile across the demographics is more similar to *W* than to *H* (in Chapter 21 we will see that Canada, for example, has a high percentage of non-responses).

Limitations in interpreting analysis of stacked tables

It is important to realize that Exhibit 17.2 is showing the separate associations between the demographic variables and the question responses, and not the relationships amongst the demographic variables. There is no information in the stacked tables about the relationship between age, education and country; for example, the fact that the youngest age group A1, the highest education group E7, and the countries Canada, USA and Israel all lie on the left-hand side does not mean that these countries have predominantly younger, more highly educated respondents. Since the variables are being related separately to the question responses, the interpretation is that the youngest age group, the highest education group and these countries all have a predominant, higher than average percentage of *W* (work full-time) responses. To confirm any relationships between the demographic variables, cross-tabulations between them would need to be made and analysed.

Decomposition of inertia in stacked tables

A very useful result here and in future chapters is the fact that when the same individuals are cross-tabulated and stacked, as in Exhibit 17.1, the total inertia in the stacked CA is the average of the inertias in the individual CAs. This result is illustrated by calculating the inertias in each of the five cross-tabulations shown in Exhibit 17.1:

Table	Inertia
Country	0.14558
Gender	0.00452
Age	0.04216
Marital Status	0.02675
Education	0.04221
Average	*0.05224*

The total inertia of the stacked analysis is 0.05271, slightly higher than the above figure, because there are some missing data for some of the demographics, which introduces some additional inertia into the stacked analysis. The totals in each table vary from 30471 for education (the whole Spanish sample, for example, has education coded as "not available") to 33590 (the full sample) for age and country. The effect of the different totals is to increase the total inertia in the stacked analysis, but only slightly since there are small differences in the column totals of each table. For the above decomposition to hold exactly each table must have the same grand total and thus exactly the same column marginal totals. Looking at the above table of inertias also shows how much more inertia there is between countries than between categories of the other variables; hence the relationship of the question responses with countries must dominate the results.

Stacking tables row- and columnwise

The idea of stacking can be broadened to include additional questions which are cross-tabulated with the demographics. In the International Social Survey Programme (ISSP) survey from which these data are taken, there were in fact four questions relating to attitudes about women working, each with the same set of four responses: work full-time, work part-time, stay at home and a category gathering the various non-responses. The respondents were asked about women working or not when they were (1) married with no children, (2) with a pre-school child at home, (3) with a schoolchild living at home (the question we have been analysing up to now), and (4) when all children are no longer living at home. Each of the five demographic variables can be cross-tabulated with each of these four questions, leading to 20 contingency tables which can be stacked row- and columnwise as shown schematically in Exhibit 17.3.

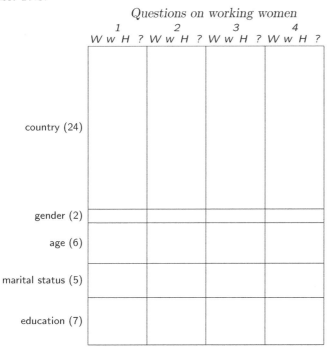

Questions on working women

Exhibit 17.3:
Stacking of contingency tables which separately cross-tabulate five demographic variables with the responses to the question on women working (W=full-time, w=part-time, H=stay at home, ?=don't know/ non-response).

Applying CA to the 20 tables stacked in this block format of five rows and four columns leads to the map in Exhibit 17.4, where each category is represented by a point. The following features of the map are of special interest:

CA of row- and columnwise stacked tables

- The 16 column categories form an even clearer arch pattern, stretching from *2W* and *3W* at top left down to *3w*, *4w* and *2H* at the bottom and up to *4H* and *1H* at top right. This is a typical result in CA where there is what ecologists call a *gradient* in the data, the gradient here being the liberal to traditional spread of attitudes. More or less one can order the categories along this curved gradient as follows (omitting the non-responses (*?*) from the discussion for the moment):

  2W – 3W – 2w & 4W – 1W – 3w – 4w – 2H – 1w – 3H – 4H – 1H

  which shows how the categories line up from extreme liberal attitudes on

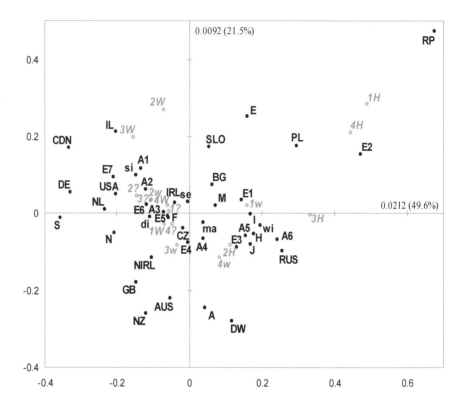

Exhibit 17.4:
*Symmetric CA map
of 20 stacked
contingency tables
shown schematically
in Exhibit 17.3;
total inertia =
0.04273, percentage
inertia in map:
71.1%.*

the left (women should work full-time even when they have children at home) to extreme traditional ones on the right (women should stay at home even though there are no children at home).

- Most of the demographic points lie along this curve, but there is a substantial spread along the second dimension which opposes groups with a polarized opinion (upper part of map, especially Spain) with groups that have higher than average percentages of the intermediate categories *3w*, *4w*, *2H* (lower part of the map, with New Zealand, Australia, Austria and West Germany). Notice that staying at home with a pre-school child (*2H*) is grouped in this intermediate part of the attitude gradient.

- The four non-response points are all in a small bunch just left of the average — in fact, these points are better represented on the third dimension of this analysis; in other words they should be imagined coming out of the page towards you, which means that the third dimension is mostly a dimension which will line up the demographic groups in terms of their non-response percentages over the four questions.

*Partitioning of
the inertia over all
subtables*

Again, the result mentioned previously about decomposition of inertia will apply here. First, the exact result is that if every contingency table in the stacked table is the cross-tabulation of exactly the same number of respon-

Variable	Qu. 1	Qu. 2	Qu. 3	Qu. 4	Average
Country	0.15268	0.12834	0.14558	0.13410	*0.14018*
Gender	0.00821	0.00336	0.00452	0.00484	*0.00523*
Age	0.01033	0.03359	0.04216	0.01266	*0.02469*
Marital Status	0.00529	0.01341	0.02675	0.00869	*0.01354*
Education	0.02306	0.02380	0.04221	0.02430	*0.02834*
Average	*0.03991*	*0.04050*	*0.05224*	*0.03692*	*0.04239*

Exhibit 17.5:
Inertias of 20 contingency tables in the stacked table analysed in Exhibit 17.4; averages of the rows and columns are given as well as the overall average.

dents, then the inertia of the stacked table is the average of the inertias of the individual contingency tables. Let us state this a little more formally, since we will be using this result again in the next chapter. Suppose $\mathbf{N}_{qs}$, $q = 1, \ldots, Q$, $s = 1, \ldots, S$ are contingency tables cross-tabulating Q categorical variables pairwise with another set of S categorical variables for the same n individuals (in our example, $Q = 5$ and $S = 4$). Let $\mathbf{N}$ be the stacked table formed by stacking row- and columnwise the $Q \times S$ tables. Then

$$\text{inertia}(\mathbf{N}) = \frac{1}{QS} \sum_{q=1}^{Q} \sum_{s=1}^{S} \text{inertia}(\mathbf{N}_{qs}) \tag{17.1}$$

This result holds approximately if there is a loss of data in some of the contingency tables owing to missing values. In the present example we have combined the missing values for the four questions about women working with other responses such as "don't know" in their respective *?* categories, so their "'missingness" is specifically coded. But there are some missing values for the demographic variables as a result of the data collection in different countries, which will affect the result in (17.1), which then doeds not hold exactly. In fact, the inertia of the stacked table $\mathbf{N}$ (left-hand side of (17.1)) will increase by a small amount ϵ because of differences between the marginal frequencies, so that (17.1) becomes:

$$\text{inertia}(\mathbf{N}) = \frac{1}{QS} \sum_{q=1}^{Q} \sum_{s=1}^{S} \text{inertia}(\mathbf{N}_{qs}) + \epsilon \tag{17.2}$$

Exhibit 17.5 reports the inertias of all the contingency tables, as well as row and column averages and overall average: as expected, the total inertia in the stacked analysis is slightly higher (0.04273) than the average of the tables (0.04239), which gives $\epsilon = 0.00034$, i.e. an increase of 0.8%. Exhibit 17.6 expresses the 20 inertia values in Exhibit 17.5 (i.e. values of inertia($\mathbf{N}_{qs}$)) as permills of 0.04273×20 (the left-hand side of (17.2) multiplied by $QS = 20$) to be able to judge the quantities more easily, just as was done in Chapter 11 to facilitate the interpretation of the numerical contributions. This shows that on average the countries account for 65.6% of the inertia in the stacked analysis, followed by education (13.3%) and age (11.6%). On question 3 the inertias are generally higher (30.6% of total inertia) — i.e. there are more differences between the demographic groups for question 3 — while on question 4 they are

Exhibit 17.6:
Permill contribu-
tions of 20 tables to
the inertia of the
stacked table; the
remaining 8 permills
(0.8%) is the extra
inertia caused by
the differing
marginal totals due
to missing data.

Variable	Qu. 1	Qu. 2	Qu. 3	Qu. 4	Total
Country	179	150	170	157	*656*
Gender	10	4	5	6	*24*
Age	12	39	49	15	*116*
Marital Status	6	16	31	10	*63*
Education	27	28	49	28	*133*
Total	*234*	*237*	*306*	*216*	*992*

generally lower (21.6% of total inertia). The total of 992 for this table shows, as we have already remarked above, that 0.8% of the inertia is accounted for by the small disparities between the margins of the 20 contingency tables owing to missing data, i.e. the contribution of ϵ to (17.2).

Only
"between"
associations
displayed, not
"within"

Once again we stress the limits of our interpretation of a map such as Exhibit 17.4. When it comes to the four questions, it should be remembered that we are not analysing the associations *within* this set of questions, but rather the associations *between* them and the demographic variables. Analysing associations within a set of variables is the subject of the next chapter on multiple correspondence analysis.

SUMMARY:
Stacked Tables

1. An approach to analysing the responses to several questions and their relationships to demographic variables is to concatenate all the contingency tables that cross-tabulate the two sets of variables and to analyse this *stacked table* by regular CA.

2. The interpretation of the CA map of a stacked table is always made bearing in mind that the information being analysed is the set of pairwise relationships between each question and each demographic variable. There is no specific information being mapped about relationships amongst the questions or amongst the demographics.

3. The analysis of a stacked table can be thought of as a consensus or average map from all the CAs of the individual contingency tables.

4. The total inertia of the stacked table is the average of the inertias of each subtable, when the row margins in each row of subtables and the column margins in each column of subtables are identical (this is true when the same number of individuals are cross-tabulated in each subtable). When there is some loss of individuals in some subtables due to missing data, this result is approximate and the total inertia of the stacked table will be slightly higher than the average of the individual subtable inertias.

Multiple Correspondence Analysis

## 18

Up to now we have analysed the association between two categorical variables or between two sets of categorical variables where the row variables have a different substantive "status" compared to the column variables, e.g. demographics versus survey questions. In this and the following two chapters we turn our attention to the association *within* one set of variables of similar status, where interest is in how strongly and in which way these variables are interrelated. In the present chapter we will concentrate on the two classic ways to approach this problem, called *multiple correspondence analysis* (MCA). One way is to think of MCA as the analysis of the whole data set coded in the form of dummy variables, called the *indicator matrix*, while the other way is to think of it as analysing all two-way cross-tabulations amongst the variables. These two ways are very closely connected, but suffer from some deficiencies that will be resolved in Chapter 19, where some improved versions of MCA are presented.

Contents

In this chapter we are concerned with a single set of variables, usually in the context of a specific phenomenon of interest. For example, the four variables used in Chapter 17, on whether women should work or not, could be such a set of interest, or a set of questions about people's attitudes to science, or a set of categorical variables describing environmental conditions at several terrestrial locations. The set of variables is "homogeneous" in that the variables are of the same substantive type, and often measured on the same scale.

A single set of "homogeneous" categorical variables

As an example let us consider the four questions on working women analysed in Chapter 17. The explanation is simplified by avoiding the large range of cross-cultural differences seen in previous analyses, using only the data from

Indicator matrix

Exhibit 18.1:
*Raw data and the
indicator (dummy
variable) coding, for
the first six
respondents out of
N = 3418.*

| Questions | | | | Qu. 1 | | | | Qu. 2 | | | | Qu. 3 | | | | Qu. 4 | | | |
1	2	3	4	W	w	H	?	W	w	H	?	W	w	H	?	W	w	H	?
1	3	2	2	1	0	0	0	0	0	1	0	0	1	0	0	0	1	0	0
2	3	3	2	0	1	0	0	0	0	1	0	0	0	1	0	0	1	0	0
4	3	3	2	0	0	0	1	0	0	1	0	0	0	1	0	0	1	0	0
4	4	4	4	0	0	0	1	0	0	0	1	0	0	0	1	0	0	0	1
4	4	4	4	0	0	0	1	0	0	0	1	0	0	0	1	0	0	0	1
1	3	2	1	1	0	0	0	0	0	1	0	0	1	0	0	1	0	0	0

... and so on for 3418 rows

one country, Germany, but including both the West and East German samples for comparison, totalling 3418 respondents (three cases with some missing demographic information were omitted from the original samples — see Computational Appendix, page 272). As before, these four variables are labelled 1 to 4, and each has four categories of response, labelled: *W* (work full-time), *w* (work part-time), *H* (stay at home) and *?* (don't know/non-response). The *indicator matrix* is the 3418 × 16 matrix which codes all responses as dummy variables, where the rows are the respondents and the 16 columns correspond to the 16 possible response categories. Exhibit 18.1 illustrates this coding for the first six rows: for example, the first respondent has responses 1, 3, 2 and 2 to the four questions, which are then coded as [1 0 0 0] indicating the response 1 (*W*) to question 1, [0 0 1 0] indicating the response 3 (*H*) to question 2, and [0 1 0 0] indicating the responses 2 (*w*) to both questions 3 and 4.

*MCA definition
number 1: CA of
the indicator
matrix*

The most common definition of MCA is that it is simple CA applied to this indicator matrix. This would provide coordinates for all 3418 rows and 16 columns, but it is mainly the positions of the 16 category points that are of interest for the moment, shown in Exhibit 18.2. The first principal axis shows all four non-response categories together, opposing all the substantive responses. In the previous analysis of these questions (see Exhibit 17.4) where the responses were related to demographic variables, the non-response points were not prominent on the first two axes. But here, because we are looking at relationships within the four questions, this is the most important feature: people who do not respond to one question tend to do the same for the others — for example, amongst the first six respondents in Exhibit 18.1 there are already two respondents who have non-responses for all four questions. On the second axis of Exhibit 18.2, we have the line-up of substantive categories from traditional attitudes at the bottom to liberal attitudes on top. Exhibit 18.3 shows the second and third dimensions of the map, which effectively partials out most of the effect of the non-response points. The positions of the points are now strikingly similar to those in Exhibit 17.4, apart from the fact that

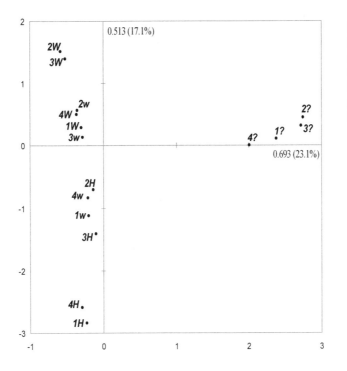

Exhibit 18.2:
MCA map of four questions on women working; total inertia = 3, percentage inertia in map: 40.2%.

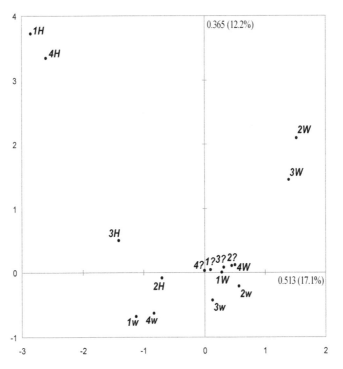

Exhibit 18.3:
MCA map of four questions on women working, showing second and third dimensions; total inertia = 3, percentage inertia in map: 29.3%.

the liberal attitude is now on the right (positive side) of the second dimension, compared to the first dimension of Exhibit 17.4. This flipping of a dimension is of no consequence to the interpretation; in fact, it is always possible to reverse a dimension by multiplying all its coordinates by -1.

Inertia of indicator matrix

The total inertia of an indicator matrix takes on a particularly simple form, depending only on the number of questions and number of response categories and not on the actual data. Suppose there are Q variables, and each variable q has J_q categories, with J denoting the total number of categories: $J = \sum_q J_q$ (in the present example, $Q = 4$, $J_q = 4$, $q = 1, \ldots, Q$, so $J = 16$). The indicator matrix, denoted by $\mathbf{Z}$, with J columns, is composed of a set of subtables $\mathbf{Z}_q$ stacked side by side, one for each variable, and the row margins of each subtable are the same, equal to a column of ones. Thus the result (17.1) in Chapter 17 applies: the total inertia of the indicator matrix is equal to the average of the inertias of the subtables. Each subtable $\mathbf{Z}_q$ has a single one in each row, otherwise zeros, so this is an example of a matrix where all the row profiles lie at the vertices, the most extreme association possible between rows and columns; hence the inertias are 1 on each principal axis of the subtable, and the total inertia of subtable $\mathbf{Z}_q$ is equal to its dimensionality, which is $J_q - 1$. Thus the inertia of $\mathbf{Z}$ is the average of the inertias of its subtables:

$$\text{inertia}(\mathbf{Z}) = \frac{1}{Q} \sum_q \text{inertia}(\mathbf{Z}_q) = \frac{1}{Q} \sum_q (J_q - 1) = \frac{J - Q}{Q} \qquad (18.1)$$

Since $J - Q$ is the dimensionality of $\mathbf{Z}$, the average inertia per dimension is $1/Q$. Notice that the first three dimensions that were interpreted in Exhibits 18.2 and 18.3 have principal inertias 0.693, 0.513 and 0.365, all above the average of $1/4 = 0.25$. The value $1/Q$ serves as a threshold for deciding which axes are worth interpreting in MCA (analogous to the average variance threshold of 1 for the eigenvalues in principal component analysis).

Burt matrix

An alternative data structure for MCA is the set of all two-way cross-tabulations of the set of variables being analysed. The complete set of pairwise cross-tabulations is called the *Burt matrix*, shown in Exhibit 18.4 for the present example. The Burt matrix is a 4×4 block matrix, with 16 subtables. Each of the 12 off-diagonal subtables is a contingency table cross-tabulating the 3418 respondents on a pair of variables. The Burt matrix is symmetric so there are only 6 unique cross-tabulations, which are transposed on either side of the diagonal blocks. The diagonal subtables (by which we mean the tables on the block diagonal) are cross-tabulations of each variable with itself — these are diagonal matrices with the marginal frequencies of the variables down their diagonals. For example, the marginal frequencies for question 1 are 2501 *W* responses, 476 *w*s, 79 *H*s and 362 *?*s. The Burt matrix, denoted by $\mathbf{B}$, is simply related to the indicator matrix $\mathbf{Z}$ as follows:

$$\mathbf{B} = \mathbf{Z}^\mathsf{T}\mathbf{Z} \qquad (18.2)$$

	1W	1w	1H	1?	2W	2w	2H	2?	3W	3w	3H	3?	4W	4w	4H	4?
1W	2501	0	0	0	172	1107	1131	91	355	1710	345	91	1766	538	40	157
1w	0	476	0	0	7	129	335	5	16	261	181	18	128	293	17	38
1H	0	0	79	0	1	6	72	0	1	17	61	0	14	21	38	6
1?	0	0	0	362	1	57	108	196	7	96	55	204	51	45	2	264
2W	172	7	1	1	181	0	0	0	127	48	4	2	165	15	0	1
2w	1107	129	6	57	0	1299	0	0	219	997	61	22	972	239	13	75
2H	1131	335	72	108	0	0	1646	0	24	989	573	60	760	616	84	186
2?	91	5	0	196	0	0	0	292	9	50	4	229	62	27	0	203
3W	355	16	1	7	127	219	24	9	379	0	0	0	360	14	1	4
3w	1710	261	17	96	48	997	989	50	0	2084	0	0	1348	567	23	146
3H	345	181	61	55	4	61	573	4	0	0	642	0	202	286	73	81
3?	91	18	0	204	2	22	60	229	0	0	0	313	49	30	0	234
4W	1766	128	14	51	165	972	760	62	360	1348	202	49	1959	0	0	0
4w	538	293	21	45	15	239	616	27	14	567	286	30	0	897	0	0
4H	40	17	38	2	0	13	84	0	1	23	73	0	0	0	97	0
4?	157	38	6	264	1	75	186	203	4	146	81	234	0	0	0	465

Exhibit 18.4:
Burt matrix of all two-way cross-tabulations of the four variables of the example on attitudes to women working. Down the diagonal are the cross-tabulations of each variable with itself.

The other "classic" way of defining MCA is the application of CA to the Burt matrix **B**. Since **B** is a symmetric matrix, the row and column solutions are identical, so only one set of points is shown — see Exhibit 18.5. Because of the direct relationship (18.2), it is no surprise that the solutions are related, in fact at first glance Exhibit 18.5 looks identical to Exhibit 18.2, only the scale has changed slightly on the two axes. This is the only difference between the two analyses — the Burt version of MCA gives principal coordinates which are reduced in scale compared to the indicator version, where the reduction is relatively more on the second axis compared to the first.

MCA definition number 2: CA of the Burt matrix

The CAs of the indicator matrix and the Burt matrix are almost identical with respect to the category points, specifically:

Comparison of MCA based on indicator and Burt matrices

- In both analyses the standard coordinates of the category points are identical — this is a direct result of the relationship (18.2).

- Also as a result of (18.2), the principal inertias of the Burt analysis are the squares of those of the indicator matrix. Since the principal inertias are less than 1, squaring them makes them smaller in value (and the lower principal inertias relatively smaller still).

- Consequently, the percentages of inertia are always higher in the Burt analysis.

- The principal coordinates are the standard coordinates multiplied by the square roots of the principal inertias, which accounts for the reduction in scale in Exhibit 18.5 compared to Exhibit 18.2. Apart from the overall reduction in scale the dispersion along the first axis of the Burt analysis (Exhibit 18.5) is relatively higher than on the second, as if the display has been squashed down slightly on the vertical axis.

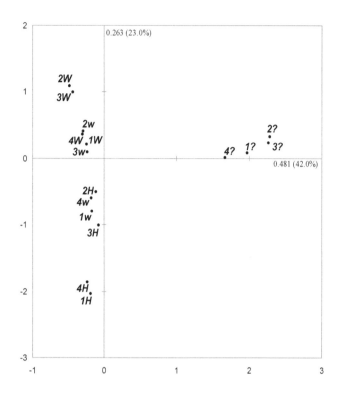

The subtables of the Burt matrix have the same row margins in each set of horizontal tables and the same column margins in each set of vertical tables, so the result (17.1) applies exactly: the inertia of **B** will be the average of the inertias of the subtables $\mathbf{B}_{qs}$. Exhibit 18.6 shows the 16 individual inertias of the Burt matrix, and their row and column averages. The overall average is equal to the total inertia 1.145 of **B**. In this table the inertias of the diagonal blocks are exactly 3; in fact their inertias have the same definition (18.1) as the inertias of the subtables of the indicator matrix — they are $J_q \times J_q$ tables of dimensionality $J_q - 1$ with perfect row–column association, and so have maximal inertia equal to the number of dimensions. These high values on the diagonal of Exhibit 18.6 demonstrate why the total inertia of the Burt matrix

QUESTIONS	Qu. 1	Qu. 2	Qu. 3	Qu. 4	Average
Qu. 1	3.0000	0.3657	0.4262	0.6457	1.1094
Qu. 2	0.3657	3.0000	0.8942	0.3477	1.1519
Qu. 3	0.4262	0.8942	3.0000	0.4823	1.2007
Qu. 4	0.6457	0.3477	0.4823	3.0000	1.1189
Average	1.1094	1.1519	1.2007	1.1189	1.1452

is so high, which is the cause of the low percentages of inertia on the axes. We return to this topic in the next chapter.

Suppose we wish to relate the demographic variables gender, age, etc. to the patterns of association shown in the MCA maps. There are two ways of doing this, highly related, but one of these has some advantages. The first way is to code these as additional dummy variables and add them as supplementary columns of the indicator matrix. The second way is to cross-tabulate the demographics with the four questions, as we did in the stacked analysis of Chapter 17, and add these cross-tables as supplementary rows of the indicator matrix or as supplementary rows (or columns) of the Burt matrix. The latter strategy is preferred because it can be used in both forms of MCA as well as in the improved versions that we present in the next chapter. In the indicator matrix version of CA each respondent has a position in the display, usually not shown when there are thousands of respondents. The respondents can be similarly added as supplementary points in the Burt matrix version of CA — these have the same positions as for the indicator matrix. Thus the supplementary points for demographic groups have the same locations in both MCA versions and are the average positions of those respondent points belonging to the particular demographic category. Exhibit 18.7 shows the positions of five of the demographic variables that were used previously, which can be superimposed on either of the maps in Exhibit 18.2 or 18.5. Notice that the scale in Exhibit 18.7 is quite reduced — these average points lie quite close to the centre of the maps, since the differences between the groups are relatively small. Remember that MCA is showing the differences between the response categories, not between the demographics (cf. Chapter 17, where demographic differences across the response categories were displayed).

Positioning supplementary categories in the map

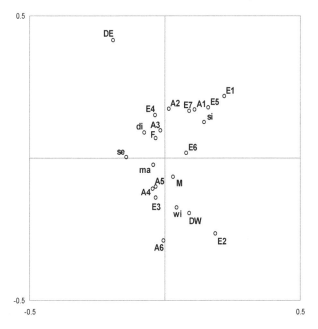

Exhibit 18.7: *Supplementary demographic category points, based on (row) profiles of responses aggregated into each category, with respect to first two principal axes. These points should be superimposed on the maps of Exhibits 18.2 or 18.5, where they occupy a small area at the centre of the map (note the scale).*

*Interpretation
of supplementary
points*

Based on the positions of the response categories on the first two dimensions of Exhibit 18.2 (similarly, Exhibit 18.5), the farther a demographic category is to the right, the higher will be the frequency of non-responses. The higher up a category is, the more liberal the attitude, and the lower down it is, the more traditional the attitude. Hence West Germany (DW) has more traditional attitudes and more non-responses than East Germany (DE), a pattern that is mimicked almost identically by the male–female (M–F) contrast but not as much as the difference between the two German regions. The age groups show the same trend as before, from young (A1) at the top (liberal) to old (A6) at the bottom (traditional). The lowest education groups E1 and E2 have the highest frequency of non-response. The highest education groups tend to have more liberal attitudes, but so does the lowest education group E1 (but note that this is a very small group). Amongst the marital status groups, single (si) respondents have higher than average non-response and liberal attitudes, opposing separated (se) respondents who have the least non-response, but are otherwise average on the liberal–traditional dimension. The widowed group wi is clearly on the traditional side, but this is surely related to the older age groups.

*SUMMARY:
Multiple
Correspondence
Analysis*

1. MCA is concerned with relationships amongst a set of variables that are usually homogeneous in the sense that they revolve around one particular substantive issue, and often the response scales are the same.

2. The variables can be recoded as dummy variables in an *indicator matrix*, which has as many rows as cases and as many columns as categories of response. The data in each row are 0s apart from the 1s that indicate the particular category of each variable corresponding to the individual case.

3. An alternative coding of such data is as a *Burt matrix*, a square symmetric categories-by-categories matrix formed from all two-way contingency tables of pairs of variables, including on the block diagonal the cross-tabulations of each variable with itself.

4. The two alternative definitions of MCA, applying CA to the indicator matrix or to the Burt matrix, are almost equivalent. Both yield identical standard coordinates for the category points.

5. The difference between the two definitions is in the principal inertias: those of the Burt matrix are the squares of those of the indicator matrix. As a result, the percentages of inertia in the Burt analysis are always more optimistic than those in the indicator analysis.

6. In both approaches, however, the percentages of inertia are artificially low, due to the coding, and underestimate the true quality of the maps as representations of the data.

7. Each respondent has a position in either version of MCA, and additional categorical variables (e.g. demographics) can be displayed as supplementary points at positions of respondent category averages.

Joint Correspondence Analysis

## 19

Extending simple correspondence analysis (CA) of a two-way table to many variables is not so straightforward. The usual strategy is to apply CA to the indicator or Burt matrices, but we have seen that the geometry is not so clear any more — for example, the total inertia and principal inertias change depending on the matrix analysed, and percentages of inertia explained are low. The Burt matrix version of multiple correspondence analysis (MCA) shows that the problem lies in trying to visualize the whole matrix, whereas we are really interested only in the off-diagonal contingency tables which cross-tabulate distinct pairs of variables. Joint correspondence analysis (JCA) concentrates on these tables, ignoring those on the diagonal, resulting in improved measures of total inertia and much better data representation in the maps.

Contents

Exhibit 18.6 shows the inertias in each subtable of the Burt matrix, and their average which is the total inertia of the Burt matrix. This average is inflated by the inertias on the diagonal, which are fixed values equal to the number of categories minus 1 of the corresponding variable (e.g. $4-1 = 3$ in this example, with four categories for each variable). Since the analysis tries to explain the inertia in the whole table, the high inertias on the diagonal are going to seriously affect the fit to the whole table. For example, we can evaluate from the results of the MCA how much inertia for each subtable is explained by the two-dimensional MCA map — see Exhibit 19.1. Although the MCA reports that 65.0% of the total inertia is explained, we can see that the off-diagonal tables are explained much better than that, and the tables on the diagonal much worse. By summing the parts explained in the off-diagonal tables and expressing this sum relative to the sum of their total inertias, it turns out that

MCA gives bad fit because the total inertia is inflated

Exhibit 19.1:
*Percentage of inertia
explained in each
subtable of the Burt
matrix, based on
two-dimensional
MCA of the Burt
matrix.*

QUESTIONS	Qu. 1	Qu. 2	Qu. 3	Qu. 4	Per question
Qu. 1	51.9	78.4	82.5	80.4	61.2
Qu. 2	78.4	55.5	88.2	76.6	65.3
Qu. 3	82.5	88.2	59.6	86.6	69.7
Qu. 4	80.4	76.6	86.6	54.6	63.5

83.2% of the off-diagonal tables is explained by the MCA solution (the parts explained in each subtable can be recovered using the percentages in Exhibit 19.1 and the total inertias in Exhibit 18.6). Similarly, we can calculate that 55.4% of the inertia on the diagonal tables is explained (this is a simple average of the diagonal of Exhibit 19.1 because the total inertias for these tables are constant). So, since we are really interested in only the off-diagonal tables, we should report a figure such as 83.2% explained rather than 65.0%. But, as will be shown now, the fit is even better than 83.2%.

*Ignoring the
diagonal blocks —
joint CA*

It is clear that the inclusion of the tables on the diagonal of the Burt matrix degrades the whole MCA solution. The method is trying to visualize these tables unnecessarily, and moreover these are tables with extremely high inertias, in fact the highest possible inertias attainable. It is possible to improve the calculation of explained inertia by completely ignoring the diagonal blocks in the search for an optimal solution. To do this we need a special algorithm to solve the problem, called *joint correspondence analysis* (JCA). This is an iterative algorithm that performs CA on the Burt matrix, treating the diagonal subtables as missing values, so that attention is focused on optimizing the fit to the off-diagonal subtables only. The method uses an imputation algorithm for missing values, starting from the MCA solution and then replacing the diagonal subtables with values estimated from the solution itself, calculated by the reconstitution formula (13.4). Since there is only one set of coordinates and masses for the rows and columns of the symmetric Burt matrix, this formula takes the following form, for a two-dimensional solution, say:

$$\hat{p}_{jj'} = c_j c_{j'} (1 + \sqrt{\lambda_1} \gamma_{j1} \gamma_{j'1} + \sqrt{\lambda_2} \gamma_{j2} \gamma_{j'2}) \tag{19.1}$$

where $\hat{p}_{jj'}$ is the estimated value of the relative frequency in the (j, j')-th cell of the Burt matrix, and the γ_{jk}s are the standard column coordinates. Using this formula the diagonal subtables of the Burt matrix are replaced with these estimated values, giving a *modified Burt matrix*. CA is then performed on the modified Burt matrix to get a new solution, from which the diagonal subtables are replaced again with estimates from the new solution to update the modified Burt matrix. This process is repeated several times until convergence, and at each iteration the fit to the off-diagonal subtables is improved.

Results of JCA

Applying JCA to the four-variable data set on women working leads to the following results: 90.2% inertia explained, and percentages for individual tables as shown in Exhibit 19.2. The results are clearly much better than before;

QUESTIONS	Qu. 1	Qu. 2	Qu. 3	Qu. 4	Per question
Qu. 1	—	97.8	95.8	77.8	*88.2*
Qu. 2	97.8	—	87.4	97.0	*91.8*
Qu. 3	95.8	87.4	—	96.7	*91.9*
Qu. 4	77.8	97.0	96.7	—	*88.5*

Exhibit 19.2:
Percentage of inertia explained in each of the 12 off-diagonal subtables of the Burt matrix, based on results of JCA of the Burt matrix.

all subtables are very well represented, the worst being the cross-tabulation of question 1 with question 4, where the explained inertia is 77.8%. Exhibit 19.3 shows the JCA map, where the scale is intentionally kept identical to that of Exhibits 18.2 and 18.5 for purposes of comparison. It is clear that the solution is practically identical apart from a contraction of the points in scale. In Exhibit 18.5 the principal inertias along the first two axes were 0.481 and 0.263, and here they are 0.353 and 0.128 respectively. So once again there has been a contraction but more of a contraction on the second axis than on the first. This also happened when passing from the MCA of the indicator matrix in Exhibit 18.2 to that of the Burt matrix in Exhibit 18.5, but in that case the standard coordinates of the two analyses were identical — the JCA solution here is different from the MCA solution, with slight rearrangement of the category points, as can be seen by closer inspection of Exhibit 19.3 and comparison with the MCA maps in Exhibits 18.2 and 18.5.

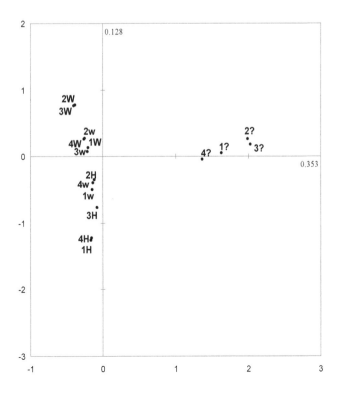

Exhibit 19.3:
JCA map of Burt matrix of four questions on women working; percentage of inertia in map: 90.2%. The percentage of inertia is the sum of the parts explained in each subtable (obtained from Exhibits 19.2 and 18.6) expressed as a percentage of the sum of off-diagonal inertias — see also the Theoretical Appendix, pages 247–248).

JCA results are
not nested

The principal axes in JCA are not nested as in the MCA analyses — that is, the solution in two dimensions does not exactly contain the best one-dimensional solution as its first axis, although in practice the nesting is approximate. This is why no percentages of inertia are reported along the axes in Exhibit 19.3 — it is possible to report only a percentage of inertia explained for the solution as a whole, in this case 90.2%. This will affect the reporting of inertia contributions as well: each category point has a certain quality of representation in the map, but cannot be decomposed into parts for each axis.

Adjusting the
results of MCA to
fit the off-diagonal
tables

The similarity between the JCA solution and the MCA solution occurs in almost all examples in our experience. This suggests that it is mainly the scale change in the solution that distinguishes JCA from MCA; hence, as an alternative, we can investigate simple scale changes of the MCA solution to improve the fit. Given the standard coordinates of the MCA solution, therefore, the question is how the solution should be rescaled (i.e. how to define principal coordinates) so that the data in the off-diagonal subtables of the Burt matrix are optimally reconstructed. This turns out to be a regression problem, again using the reconstitution formula (19.1), but considering the scale factors (the square roots of the principal inertias) as unknown regression coefficients β_1 and β_2, for example, in a two-dimensional solution:

$$\frac{p_{jj'}}{c_j c_{j'}} - 1 = \beta_1 \gamma_{j1} \gamma_{j'1} + \beta_2 \gamma_{j2} \gamma_{j'2} + e_{jj'} \qquad (19.2)$$

The regression is performed by stringing out all the values on the left-hand side of (19.2) in a vector, just for the cells in the off-diagonal tables, constituting the "response variable" of the regression — in the present four-variable example, with six 4×4 tables off the diagonal, there will be $6 \times 16 = 96$ values in this vector. As "explanatory variables" we have the corresponding products $\gamma_{j1} \gamma_{j'1}$ and $\gamma_{j2} \gamma_{j'2}$, also strung out as vectors. A weighted least-squares regression with no constant is performed with weights $c_j c_{j'}$ on the respective values — in this example this gives coefficient estimates $\hat{\beta}_1 = 0.5922$ and $\hat{\beta}_2 = 0.3532$. Squaring these values gives the optimal values 0.351 and 0.125 respectively for the "principal inertias", for which the explained inertia is 89.9% (this is the coefficient of determination R^2 of the regression). This is the best we can do with the MCA solution — notice how close these are to the principal inertias in the JCA of Exhibit 19.3, which were 0.353 and 0.128. For mapping the categories, the principal coordinates are calculated as the MCA standard coordinates on the first two axes multiplied by the scaling factors $\hat{\beta}_1$ and $\hat{\beta}_2$. But once again the solution is not nested and depends on the dimensionality of the solution — if we perform the same calculation for the three-dimensional solution the first two regression coefficients will not be exactly those obtained above. The nested property will hold only if the "explanatory variables" in (19.2) are uncorrelated. By simply ignoring their correlations, we obtain a simpler (but sub-optimal) way to adjust the solution, which is indeed nested, as described in the following section.

We now describe a simpler adjustment of the principal inertias, which has the nested property; in our experience it gives a solution that is usually very close to optimal. It is also quite easy to compute, involving (i) a recomputation of the total inertia, just for the off-diagonal subtables, and (ii) a simple adjustment of the principal inertias emanating from MCA. Principal coordinates are then calculated in the usual way, as are percentages of inertia.

A simple adjustment of the MCA solution

In MCA of the Burt matrix $\mathbf{B}$, total inertia is the average of the inertias of all subtables, including the problematic ones on the diagonal. But now, as in JCA, the total inertia is the average of the inertias of the off-diagonal subtables. This is easily calculated from the total inertia of $\mathbf{B}$ because we know exactly what the values are of the inertias of the diagonal subtables: $J_q - 1$, where J_q is the number of categories of the q-th variable. Hence

Adjusted inertia = average inertia in off-diagonal blocks

$$\text{sum of inertias of } Q \text{ diagonal subtables} = J - Q \quad (19.3)$$

while

$$\text{sum of inertias of all two-way tables} = Q^2 \times \text{inertia}(\mathbf{B}) \quad (19.4)$$

Subtracting (19.3) from (19.4) to obtain the sum of inertias in the off-diagonal blocks and then dividing by $Q(Q-1)$ to obtain the average, leads to:

$$\text{average off-diagonal inertia} = \frac{Q}{Q-1} \times \left(\text{inertia}(\mathbf{B}) - \frac{J-Q}{Q^2} \right) \quad (19.5)$$

Using our data on women working as an example:

$$\text{average off-diagonal inertia} = \frac{4}{3} \times \left(1.1452 - \frac{12}{16} \right) = 0.5270$$

Another way to compute this value is by directly averaging the inertias of each subtable, which are given in Exhibit 18.6. This needs to be done on only one triangle of the symmetric Burt matrix, which contains $\frac{1}{2}Q(Q-1) = 6$ pairwise cross-tables:

$$\frac{1}{6}(0.3657 + 0.4262 + 0.6457 + 0.8942 + 0.3477 + 0.4823) = 0.5270$$

Suppose that the principal inertias (eigenvalues) in the MCA of the Burt matrix $\mathbf{B}$ are denoted by λ_k, for $k = 1, 2$, etc. The adjusted principal inertias λ_k^{adj} are calculated as follows:

Adjusting each principal inertia

$$\lambda_k^{\text{adj}} = \left(\frac{Q}{Q-1} \right)^2 \times \left(\sqrt{\lambda_k} - \frac{1}{Q} \right)^2, \quad k = 1, 2, \dots \quad (19.6)$$

In our example the first two adjusted inertias are:

$$\frac{16}{9} \times (0.6934 - \frac{1}{4})^2 = 0.3495 \ \text{ and}$$

$$\frac{16}{9} \times (0.5132 - \frac{1}{4})^2 = 0.1232$$

(notice again how close these are to the optimal ones from the regression, which were given on the previous page as 0.351 and 0.125).

The adjusted inertias are then expressed relative to the adjusted total to give percentages of inertia along each principal axis:

$$100 \times \frac{0.3495}{0.5269} = 66.3\% \quad \text{and} \quad 100 \times \frac{0.1232}{0.5269} = 23.4\%$$

The percentage of inertia in the two-dimensional adjusted solution (which is nested) is thus 89.7%, only 0.2% less than the optimal adjustment from the regression (which is not nested) and 0.5% less than the JCA solution. It has been proved that the percentages calculated according to these simple adjustments give an overall percentage for the solution which is a lower bound for the optimal percentage explained in a JCA solution, as illustrated in this example. Hence, when reporting an MCA, the best way to express measure of fit is as above, and then the square roots of the adjusted principal inertias should be used to scale the standard coordinates to obtain principal coordinates for mapping. We do not give the map here since the relative positions of the points along the two axes are identical to those in Exhibits 18.2 and 18.5 — just the scales are different, more like the scale of Exhibit 19.3.

As another example of JCA, and also to illustrate supplementary points for JCA, consider a large data set from the 2005 Eurobarometer survey on interest in science. As part of this survey each respondent was asked how interested he or she was in the following six news issues: sports news (S), politics (P), new medical discoveries (M), environmental pollution (E), new inventions and technologies (T), and new scientific discoveries (D). The response scale was "very interested" (++), "moderately interested" (+) and "not at all interested" (0). Hence the response categories are depicted by, for example, E+ for "moderately interested in environmental pollution", or P0 for "not at all interested in politics". In order to avoid the usual phenomenon of non-responses that strongly affect the results, as in the previous example, respondents with any "don't know" and missing responses were omitted, which reduced the sample size from 33190 to 29652, a reduction of 10.7% — we shall deal with non-response issues specifically in Chapter 21. The adjusted MCA map of these data is shown in Exhibit 19.4. The map shows the "no interest" points forming a diagonal spread of their own to the right and the "interest" points spreading from "moderately interested" at the bottom to "very interested" top left. This is an example of a map whose axes could be rotated, if one wanted these two lines of dispersion to coincide more with the principal axes. But a rotation would destroy the property of the horizontal first axis to capture a dimension of general interest in news issues — see the remark about rotations in CA in the Epilogue.

This first axis accounts for 67.0% of the inertia. The second axis (22.0%) shows the interest in scientific discovery (D) and technological innovation (T) at the extremes, indicating high correlation between these two. The two points for interest in sports, however, are near the centre of this spread of points, which indicates that high and moderate interest in sport (S++ and S+), for

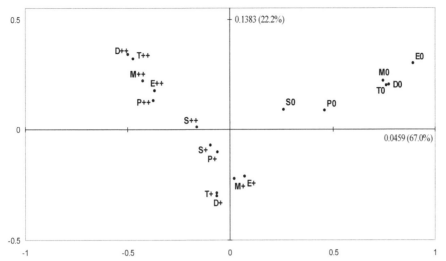

Exhibit 19.4:
*Adjusted MCA map
of news interest
data. Percentage
inertia in map:
89.2%. (If the MCA
of the indicator
matrix were
performed, the
percentage of
explained inertia
would be only
41.1%.)*

example, must be also associated with moderate interest in the other issues, and vice versa. Remember that what is being visualized is the association of each category of a particular variable with the categories of all the other variables.

Even though we do not show the positions of the 29652 cases in this data set, they can be imagined in the map as supplementary points. That is, if we added the huge 29652×18 indicator matrix to the Burt matrix as supplementary rows, each respondent would have a position in the MCA map (but notice that there are only $3^6 = 729$ unique response patterns, so the respondents would pile up at the points representing each response pattern). Since the standard coordinates in the three versions of MCA (indicator, Burt and adjusted) are the same, the principal coordinate positions of the respondents would be the same in all three. As stated in Chapter 18, the way to show supplementary categories is to add their cross-tabulations with the active variables as supplementary rows of the Burt matrix. For example, in this data set there are samples from 34 European countries. Each respondent from a particular country has a position in the map and the row of frequencies in the aggregation of responses for a particular country has a profile exactly at the average position of that country's respondents. Exhibit 19.5 shows the positions of these average country points, labelled by their local names — this display should be imagined overlaid on Exhibit 19.4. Of all the countries TURKIYE (Turkey) is the most in the "no interest" direction — about 40% of Turkish respondents express very little interest on all issues except environmental pollution (22%); whereas KYPROS (Cyprus), ELLADA (Greece) and MALTA seem to be the most interested — for example, Cyprus has the highest percentages of "very interested" responses in issues of environmental pollution (75%), medical discoveries (62%), technological innovation (53%) and scientific discoveries (55%).

*Supplementary
points in adjusted
MCA and JCA*

Exhibit 19.5:
*Supplementary
country points in
the MCA space of
the data set of news
interest. The
original country
names are used as
labels, as given in
the Eurobarometer.*

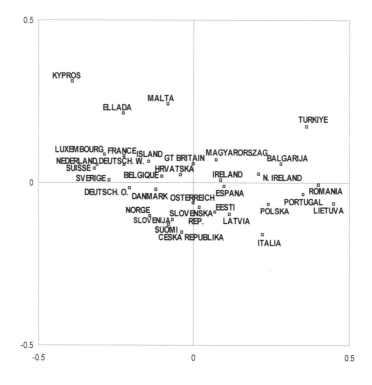

SUMMARY:
Joint
Correspondence
Analysis

1. One way of defining MCA is as the CA of the Burt matrix of all two-way cross-tabulations of a set of variables, including the cross-tables of each variable with itself, which inflate the total inertia.

2. *Joint correspondence analysis* (JCA) finds a map which best explains the cross-tabulations of all pairs of variables, ignoring those on the block diagonal of the Burt matrix. This requires a different iterative algorithm, but in the optimal solution the axes are not nested.

3. The total inertia to be explained is now the average of all the inertias in the off-diagonal tables of the Burt matrix.

4. An intermediate solution is to condition on the standard coordinates of the MCA solution and find the best weighted least-squares fit to the cross-tables of interest, using regression analysis. However, this solution is again not nested.

5. A simple solution, called *adjusted* MCA, which is nested and thus conserves all the good properties of MCA while approximating the JCA solution, is to apply certain adjustments to the MCA principal inertias of the axes and to the total inertia.

6. Supplementary categories are represented as in all forms of MCA, i.e. by cross-tabulating them with the active variables and then adding them as supplementary rows of the Burt matrix (or modified Burt matrix in JCA).

Scaling Properties of MCA

## 20

As was shown in Chapters 7 and 8, there are several alternative definitions of correspondence analysis (CA) and different ways of thinking about the method. In this book Benzécri's geometric approach has been emphasised, leading to data visualization. In Chapters 18 and 19 it was clear that the passage from simple two-variable CA to the multivariate form of the analysis is not straightforward, especially if one tries to generalize the geometric interpretation. An alternative approach to the multivariate case, which relies on exactly the same mathematics as multiple correspondence analysis (MCA), is to see the method as a way of quantifying categorical data, generalizing the optimal scaling ideas of Chapter 7. As before, there are several equivalent ways to think about MCA as a scaling technique, and these different approaches enrich our understanding of the method's properties. The optimal scaling approach to MCA is often referred to in the literature as *homogeneity analysis*.

Contents

This data set is taken from the multinational International Social Survey Program (ISSP) survey on environment in 1993. We are going to look specifically at $Q = 4$ questions on attitudes towards the role of science. Respondents were asked if they agreed or disagreed with the following statements:

Data set 11: Attitudes to science and environment

A We believe too often in science, and not enough in feelings and faith.

B Overall, modern science does more harm than good.

C Any change humans cause in nature, no matter how scientific, is likely to make things worse.

D Modern science will solve our environmental problems with little change to our way of life.

There were five possible response categories: *1.* strongly agree, *2.* somewhat agree, *3.* neither agree nor disagree, *4.* somewhat disagree, *5.* strongly disagree. For simplicity data for the West German sample only are used and cases with missing values on any one of the four questions are omitted, reducing the sample size to $N = 871$ (these data are provided with the **ca** package in R — see the Computational Appendix).

Category quantification as a goal

In Chapter 7 CA was defined as the search for quantifications of the categories of the column variable, say, which lead to the greatest possible differentiation, or discrimination, between the categories of the row variable, or vice versa. This is what we would call an "asymmetric" definition because the rows and columns play different roles in the definition and the results reflect this too; for example, the column solution turns out to be in standard coordinates and the row solution in principal coordinates. In Chapter 8 CA was defined "symmetrically" as the search for new scale values which lead to the highest correlation between the row and column variables. Here the rows and columns have an identical role in the definition. These scaling objectives do not include any specific geometric concepts and, in particular, make no mention of a full space in which the data are imagined to lie, which is an important concept in the geometric approach for measuring total inertia and percentages of inertia explained in lower-dimensional subspaces.

MCA as a principal component analysis of the indicator matrix

The asymmetric definition of optimal scaling, when applied to an indicator matrix, resembles closely principal component analysis (PCA). PCA is usually applied to matrices of continuous-scale data, and has close theoretical and computational links to CA — in fact, one could say that CA is a variant of PCA for categorical data. In the PCA of a data set where the rows are cases and the columns variables (m variables, say, $x_1,...,x_m$), coefficients $\alpha_1,...,\alpha_m$ (to be estimated) are assigned to the columns, leading to linear combinations for the rows (cases) of the form $\alpha_1 x_1 + \cdots + \alpha_m x_m$, called *scores*. The coefficients are then calculated to maximize the variance of the row scores. As before, identification conditions are required to solve the problem, and in PCA this is usually that the sum of squares of the coefficients is 1: $\sum_j \alpha_j^2 = 1$. Applying this idea to the indicator matrix, which consists of zeros and ones only, assigning coefficients $\alpha_1,...,\alpha_J$ to the J dummy variables and then calculating linear combinations for the rows, simply means adding up the coefficients (i.e. scale values) for each case. Then maximizing the variance for each case sounds just like the optimal scaling procedure of Chapter 7 (maximizing discrimination between the rows); in fact this is almost the same except for one aspect, namely the identification conditions. In optimal scaling the identification conditions would be that the weighted variance (inertia) of the coefficients (not the simple sum of squares) be equal to 1: $\sum_j c_j \alpha_j^2 = 1$. Here the weights c_j are the column masses, i.e. the column sums of the indicator matrix divided by the grand total NQ of the indicator matrix — thus each set of c_js for one categorical variable adds up to $1/Q$. So with this change in the identification condition, MCA could be called the PCA of categorical data, maximizing the variance across cases. The coefficients are the standard coordinates of the column categories, while the MCA principal coordinate of a case is the average of that case's scale values, i.e. $1/Q$ times the sum that was called the score before. The first dimension of MCA maximizes the variance (first principal inertia), the second dimension maximizes the variance subject to the scores being uncorrelated with those on the first dimension, and so on.

MCA as a scaling technique, usually called *homogeneity analysis*, is more commonly seen as a generalization of the correlation approach of Chapter 8. In Equation (8.1) on page 63, an alternative way of optimizing correlation between two categorical variables was described, which is easy to generalize to more than two variables. Again we shall use a pragmatic notation for this specific four-variable example, i.e. $Q = 4$, with total number of categories $J = 20$, but the idea easily extends to Q variables with any number of categories. Suppose that the four variables have (unknown) scale values a_1 to a_5, b_1 to b_5, c_1 to c_5, and d_1 to d_5. Assigning four of these values a_i, b_j, c_k and d_l to each respondent according to his or her set of responses leads to the quantified responses for the whole sample, which we denote by a, b, c and d (i.e. the vector a denotes all 871 quantified responses to question A, etc.). Each respondent has a score $a_i + b_j + c_k + d_l$ which is the sum of the scale values, so the scores for the whole sample are denoted by $a + b + c + d$. In this context the variables are often referred to as *items* and we talk of the values in a to d as *item scores* and those in $a + b + c + d$ as the *summated scores*. The criterion for finding optimal scale values is thus to maximize the average squared correlation between the item scores and the summated score:

Maximizing inter-item correlation

$$\text{ave. squared correlation} = \frac{1}{4}[\text{cor}^2(a, a + b + c + d) + \text{cor}^2(b, a + b + c + d)$$
$$+ \text{cor}^2(c, a + b + c + d) + \text{cor}^2(d, a + b + c + d)] \quad (20.1)$$

(cf. the two-variable case, (8.1) on page 63). Again, identification conditions are required, and it is convenient to use the mean 0 and variance 1 conditions on the summated scores: $\text{mean}(a + b + c + d) = 0$, $\text{var}(a + b + c + d) = 1$. The solution to this maximization problem is then given exactly by the standard coordinates of the item categories on the first principal axis of MCA, and the maximized average squared correlation of (20.1) is exactly the first principal inertia of the indicator matrix version of MCA.

Exhibit 20.1 shows the two-dimensional MCA map based on the indicator matrix, showing again the very low percentages of inertia (the percentages based on adjusted inertias are 44.9% and 34.2% respectively). But in this case the percentages should be ignored, since it is the values of the principal inertias that are of interest *per se*, being average squared correlations. The maximum value of (20.1) is thus 0.457. The second principal inertia, 0.431, is found by looking for a new set of scale values that lead to a set of scores uncorrelated with those obtained previously, and which maximize (20.1) —

MCA of scientific attitudes example

Exhibit 20.1:
*MCA map
(indicator matrix
version) of science
attitudes, showing
category points in
principal coordi-
nates. Since the
principal inertias
differ only slightly
(and even less in
their square roots),
these principal
coordinates are
almost the same
scale contraction
(approximately 2/3,
close to the square
roots of 0.457 and
0.431) of the
standard coordi-
nates on both axes.*

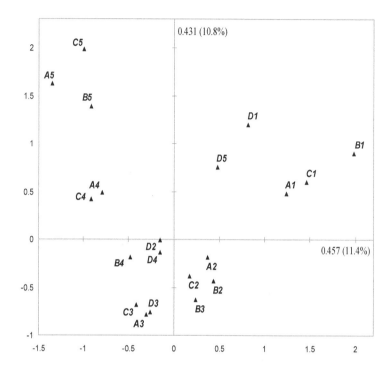

this maximum is the value 0.431. And so it would continue for solutions on subsequent axes, always uncorrelated with the ones already found. Looking at the map in Exhibit 20.1, questions *A*, *B* and *C* can be seen to follow a very similar pattern, with strong disagreements on the left to strong agreements on the right, in a wedge-shaped curved pattern (the arch, or "horseshoe"). Question *D*, however, has a completely different trajectory, with the two poles of the scale very close together. Now the first three questions were all worded negatively towards science whereas question *D* was worded positively, so we would have expected *D5* to lie towards *A1*, *B1* and *C1*, and *D1* on the side of *A5*, *B5* and *C5*. The fact that *D1* and *D5* lie close together inside the horseshoe means that they are both associated with the extremes of the other three questions — the most likely explanation is that some respondents are misinterpreting the change of direction of the fourth question.

*Individual
squared
correlations*

Knowing the values of the individual squared correlations in (20.1) is also interesting information. These can be obtained adding up the individual inertia contributions to the first principal inertia for each question. The results of an MCA usually give these expressed in proportions or permills, so we show these as permills in Exhibit 20.2 as an illustration of how to recover these correlations. Questions *A* to *D* thus contribute proportions 0.279, 0.317, 0.343 and 0.062 of the principal inertia of 0.457. Since 0.457 is the average of the four squared correlations, the squared correlations and thus the correlations are:

	QUESTIONS				
CATEGORIES	A	B	C	D	Sum
1 "strongly agree"	115	174	203	25	518
2 "somewhat agree"	28	21	6	3	57
3 "neither–nor"	12	7	22	9	49
4 "somewhat disagree"	69	41	80	3	194
5 "strongly disagree"	55	74	32	22	182
Sum	279	317	343	62	1000

Exhibit 20.2: Permill (‰) contributions to inertia of first principal axis of MCA (indicator matrix version) of data on science and environment.

A: $0.279 \times 0.457 \times 4 = 0.510$ correlation $= \sqrt{0.510} = 0.714$

B: $0.317 \times 0.457 \times 4 = 0.579$ correlation $= \sqrt{0.579} = 0.761$

C: $0.343 \times 0.457 \times 4 = 0.627$ correlation $= \sqrt{0.627} = 0.792$

D: $0.062 \times 0.457 \times 4 = 0.113$ correlation $= \sqrt{0.113} = 0.337$

This calculation shows how much lower the correlation is of question D with the summated score on the first dimension. Notice that, although the MCA of the indicator matrix was the worst from the usual CA geometric point of view of χ^2-distances, total inertia, etc., the principal inertias and the contributions to the principal inertias do have a very interesting interpretation by themselves. In the approach called *homogeneity analysis*, which is theoretically equivalent to the MCA of the indicator matrix but which interprets the method from a scaling viewpoint, the squared correlations 0.510, 0.579, 0.627 and 0.113 are called *discrimination measures*.

In homogeneity analysis the objective function (8.3) (see Chapter 8, page 63) is generalized to many variables. Using the notation above for the present four-variable example, the average score $\frac{1}{4}(a_i + b_j + c_k + d_l)$ of the item scores can be calculated for each respondent and then the respondent's measure of variance within his or her set of quantified responses is:

Loss of homogeneity

$$
\begin{aligned}
\text{variance (for one case)} = \frac{1}{4}\big(\ & [a_i - \tfrac{1}{4}(a_i + b_j + c_k + d_l)]^2 \\
& + [b_j - \tfrac{1}{4}(a_i + b_j + c_k + d_l)]^2 \\
& + [c_k - \tfrac{1}{4}(a_i + b_j + c_k + d_l)]^2 \\
& + [d_l - \tfrac{1}{4}(a_i + b_j + c_k + d_l)]^2 \ \big) \qquad (20.2)
\end{aligned}
$$

The average of all these values over the N cases is then calculated, called the *loss of homogeneity*, and the objective is to minimize this loss. Again the MCA (indicator matrix version) solves this problem and the minimized loss is 1 minus the first principal inertia; i.e. $1 - 0.457 = 0.543$. Minimizing the loss is equivalent to maximizing the correlation measure defined previously.

Geometry of loss function in homogeneity analysis

The objective of minimizing loss has a very attractive geometric interpretation which is closely connected to the row-to-column distance definition of CA discussed in Chapter 7. In fact, the homogeneity loss function is exactly the weighted distance function (7.6) on page 55, applied to the indicator matrix. Exhibit 20.3 shows the asymmetric MCA map of all $N = 871$ respondents (in principal coordinates) and the $J = 20$ category points (in standard coordinates), which means that the respondents lie at the centroids of the categories, where the weights are the relative values in the rows of the indicator matrix. Each respondent has a profile consisting of zeros apart from values of $\frac{1}{4}$ in

Exhibit 20.3:
Asymmetric MCA map (indicator matrix version) of science attitudes, showing respondents in principal coordinates and categories in standard coordinates. Each respondent is at the average of the four categories given as responses. MCA minimizes the sum of squared distances between category points and respondents.

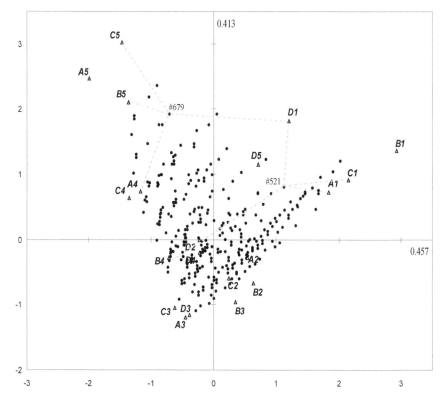

the positions of the four responses; hence each repondent point lies at the ordinary average position of his or her responses. Two respondents, #679 and #521, are labelled in Exhibit 20.3. Respondent #679 chose the categories $(A4,B5,C5,D1)$, disagreeing with the first three questions and agreeing to the fourth — those categories are linked to the respondent point on the upper left-hand side of the display. This is a strong and consistent position in favour of science. Respondent #521, however, has a mixed opinion: $(A1,B4,C1,D1)$, strongly agreeing that we believe too much in science and that human interference in nature will make things worse, but at the same time strongly agreeing that science will solve our environmental problems while disagreeing that science does more harm than good. This shows one of the reasons why *D1* has been pulled to the middle between the two extremes of opinion. Every respondent is at the average of the four categories in his or her set of answers. For any configuration of category points, the respondents could be located at average positions, but the result of Exhibit 20.3 is optimal in the sense that the lines linking the respondents to the category points are the shortest possible (in terms of sum of squared distances). Showing all the links between respondents and their response categories has been called a *star plot*, so the objective of MCA can be seen as obtaining the star plot with the shortest links in the least-squares sense. The number of links between the N respondent points and the corresponding Q category points is NQ. The value of the loss is actually the average of the squares of the links (for example, in (20.2) where $Q = 4$ the sum of the four squares is divided by 4, and then the average over N is calculated, so that the sum of squared values is divided by $4N$). So the average sum of squared links on the first dimension is $1 - 0.457 = 0.513$, and on the second dimension it is $1 - 0.413 = 0.587$; by Pythagoras' theorem, the average sum of squared links in the two-dimensional map of Exhibit 20.3 is $0.513 + 0.587 = 1.100$.

In the present example of the science and environment data, we saw that question D is not correlated highly with the others (see page 157). If we were trying to derive an overall measure of attitude towards science in this context, we would say that these results show us that question D has degraded the *reliability* of the total score, and should preferably be removed. In reliability theory, the Q variables, or items, are supposed to be measuring one underlying construct. *Cronbach's alpha* is a standard measure of reliability, defined in general as:

$$\alpha = \frac{Q}{Q-1} \left(1 - \frac{\sum_q s_q^2}{s^2} \right) \tag{20.3}$$

where s_q^2 is the variance of the q-th item score, $q = 1, \ldots, Q$ (e.g. variances of $a + b + c + d$) and s^2 is the variance of the average score (e.g. variance of $\frac{1}{4}(a+b+c+d)$). Applying this definition to the first dimension of the MCA solution, it can be shown that Cronbach's alpha reduces to the following:

$$\alpha = \frac{Q}{Q-1} \left(1 - \frac{1}{Q\lambda_1} \right) \tag{20.4}$$

where λ_1 is the first principal inertia of the indicator matrix. Thus the higher the principal inertia, the higher the reliability. Using $Q = 4$ and $\lambda_1 = 0.4574$ (four significant digits for slightly better accuracy) we obtain:

$$\alpha = \frac{4}{3} \left(1 - \frac{1}{4 \times 0.4574} \right) = 0.6046$$

Having seen the low correlation of question D with the other questions, an option now is to remove it and recompute the solution with the three questions that are more highly intercorrelated. The results are not given here, apart from reporting that the first principal inertia of this three-variable MCA is

$\lambda_1 = 0.6018$, with an increase in reliability to $\alpha = 0.6692$ (remember to use (20.4) with $Q = 3$).

The
adjustment
threshold
rediscovered

As a final remark, it is interesting to notice that the average squared correlation of a set of random variables, with no zero pairwise correlation between them, is equal to $1/Q$, and this corresponds to a Cronbach's alpha of 0. The value $1/Q$ is exactly the threshold used in (19.6) for adjustment of the principal inertias (eigenvalues), and is also the average principal inertia in the MCA of the indicator matrix, mentioned in Chapter 18. See also the conjecture in the last paragraph of this book — page 304 of the Epilogue in Appendix E.

SUMMARY:
Scaling Properties
of MCA

1. Optimal scaling in a two-variable context was defined as the search for scale values for the categories of one variable which lead to the highest separation of groups defined by the other variable. This problem is equivalent to finding scale values for each set of categories which lead to the highest possible correlation between the row and column variables.

2. In a multivariate context, optimal scaling can be generalized as the search for scale values for the categories of all variables so as to optimize a measure of correlation between the variables and their sum (or average). Specifically, the average squared correlation is maximized between the scaled observations for each variable, called *item scores*, and their sum (or average), called simply the *score*.

3. Equivalently, a minimum can be sought for the variance between item scores within each respondent, averaged over the sample. This is the usual definition of *homogeneity analysis*.

4. The scaling approach in general, exemplified by homogeneity analysis, is a better framework for interpreting the results of MCA of an indicator matrix. The principal inertias and their breakdown into contributions are more readily interpreted as squared correlations, rather than quantities with a geometric significance as in simple CA.

5. The first principal inertia in the indicator matrix version of MCA has a monotonic functional relationship with *Cronbach's alpha* measure of reliability: the higher the principal inertia, the higher the reliability.

6. Since the standard coordinates are identical for the MCA of the indicator matrix, the Burt matrix and in the adjusted form, these scaling properties apply to all three versions of MCA.

Subset Correspondence Analysis

## 21

It is often desirable to restrict attention to part of a data matrix, leaving out either some rows or some columns or both. For example, the columns might subdivide naturally into groups and it would be interesting to analyse each group separately. Or there might be categories corresponding to missing values and one would like to exclude these from the analysis. The most obvious approach would be simply to apply correspondence analysis (CA) to the submatrix of interest. However, one or both of the margins of the submatrix would differ from those of the original data matrix, and so the profiles, masses and distances would change accordingly. The approach presented in this chapter, called *subset correspondence analysis*, fixes the original margins of the whole matrix, using these to determine the masses and χ^2-distance in the analysis of any submatrix. Subset CA has many advantages; for example, the total inertia of the original data matrix is decomposed amongst the subsets, hence the information in a data matrix can be partitioned and investigated separately.

Contents

The author data set of Exhibit 10.6 is a good example of a table with columns that naturally divide into subsets — the 26 letters of the alphabet formed by 21 consonants and 5 vowels. We have seen in Chapter 10, page 78, that the total inertia of this table is very low, 0.01873, but that there is a definite structure amongst the rows (the 12 texts by the six authors). It would be interesting to see how the results are affected if attention is restricted to the subset of consonants or the subset of vowels. One way of proceeding would be simply to analyse the two submatrices, the 12 × 21 matrix of consonant frequencies and the 12 × 5 matrix of vowel frequencies. But this means that the values in the profiles of each text would be recalculated with respect to the new margins of the submatrix. In the analysis of consonants, for example, the

Analysing the consonants and vowels of author data set

relative frequencies of *b, c, d, f,* ..., etc. for each text (row) would be calculated relative to the total number of consonants in the text, not the total number of letters. As for the consonant profiles (columns), these would remain the same as before but the χ^2-distances between them would be different because they depend on the row masses, which have changed.

Subset analysis keeps original margins fixed

An alternative approach, which has many advantages, is to analyse the submatrix but keep the original margins of the table fixed for all calculations of mass and distance. Algorithmically, this is a very simple modification of any CA program — all that needs to be done is to suppress the calculation of marginal sums that are "local" to the submatrix selected, maintaining the calculation of the row and column sums of the original complete table, and using these sums to determine the profile values, masses and distances. This method is called *subset correspondence analysis*.

Subset CA of consonants, contribution biplot

Applying subset CA to the table of consonant frequencies (see pages 265–266) gives the map of Exhibit 21.1. Here the standard CA biplot is given rather than the symmetric or asymmetric map (see Chapter 13). The texts are in principal coordinates, so their interpoint distances are approximate χ^2-distances, where the distances are based on that part of the original χ^2-distance function due only to the consonants, dropping the terms due to the vowels. The consonants are in standard coordinates multiplied by the respective square roots of the relative frequency of the consonant (i.e. relative frequency in the set of 26 letters, remembering that the marginal sums are always those of the original table). The squared lengths of the consonant vectors on each axis are proportional to their contribution to the axis, which is why the letter *y* is so prominent on the second axis (more than 50% in this case). This contribution biplot works just as well for tables with low or high inertias and is particularly useful in this example where the inertia is extremely small. Comparing this map with the asymmetric map of Exhibit 10.7, we can see that the letters are pointing in more or less the same directions and that the configuration of the texts is quite similar. The total inertia is 0.01637, and this value is exactly the sum of the inertias of the consonants in the previous full analysis in Chapter 10. On page 79 we reported the total inertia of the full table to be 0.01873; hence the consonants are responsible for 87.4% (0.01637 relative to 0.01873) of the inertia. Having realized that the consonants contribute the major part of the inertia, it is no surprise that most of the structure displayed in the full analysis of Exhibit 10.7 and the subset analysis of Exhibit 21.1 is the same.

Subset CA of the vowels, contribution biplot

The total inertia of the orginal table is decomposed as follows between the consonants and the vowels:

$$\text{total inertia} = \text{inertia of consonants} + \text{inertia of vowels}$$
$$0.01873 = 0.01637 + 0.00236$$
$$(87.4\%) \quad (12.6\%)$$

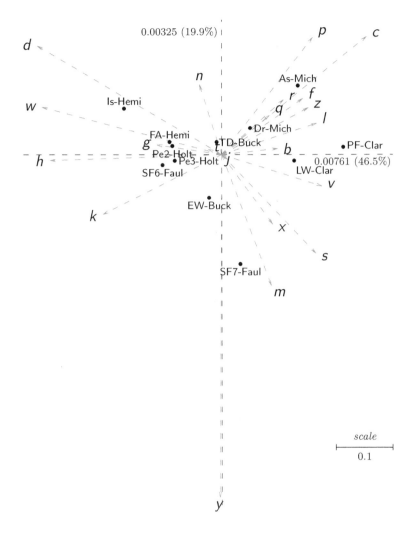

Exhibit 21.1:
Subset analysis of consonants in author example; contribution CA biplot, with rows (texts) in principal coordinates and columns (letters) in contribution coordinates, i.e. standard coordinates multiplied by the square roots of column masses.

The inertia in the vowels subtable is much smaller, only 12.6% of the original total. The vowels are, as expected, more frequent (38.3% of the letters correspond to the 5 vowels, compared to 61.7% for the 21 consonants). The subset CA of the vowels, again with contribution biplot scaling, is shown in Exhibit 21.2. The biplot is on the same scale as Exhibit 21.1 and the lower dispersion of the texts compared to the vectors for the letters is immediately apparent. However, some pairs of texts are still lying in fairly close proximity. There is an opposition of the letter *e* on the left versus the letter *o* on the right, with a corresponding opposition of the texts by Buck versus those by Faulkner. Of the six authors, the texts of Holt seem to be the most different. In Chapter 29, page 232, permutation tests for testing the significance of these results will be reported — anticipating this, it turns out the pairing of the texts is highly

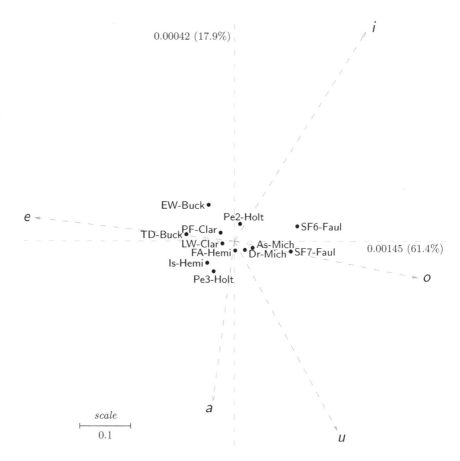

Exhibit 21.2:
Subset analysis of vowels in author example; contribution CA biplot, with rows (texts) in principal coordinates and columns (letters) in contribution coordinates, i.e. standard coordinates multiplied by the square roots of column masses.

significant for both consonants and vowels. So, even though the inertia in the vowels is much lower than the consonants, they still signficantly distinguish between the authors.

Subset MCA   The subset idea can be applied to multiple correspondence analysis (MCA) in much the same way and provides a very useful tool for investigating patterns in specific categories in multivariate categorical data. In questionnaire surveys it may be interesting to focus on a particular subset of responses, for example only the categories of agreement on a five-point agreement–disagreement scale, or all the "middle" response categories ("neither agree nor disagree") or the various non-substantive response categories ("don't know", "no response", "other", etc.). Or we might want to exclude the non-substantive response categories and focus only on the categories of substantive responses. In all these cases, a subset analysis will allow us to see more clearly how demographic variables relate to these specific response categories, which might not be so clear when all categories are analysed together. The subset option allows us to partition the variation in the data into parts for different sets of categories, which can then be visualized separately. Subset MCA is performed by applying subset CA to the appropriate parts of the indicator matrix or Burt matrix, as we illustrate now.

We return to the data set on working women introduced in Chapter 17 and analysed by MCA in Chapter 18. Each of the four questions has a response category for "don't know"/missing responses, labelled by *?*. These categories were prominent on the first principal axis of the MCA (Exhibit 18.2). A subset analysis can exclude these columns of the indicator matrix corresponding to the non-substantive responses, so the subset includes only the substantive response categories, *W* (work full-time), *w* (work part-time) and *H* (stay at home), maintaining the original row sums of the indicator matrix. Since the row sums of the indicator matrix are 4 in this case, the subset analysis maintains the equal weighting for each row (respondent) and the profile values are still zero or $\frac{1}{4}$. The respondents with four substantive answers will have four nonzero values of $\frac{1}{4}$ in their profiles, while those with three substantive answers will have three values of $\frac{1}{4}$, and so on. If we had simply omitted these columns of the indicator matrix and performed a regular CA, then there would be values of $\frac{1}{3}$ for those with three substantive responses, $\frac{1}{2}$ for those with two, and 1 for those with just one substantive response. The profile of a case with four non-substantive responses would be ill-defined, whereas in a subset analysis such a case has a set of zeros as data and is represented at the origin of the map. The total inertia of the subset of 12 categories is 2.1047. Since the total inertia of the whole indicator matrix is 3, this shows that the inertia is decomposed as 2.1047 (70.2%) for the substantive categories and 0.8953 (29.8%) for the non-substantive ones. The principal inertias and percentages of inertia for the first two dimensions of this subset analysis are 0.5133 (24.4% of the total of 2.1047) and 0.3652 (17.4%), i.e. 41.8% in the two-dimensional solution. These percentages again suffer from the problem, as in MCA, of being artificially low. As in Chapter 19, an adjustment of the scaling factors on the axes can be implemented, as will be demonstrated below.

Subset analysis on an indicator matrix

As in regular MCA, the subset analysis can also be performed on the appropriate part of the Burt matrix. To illustrate the procedure, the Burt matrix, given in Chapter 18 in Exhibit 18.4, can be rearranged so that all categories of the subset are in the top left part of the table, as shown in Exhibit 21.3. So the subset of interest is the 12×12 submatrix, itself in a block structure made up of the four sets of three substantive responses, while the four non-substantive categories are now the last rows and columns of the table. The analysis of the subset gives a total inertia of 0.6358 and principal inertias and percentages of 0.2635 (41.4%) and 0.1333 (21.0%) on the first two dimensions: as in MCA, this is an improvement over the indicator matrix version, explaining 62.4% of the inertia compared to 41.8%. Notice that the connection between the indicator and Burt versions of subset MCA is the same as in regular MCA: the principal inertias in the Burt analysis are the squares of those in the indicator version, for example $0.2635 = 0.5133^2$.

Subset analysis on a Burt matrix

Exhibit 21.3:
Burt matrix of four categorical variables of Exhibit 18.4, re-arranged so that all non-substantive response categories (?) are in the last rows and columns. All substantive responses (W, w and H) are in the upper left 12 × 12 part, while the lower right 4 × 4 corner contains the co-occurences of the non-substantive responses ("don't know/missing").

1W	1w	1H	2W	2w	2H	3W	3w	3H	4W	4w	4H	1?	2?	3?	4?
2501	0	0	172	1107	1131	355	1710	345	1766	538	40	0	91	91	157
0	476	0	7	129	335	16	261	181	128	293	17	0	5	18	38
0	0	79	1	6	72	1	17	61	14	21	38	0	0	0	6
172	7	1	181	0	0	127	48	4	165	15	0	1	0	2	1
1107	129	6	0	1299	0	219	997	61	972	239	13	57	0	22	75
1131	335	72	0	0	1646	24	989	573	760	616	84	108	0	60	186
355	16	1	127	219	24	379	0	0	360	14	1	7	9	0	4
1710	261	17	48	997	989	0	2084	0	1348	567	23	96	50	0	146
345	181	61	4	61	573	0	0	642	202	286	73	55	4	0	81
1766	128	14	165	972	760	360	1348	202	1959	0	0	51	62	49	0
538	293	21	15	239	616	14	567	286	0	897	0	45	27	30	0
40	17	38	0	13	84	1	23	73	0	0	97	2	0	0	0
0	0	0	1	57	108	7	96	55	51	45	2	362	196	204	264
91	5	0	0	0	0	9	50	4	62	27	0	196	292	229	203
91	18	0	2	22	60	0	0	0	49	30	0	204	229	313	234
157	38	6	1	75	186	4	146	81	0	0	0	264	203	234	465

Subset MCA with rescaled solution and adjusted inertias

The problem of low inertias is the same here as in MCA: in Exhibit 21.3 there are still 3 × 3 diagonal matrices on the block diagonal of the 12 × 12 submatrix which forms the subset being analysed. As described in Chapter 19, it is possible to rescale the solution by regression analysis so that the off-diagonal submatrices are optimally fitted. This involves stringing out the elements of these 6 off-diagonal matrices, each with 9 elements, as a vector of 54 elements, forming the "y"-variable of the regression. These elements should be expressed as in (19.2), as contingency ratios minus 1. The two "x"-variables (for a two-dimensional solution) are formed by the corresponding products of the standard coordinates. The optimal values for the scale factors are then found by weighted least squares, as before (see Chapter 19, page 148), giving a fit of $R^2 = 0.849$. Once again, this fit applies to the two-dimensional solution only and there is no nesting of the dimensions. However, a simple sub-optimal adjustment is also available here, which is nested, and is implemented in the **ca** package — see the Computational Appendix, page 276). For this example, the percentage explained in this adjusted solution is 82.9% (thus, only 2 percentage points lower than the optimally adjusted 84.9%), but this percentage can be decomposed on the two dimensions: 69.6% and 13.3% respectively — see Exhibit 21.4.

Supplementary points in subset CA

Displaying supplementary points depends on whether rows or columns have been subsetted. In the case of the author data, for example, where the subset of the vowels was analysed (Exhibit 21.2), the usual centring condition holds for the rows (texts), which were not subsetted, but it does not hold for the columns (vowels). If we wanted to project the letter Y onto the subset map of the vowels, we use the zero-centred row coordinates ϕ_{ik} (i.e. row vertices), so it is not necessary to centre Y's profile, and the usual weighted averaging

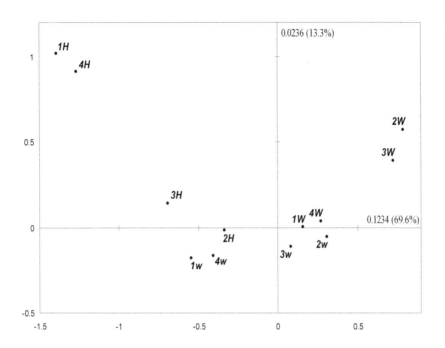

Exhibit 21.4:
*Map of subset of
response categories,
excluding the
non-substantive
categories. The
solution has been
adjusted to fit the
off-diagonal tables of
the subset matrix,
explaining 82.9% of
the inertia.*

gives the principal coordinates — see Chapter 12 and the specific transition formula (14.2) applicable to this case (for a two-dimensional solution):

$$\sum_i y_i \phi_{ik} \qquad k = 1, 2 \qquad (21.1)$$

where y_i is the i-th profile value of Y. On the other hand, to project a new text, with profile values t_j for the subset (these add up to the proportion of vowels in that text, not 1), centring has to be performed with respect to the original centroid values c_j before performing the scalar product operation with the standard column coordinates γ_{jk}:

$$\sum_j (t_j - c_j) \gamma_{jk} \qquad k = 1, 2 \qquad (21.2)$$

Notice that to situate a supplementary point in subset CA and in regular CA, this type of centring can always be done, but is not necessary when the standard coordinates satisfy $\sum_i r_i \phi_{ik} = 0$ and $\sum_j c_j \gamma_{jk} = 0$, which is the case when the summation is over the complete set.

Every respondent (row) of the indicator matrix can be represented as a supplementary point, as well as any grouping of rows into education groups, gender groups, etc. So, as in regular MCA, the categories of supplementary variables are displayed at the centroids of the respondent points that fall into the respective groups. Exhibit 21.5 shows the positions of various demographic categories with respect to the same principal axes as in Exhibit 21.4. Since

*Supplementary
points in subset
MCA*

the horizontal axis now coincides with a traditional attitude (on the left) to liberal (on the right), we see again West Germany opposing East Germany on this axis and the line-up of the other demographic categories, similar to previous solutions.

Exhibit 21.5:
Positions of supplementary points in the map of Exhibit 21.4. Some abbreviations can be found in Chapter 17, page 129; DW and DE are West and East Germany.

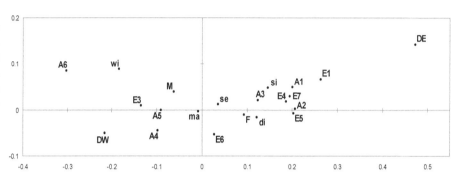

SUMMARY: Subset Correspondence Analysis

1. The idea in *subset CA* is to visualize a subset of the rows or a subset of the columns (or both) in subspaces of the same full space as the original complete set. The original centroid is maintained at the centre of the map, as well as the original masses and χ^2-distance weights.

2. Because the properties of the original space are conserved in the subset analysis, the original total inertia is decomposed exactly into parts of inertia for each subset.

3. Subset CA is implemented simply by suppressing the recomputation of the margins of the subset, and using the original margins (masses) in all the usual CA computations.

4. This idea extends to MCA as well, allowing the selection of any subset of categories, providing an analytic strategy that can be put to great advantage in the analysis of questionnaire data. For example, missing categories can be excluded, or the analysis can focus on one type of response category for all questions, visualizing the dimensions of the subset without any interference from the other categories.

5. As in regular MCA, subset MCA can be applied to the indicator matrix or Burt matrix, and the solutions can be rescaled to optimize the fit to the actual subtables of interest, which dramatically improves the percentages of explained inertia.

6. Supplementary points can also be added to a subset map. In subset MCA this facility allows demographic categories to be related to particular types of response categories.

Compositional Data Analysis

## 22

Compositional data consist of sets of osbervations that add up to a constant, such as proportions that sum to 1, or percentages that sum to 100. Data such as these, called *compositions*, are prolific in chemistry, biochemistry and geology, where samples are analytically measured for their relative contents of different substances, for example oxides in chemistry, fatty acids in biochemistry and grain sizes in geology. But compositional data are also found in the social and economic sciences, such as daily time budgets that sum to 24 hours, family expenses on different items as proportions of total expenditure, or government budgets for different public services. Methods for visualizing compositional data are intimately related to correspondence analysis, since correspondence analysis also analyses the relative values in a data set. In this chapter we will look at methods for visualizing compositional data and show their relationship with correspondence analysis.

Contents

Compositional data

Compositional data are special because they have the property of *closure*, i.e. each set of values for each sample has a constant sum. The original measurements, for example the weights in grams of different oxides in a chemical sample, do have a certain total, but this total is of no interest — it is the proportional amounts that are relevant. We have seen that correspondence analysis treats the rows and/or columns of a table after they have been *closed*, i.e. expressed relative to their respective marginal totals, but these marginal totals are also of interest and used in the analysis as weights.

Subcompositional coherence

Consider the small set of three biological samples in Exhibit 22.1(a), analysed for their proportions of four fatty acids. The column variables are often referred to as *parts*, and so each row is a four-part composition. In the sec-

(a)

Samples	16:1(n-7)	20:5(n-3)	18:4(n-3)	18:00	Sum
A	0.342	0.217	0.054	0.387	1
B	0.350	0.196	0.050	0.404	1
C	0.442	0.294	0.018	0.246	1
Average	*0.378*	*0.236*	*0.041*	*0.346*	*1*

(b)

Samples	16:1(n-7)	20:5(n-3)	18:4(n-3)	Sum
A	0.558	0.354	0.088	1
B	0.587	0.329	0.084	1
C	0.586	0.390	0.024	1
Average	*0.577*	*0.358*	*0.065*	*1*

ond table, Exhibit 22.1(b) the fourth fatty acid has been removed and the compositions recomputed, i.e. the table has been reclosed to give three-part *subcompositions*. The values for sample C do not change too much because its proportion of *18:00* is smaller than the others, whereas the values for samples A and B change more. This means that the usual statistical measures for the column variables, for example the correlation coefficient, are meaningless for such data. Even though one would not seriously compute correlations for such a small table, the correlation between *16:1(n-7)* and *20:5(n-3)* is 0.962 in the full composition (Exhibit 22.1(a)), whereas it is 0.070 in the subcomposition (Exhibit 22.1(b)) — from almost 1 to almost 0! The means and the variances of the three fatty acids in the subcomposition are also radically changed by the dropping of the fourth one. These regular summary statistics are not following what is called the principle of *subcompositional coherence* — this principle states that statistics computed on the parts should be unaffected by the presence or absence of other parts.

Ratios and log-ratios are subcompositionally coherent

One relationship between parts that remains stable, whether in a composition or subcomposition, is the ratio between them. In Exhibit 22.1(a) the ratio *16:1(n-7)/20:5(n-3)* for sample A is $0.342/0.217 = 1.58$, identical to the corresponding ratio in the subcomposition in Exhibit 22.1(b): $0.558/0.354 = 1.58$. Generally, ratios are compared multiplicatively, so the logarithmic transformation is appropriate: $\log(0.342/0.217) = 0.455$, and the logarithm of the inverse ratio is conveniently $\log(0.217/0.342) = -0.455$, which is obvious since $\log(a/b) = \log(a) - \log(b)$. This is called the *log-ratio transformation*, and it is the subcompositionally coherent transformation that is at the heart of compositional data analysis (the natural logarithm is always used). If the compositional values are the same, the ratio is 1 and the log-ratio is $\log(1) = 0$.

There are two equivalent ways of measuring the distances between samples (i.e. the rows) based on their m parts. The first involves the sum of squared differences over their full set of $m(m-1)/2$ log-ratios, while the more compact second way involves sum of squared differences using the set of m so-called *centred log-ratios*, i.e. the logarithm of each compositional part a_{ij} relative to its respective geometric mean $\bar{g}$ across the m parts:

$$d_{ii'} = \sqrt{\sum_{j<j'}\sum \frac{1}{m^2}\left[\log\left(\frac{a_{ij}}{a_{ij'}}\right) - \log\left(\frac{a_{i'j}}{a_{i'j'}}\right)\right]^2} = \sqrt{\sum_j \frac{1}{m}\left[\log\left(\frac{a_{ij}}{\bar{g}_i}\right) - \log\left(\frac{a_{i'j}}{\bar{g}_{i'}}\right)\right]^2}$$

(22.1)

where $\bar{g}_i = (\prod_j a_{ij})^{1/m}$. For the small data matrix of Exhibit 22.1(a), these *log-ratio distances* are given in Exhibit 22.2:

	A	B	C
A	0	0.0622	0.5754
B	0.0622	0	0.5735
C	0.5754	0.5735	0

Exhibit 22.2: *Log-ratio distances between samples (rows) of Exhibit 22.1(a)*

The indices i and j in the log-ratio distance function between samples in (22.1) can be interchanged to give the log-ratio distance between parts (i.e. the columns), either on pairwise log-ratios across the samples or using centred log-ratios with respect to geometric means $\bar{g}_j = (\prod_i a_{ij})^{1/n}$ of the columns:

$$d_{jj'} = \sqrt{\sum_{i<i'}\sum \frac{1}{n^2}\left[\log\left(\frac{a_{ij}}{a_{i'j}}\right) - \log\left(\frac{a_{ij'}}{a_{i'j'}}\right)\right]^2} = \sqrt{\sum_i \frac{1}{n}\left[\log\left(\frac{a_{ij}}{\bar{g}_j}\right) - \log\left(\frac{a_{ij'}}{\bar{g}_{j'}}\right)\right]^2}$$

(22.2)

Since $\log(a_{ij}/a_{i'j}) - \log(a_{ij'}/a_{i'j'}) = \log(a_{ij}/a_{ij'}) - \log(a_{i'j}/a_{i'j'})$ (notice the interchange of $a_{i'j}$ and $a_{ij'}$), and the ratios $a_{ij}/a_{ij'}$ remain constant in subcompositions for all i and j, j', it follows that the log-ratio distances between the three parts of the subcomposition in Exhibit 22.1(b) are exactly the same as those computed from the full composition — the log-ratios are subcompositionally coherent, hence also the log-ratio distances between parts.

The distance function (22.1) is more specifically called the *unweighted* log-ratio distance between samples. An alternative version, which proves to be more useful, is the *weighted* log-ratio distance between samples, which uses the component averages c_j as weights, rather than the equal weights $\frac{1}{m}$:

$$d_{ii'} = \sqrt{\sum_{j<j'}\sum c_j c_{j'}\left[\log\left(\frac{a_{ij}}{a_{ij'}}\right) - \log\left(\frac{a_{i'j}}{a_{i'j'}}\right)\right]^2} = \sqrt{\sum_j c_j\left[\log\left(\frac{a_{ij}}{\bar{g}_i}\right) - \log\left(\frac{a_{i'j}}{\bar{g}_{i'}}\right)\right]^2}$$

(22.3)

(see, for example, the averages c_j in the last rows of the tables in Exhibit 22.1). This means that the log-ratios involving rarer parts get less weight than those involving more "frequent" parts. This makes sense if one realizes that the ratios $\log(a_{ij}/a_{i'j})$ for a fixed part j will generally have higher variance when the values are on average smaller. For example, in the column for the most common fatty acid 16:1(n-7) in Exhibit 22.1(a), notice that the ratios are generally close to 1, e.g. 0.342/0.442 or 0.350/0.442, whereas in the column for the least common one 18:4(n-3) there are two large ratios, 0.054/0.018 and 0.050/0.018, in addition to the smaller one 0.054/0.050. The weighting thus serves as a normalizing factor, similar to the idea in correspondence analysis.

Data set 12:
Time budgets

To illustrate the visualization of compositional data, we use a data set on average time budgets in samples from the USA, some Western countries and Eastern countries (Exhibit 22.3). The data are in hours and the sums of every row are a constant 24 hours, hence making these data compositional.

Exhibit 22.3:
Average time budgets (in hours) for men and women in the USA, Western and Eastern countries. Abbreviations in row labels: M/F is male/female, s/m is single/married, U/W/E is USA/Western countries/Eastern countries. The three rounded zero values 0.00 have been replaced by 0.01 to make the log-ratio analysis possible.

	Prof	Tran	Hous	Kids	Shop	Pers	Eat	Slee	Tele	Leis	Sum
MsU	5.85	1.15	0.50	0.00	1.50	1.05	1.00	7.60	1.50	3.85	24
FsU	4.82	0.94	1.96	0.18	1.41	1.30	0.96	7.75	1.32	3.36	24
MmU	6.15	1.40	0.65	0.10	1.15	0.90	1.15	7.65	1.80	3.05	24
FmU	1.79	0.29	4.21	0.87	1.61	1.12	1.19	7.76	1.43	3.73	24
MsW	6.43	1.05	0.72	0.00	0.62	0.77	1.40	8.13	1.00	3.88	24
FsW	4.29	0.34	2.62	0.14	0.92	0.97	1.47	8.49	0.84	3.92	24
MmW	6.56	0.97	0.97	0.10	0.52	0.85	1.52	8.08	1.22	3.21	24
FmW	1.68	0.22	5.28	0.69	1.02	0.83	1.74	8.24	1.19	3.11	24
MsE	6.27	1.48	0.68	0.00	0.88	0.92	0.86	7.70	0.58	4.63	24
FsE	4.34	0.86	2.97	0.21	1.29	1.02	0.94	7.99	0.58	3.80	24
MmE	6.52	1.33	1.34	0.22	0.68	0.94	1.02	7.63	1.22	3.10	24
FmE	4.36	0.79	4.33	0.60	1.19	0.90	1.07	7.72	0.73	2.31	24
Means	4.92	0.90	2.19	0.26	1.07	0.96	1.19	7.90	1.12	3.50	24
Masses	.205	.038	.091	.011	.044	.040	.050	.329	.047	.146	1

Prof=Professional activity, *Tran*=Transportation linked to professional activity, *Hous*=Household occupation, *Kids*=Occupation linked to children, *Shop*=Shopping, *Pers*=Time spent for personal care, *Eat*=Eating, *Slee*=Sleeping, *Tele*=Watching television, *Leis*=Other leisure.

Log-ratio analysis

The technical details of *log-ratio analysis* (abbreviated as LRA) are given in the Theoretical Appendix, but they are similar to correspondence analysis (CA), with a few interesting variations. We will generally use the weighted form of log-ratio analysis, where the parts (in this case, the categories of daily activity) are weighted by their averages — these are exactly the column masses in a CA of this table. Since all the row sums are equal, the column masses in CA would be the usual means of the columns, divided by 24 (see

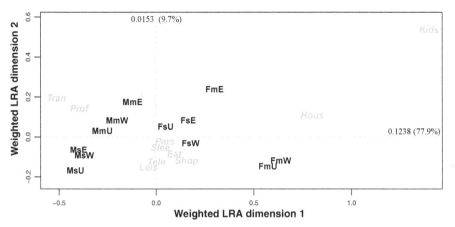

Exhibit 22.4:
Symmetric map of Exhibit 22.3, using log-ratio analysis (LRA). Variance explained is 87.6%.

the means and masses in the last two rows of Exhibit 22.3). The results of the LRA are shown in Exhibit 22.4. The interpretation is quite clear. The activities form two patterns, one from *Tran/Prof* (upper left quadrant) to *Slee/Pers/Eat/Shop/Tele/Leis* (centre) and another one from *Kids/Hous* (upper right quadrant) again to the group of points in the centre. All the male points are shifted in the direction of *Tran/Prof*, for obvious reasons. With respect to this direction they all seem to be more or less at the same position, but they differ in their positions in the direction defined by *Kids/Hous*. So married males in Eastern countries (MmE) are at one extreme, spending more time with the kids and with household occupation than single males in the USA at the other extreme (MsU). All "male single" groups have zeros for spending time with kids, which explains their positions the lowest down amongst the male groups, but single males in USA must be spending less time with household occupation to be the farthest down (see Exhibit 22.3 — their value is 0.50 hours, whereas the other single values are 0.72 and 0.68).

As for the female groups, they have a different alignment of positions. Married women in the Western countries as well as USA separate away from the professional activities and associated transportation. Married women from the Eastern countries as well as all single women occupy an intermediate position. Married females from Eastern countries separate from all single female groups in the direction of household occupation and spending time with children — in Exhibit 22.3 it can be verified that the category FmE is similar to the female single categories in terms of professional activities, but higher on household and children occupation.

An aspect that is particular to log-ratio analysis is that the lines joining pairs of parts represent the log-ratios, and these are also optimally represented. This means that if we had analysed a much wider matrix with the columns being all the $\frac{1}{2} \times 10 \times 9 = 45$ possible log-ratios, then the solution in Exhibit 22.4 would have been exactly the same in that the groups (rows) would have

Interpretation of links as estimated log-ratios

identical positions. In Exhibit 22.4 all possible lines connecting pairs of parts, called *links*, represent the directions of the respective log-ratios.

For example, in Exhibit 22.4 imagine a link in the form of an arrow connecting *Prof* (professional activity) to *Shop* (shopping), in the direction of *Shop*. This represents the log-ratio log(*Shop/Prof*), so the groups FmW and FmU are clearly the highest on this log-ratio, while all male groups are the lowest. If the arrow link pointed in the reverse direction from *Shop* to *Prof* then the link would represent the inverse log-ratio log(*Prof/Shop*), which just changes sign. If the wide matrix of all 45 log-ratios had been analysed, then the log-ratio log(*Prof/Shop*) would be exactly the same link but transferred so the link vector is anchored at the origin of the display. Thus, instead of having 45 points, anchored at the origin, that represent all the log-ratios, Exhibit 22.4 only has 10 points, with their 45 pairwise links representing all unique log-ratios (or their inverses, depending in which direction the links are considered).

Notice that this is not the case of related methods such as principal component analysis (PCA) and CA. For example, in a PCA each variable has a location in the graphical display, and we can certainly link pairs of variables and interpret them as difference vectors, assuming the difference makes sense (for variables on the same scale, for example). But that difference vector is not optimally displayed, and would not be the same if we had separately analysed the wider matrix of all the difference vectors. In log-ratio analysis, however, the difference vectors are optimally displayed in the analysis, and one should think of the method as exactly that, namely the optimal display of the log-ratios in the form of the links in the graphical result.

Diagnosing power models   Another particular property of LRA is that any lining up of points for the parts in the display indicates a power relationship between the corresponding parts. In Exhibit 22.4, for example, there is lining up of the parts *Kids*, *House* and *Leis*&*Tele* together. The log-ratio distance between *Leis* and *Tele* being so small, we can combine these two activities into one, which we denote by *LeTe*: i.e. *LeTe* = *Leis* + *Tele*. As in CA, the new merged point *LeTe* would be located somewhere between *Leis* and *Tele*. The lining up in Exhibit 22.4 of the log-ratios log(*LeTe/Kids*) and log(*LeTe/Hous*), and thus their linear relationship, can be plotted in the scatterplot of Exhibit 22.5. The dashed line shown is not the least-square linear regression line, but rather the first principal axis through the points, with the following equation:

$$\log(\textit{LeTe}/\textit{Kids}) = 1.609 + 1.982\log(\textit{LeTe}/\textit{Hous}) \qquad (22.4)$$

which, after exponentiating both sides, becomes:

$$\textit{LeTe}/\textit{Kids} = 4.998\,(\textit{LeTe}/\textit{Hous})^{1.982}$$

Since $\exp(1.609) = 4.998$ is almost exactly 5, and 1.982 very close to 2, we can simplify (22.4) to obtain the *power model*

$$\frac{(\textit{Hous})^2}{(\textit{LeTe})\,(\textit{Kids})} = 5 \qquad (22.5)$$

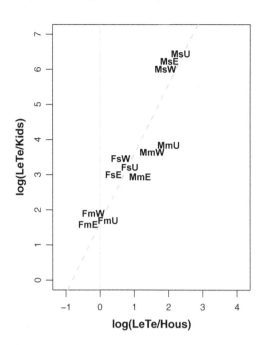

Interestingly, (22.5) shows that, for a fixed time of the "household" occupation (*Hous*), there is a trade-off between "time with kids" (*Kids*) and "other leisure" & "watching television" (*LeTe*). It also shows demographic clusters: married females with proportionally higher *Kids* and *Hous* relative to *LeTe* at one extreme and single males at the other extreme with proportionally less.

A table of compositional data can usually be analysed quite successfully using CA as well (Exhibit 22.6). Since the rows have constant sums, the rows are equally weighted and the columns are weighted proportionally to the column means, just like in (weighted) LRA. So the only difference is the distance function: chi-square distance for CA and log-ratio distance for LRA. Each method has its advantages: CA can handle zeros without any need for introducing the additive constant for the log-transformation, but the χ^2-distance lacks the LRA property of subcompositional coherence. The overall interpretation of Exhibit 22.6 is the same as Exhibit 22.4, but any lining up of the activities has no model diagnostic implications as in LRA. There is, nevertheless, an intimate theoretical connection between LRA and CA. To explain this we recall a well-known data transformation in statistics, the *Box-Cox transformation*:

Correspondence analysis and log-ratio analysis

$$f(x) = \frac{1}{\alpha}(x^{\alpha} - 1) \quad \text{for } 0 < \alpha \leq 1$$
$$= \log(x) \qquad \text{for } \alpha = 0 \tag{22.6}$$

This is a power transformation that has the log-transformation as its limit when the power α tends to 0. Using this transformation gives a continuous family of methods applicable to the contingency ratios transformed as $(1/\alpha)[p_{ij}/(r_i c_j)]^{\alpha}$ (the -1 of (22.6) can be omitted, since any additive con-

Exhibit 22.6:
Symmetric map of Exhibit 22.3, using CA. Inertia explained is 87.3%.

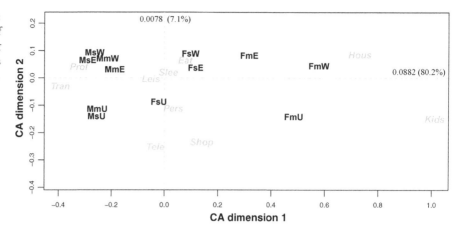

Exhibit 22.6:
Symmetric map of Exhibit 22.3, using CA. Inertia explained is 87.3%.

stant is removed by the double-centring inherent in CA and LRA). When $\alpha = 1$, the matrix elements are just the contingency ratios, hence we obtain CA. As α is reduced and approaches 0, the CA of the power-transformed matrix tends to the analysis of the log-ratios. For α close to — but not exactly equal to — 0, CA will produce a result almost identical to LRA. This also hints at the fact that CA is not far from compositional coherence.

SUMMARY:
Compositional
Data Analysis

1. Compositional data have the property that the multivariate observations on their variables (called *parts*) are nonnegative and have constant sums: 1, or 100% or any other fixed total.

2. *Subcompositional coherence* of any statistic or method means that the results on the parts of the composition remain the same if some more parts are included or some are removed. Most regular statistics, e.g. the correlation, do not have subcompositional coherence.

3. Ratios between parts are subcompositionally coherent: ratios are usually log-transformed to *log-ratios*. *Log-ratio analysis* is a method similar to principal component analysis and correspondence analysis, but measures differences between multivariate samples and between their parts by the *log-ratio distance*.

4. In a LRA map the lines connecting pairs of parts in a LRA map, or *links*, optimally display the corresponding log-ratios.

5. CA can also be applied to compositional data, but lacks the property of subcompositional coherence. However, there is a close relationship between CA and LRA thanks to the Box-Cox power transformation. The CA of power-transformed contingency ratios tends to LRA as the power of the transformation tends to 0.

6. CA is generally close to being subcompositionally coherent, and thus forms a good alternative to LRA, especially when there are a lot of data zeros.

Analysis of Matched Matrices

## 23

In Chapters 16 and 17 the analysis of concatenated tables was considered, either when two categorical variables were combined interactively (e.g. country and gender in Exhibit 16.6), or several variables were stacked (e.g. Exhibit 17.1). It often occurs that we want to compare two tables of the same size, with the same row and column entities, in order to understand their similarities and differences. The gender comparison is a classic example, where we split a cross-tabulation into two tables, one for males and another for females. But we could be comparing other groups such as Western versus Eastern Europe, urban versus rural, treatment versus control, or first time period versus second time period. Two such binary factors could also classify the data, giving four tables, e.g. males versus females in Western Europe, to be compared with males versus females in Eastern Europe. Correspondence analysis (CA) can be used to display variation common to these tables, and variation accounting for their differences.

Contents

In Exhibit 16.6 the categorical variables "country" and "gender" were inter- *Matched matrices*
actively coded to give a row for each country–gender combination. That table shows the male–female groups alternating for each country, but it is equivalent to stack the table for males on the table for females, as was in fact done in Exhibit 16.2 for the health assessment data set. The male and female tables are *matched matrices*: in the one example, a 24×4 country-by-attitude table for each gender, and in the other, a 7×5 age group-by-health category table. The male and female tables have the same row and column entities and the idea is to compare them to see on which aspects males and females tend to agree or differ the most. The correspondence analyses in Exhibits 16.3 and 16.7 showed points for each male and female group within the categories of the corresponding demographic variable, and each difference between the male and female points could be interpreted for its direction and magnitude. In these

plots the male and female points were optimally displayed, but this does not necessarily mean that the male–female differences were optimally displayed. This is reminiscent of what we said in the previous chapter about the special property of log-ratio analysis, that it optimally displays differences between parts in the form of links between pairs of points, whereas methods such as principal component analysis (PCA) and CA do not. In order to specifically optimize the display of male–female differences in tables such as Exhibits 16.2 and 16.6, we would need to compute the actual table of differences between males and females and visualize these differences directly. We can do this with a special version of CA that splits the inertia in two matched matrices into inertias for the average and the difference components, and then visualizes these parts separately.

Between- and within-groups inertia

As in Chapter 16, we are going to reweight the data for gender, since females are over-represented in the samples. This is achieved by expressing each row of the country–gender table as percentages of its respective sample size. In so doing we are assigning equal masses to each male and female point but also equal masses to every country. We could make the country masses proportional to their overall sample sizes, but there is no clear reason to do this, since the sample sizes were not related to any substantive feature of the country itself. In the analysis of the percentaged data, shown in Exhibit 16.7, the first two axes explain 0.08106 (53.4%) and 0.05030 (33.1%) of the total inertia of 0.15181. The total inertia can be decomposed into two parts: a "between-country" part and a "within-country" part. This is the same decomposition that we had in Chapter 15 when we looked at between- and within-groups inertia in cluster analysis — see Equation (15.1). Here the "groups" are defined by the 24 countries, each consisting of two elements, the male and female points, so that the within-country inertia consists of the inertias of the male–female differences in all the countries. It is easy to obtain the "between" and "within" inertia amounts — the simplest is to perform the CA on the 24×4 table where the male and female samples have been combined (because we are analysing the percentages, this aggregation is just the average of the male and female percentages). This analysis gives a total inertia of 0.14288, which is the between-country inertia (94.1% of the total), hence the part due to male–female differences that is "lost" in going from the original 48×4 table to the gender-averaged 24×4 table is $0.15181 - 0.14288 = 0.00893$ (5.9%).

One analysis that splits the "between" and "within" inertias

Let's denote the two 24×4 matrices of percentaged data for the males and females respectively by $\mathbf{A}$ and $\mathbf{B}$. In the stacked analysis, CA was applied to the 48×4 matrix of $\mathbf{A}$ stacked on top of $\mathbf{B}$, obtaining the total inertia of 0.1518. Now a new concatenated matrix is set up in the following format:

$$\begin{bmatrix} \mathbf{A} & \mathbf{B} \\ \mathbf{B} & \mathbf{A} \end{bmatrix} \qquad (23.1)$$

referred to as the "ABBA matrix" (technically, this is called a *block circulant matrix*). For this data set the ABBA matrix is 48×8, repeating the rows and columns twice. Having 8 columns, the table would be 7-dimensional in a CA, but each set of four columns has the same marginal row sums of 1, so the dimensionality is further reduced by one. The resultant six dimensions turn out to split exactly into two sets, one set for the between-country effect and one set for the within-country male–female differences. In order to identify which axes correspond to which set, it is necessary to look at the pattern of signs in the coordinates of the points, either the row points or the column points. For example, all principal coordinates of the eight column points in the ABBA analysis are given in Exhibit 23.1.

	Dim 1	Dim 2	Dim 3	Dim 4	Dim 5	Dim 6
W	−0.050	0.383	−0.102	−0.017	0.037	−0.048
w	−0.153	−0.160	−0.028	−0.052	−0.023	0.012
H	0.588	−0.067	0.014	0.163	−0.029	−0.001
?	−0.142	0.181	0.464	0.037	0.124	0.057
W	−0.050	0.383	−0.102	0.017	−0.037	0.048
w	−0.153	−0.160	−0.028	0.052	0.023	−0.012
H	0.588	−0.067	0.014	−0.163	0.029	0.001
?	−0.142	0.181	0.464	−0.037	−0.124	−0.057
sign	+	+	+	+	+	+
pattern	+	+	+	−	−	−
inertias (sum)	0.07516	0.04886	0.01886			
inertias (diff.)				0.00624	0.00186	0.00083

Exhibit 23.1: *Principal coordinates of columns in the CA of the ABBA matrix of (23.1). There are two sets of identical coordinates up to possible sign changes. The last three dimensions with a sign change correspond to the male–female differences.*

The coordinate values in the upper set of the four columns are identical to those in the lower set, but some have a sign change, specifically the last three dimensions where the lower set is the negative of the upper set (note that, in general, the dimensions with a sign change can occur on any of the dimensions, not necessarily in a contiguous group as in this example). This is the indication that the first three dimensions correspond to the sum (or, equivalently, the average) $\mathbf{A} + \mathbf{B}$, i.e. the between-country table, and the last three dimensions to the difference $\mathbf{A} - \mathbf{B}$, i.e. the within-country table of gender differences. Summing the principal inertias for these two sets of dimensions, the total of the three corresponding to the sum is $0.07516 + 0.04886 + 0.01886 = 0.14288$, the between-country inertia; and the total of the three corresponding to the difference: $0.00624 + 0.00186 + 0.00083 = 0.00893$, the within-country gender difference inertia. Exactly the same inertia components were computed in a different way in the previous section.

The beauty of the ABBA analysis is that it separates the "between" and "within" inertia components (equivalently, "sum" and "difference" components) on different dimensions of the solution space. Exhibit 23.2 shows the

Display of the sum and difference components

Exhibit 23.2:
CA maps of (a)
between-country
and (b) within-
country (i.e. gender
difference) compon-
ents of the matrices
A *and* **B** *for the*
working women data
in 1994. In map (b)
the male–female
differences are
plotted, with the
female points
anchored at the
origin.

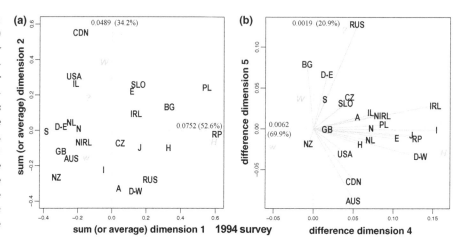

two solutions plotted in symmetric maps side by side. The coordinates for these maps are obtained from the first set of row and column principal co-ordinates of the ABBA analysis: dimensions 1 and 2 for the sum component and dimensions 4 and 5 for the difference component. In each case the per-centages indicated on the axes are those of the inertias relative to the total of the respective component.

Notice that the map of Exhibit 23.2(a) strongly resembles those of Exhibits 16.5 and 16.7 — in the latter case, one should imagine a country point mid-way between its coresponding male and female points. In Exhibit 23.2(a) 86.8% (86.80% to two decimals) of the between-country inertia is displayed, while in Exhibit 23.2(b) 90.8% of the within-country (gender-difference) inertia is displayed. Both these percentages are necessarily as good or better than the corresponding percentages that can be computed in Exhibit 16.7, where both between- and within-country inertias were accounted for together in a single map: there these percentages were 86.8% (86.78% to two decimals) and 82.6% respectively. Thus the gain in Exhibit 23.2(a), compared to Exhibit 16.7, is only 0.02 percentage points for the explanation of the between-country inertia — remember that the between-country inertia dominates the total inertia, being 94.1% of the total, so it is no surprise that the map of Exhibit 16.7 concentrates almost exclusively on this aspect. The gender difference (within-country) component, on the other hand, is noticeably better explained in Exhibit 23.2(b) — 90.8% compared to 82.6% in Exhibit 16.7.

Interpretation
of the difference
map

Apart from this improvement in explained inertia for the difference map, the gender differences are much easier to interpret in Exhibit 23.2(b) when there is just one point displaying each country. Firstly, the countries close to the origin have short lines indicating small differences between male and female attitudes. Secondly, the horizontal dimension is still a "work" versus "stay at home" one, like in the between-country analysis, but the second dimension is more clearly defined by the missing response category. Remember that

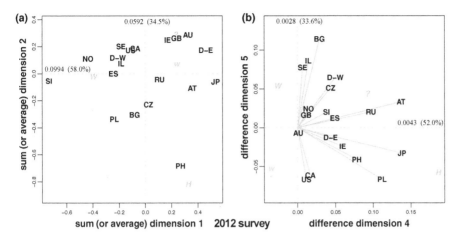

the origin in Exhibit 23.2(b) is not the average male–female difference, but the point representing zero difference. Only Bulgaria and New Zealand point slightly to the left of the origin — these were the only countries in Exhibit 16.7 that showed a slightly different direction in between-gender attitudes, whereas all the other countries point to the right, indicating the consistently more conservative attitude of the males, especially for Italy, Ireland, Japan, Philippines, former West Germany and Spain. Countries such as Russia and Bulgaria show male differences still more conservative but also giving more missing responses, whereas in Australia and Canada it is the females who give more missing responses.

The same approach can be adopted when two data sets are available at different time points, in which case the difference map would show the different trends amongst the countries. It is then even more interesting if one wants to compare trends in male-female differences over time. The data set used up to now has been from the International Social Survey (ISSP) survey in 1994, from the second Family & Changing Gender Roles survey. More recently in 2012, the fourth survey on the same theme was conducted. The same set of countries was not present in this more recent survey: absent were Italy, Hungary, the Netherlands, New Zealand and Northern Ireland, so the common set between the two surveys consists of only 19 countries. The same analysis as was performed in Exhibit 23.2 for the 1994 survey is repeated in Exhibit 23.3 for the 2012 data, where the tables of percentages for the male and female samples across the 19 countries are denoted by **C** and **D** respectively, and in the format of (23.1) we call this the CDDC matrix. It again turns out that dimensions 1 to 3 coincide with the between-country differences and dimensions 4 to 6 with the within-country gender differences. The configurations in Exhibit 23.3 are different from those in Exhibit 23.2: notably the first axis in Exhibit 23.3(a) opposes working full-time (*W*) and the other categories, and the second axis resembles more the first axis in Exhibit 23.2(a), with

Data set 13: Attitudes to women working in 2012

traditional groups lower down on this axis, versus more liberal groups vertically. The positions of Slovenia (SI) and former East Germany (D-E) have changed substantially from the previous survey. The gender-difference map in Exhibit 23.3(b) is similar to Exhibit 23.2(b) as far as response categories are concerned, where staying at home (*H*) mainly accounts for the male–female contrast, with Austrian, Japanese, Polish and Russian males showing the strongest traditional contrast with their female counterparts. Notice that there are no countries here that are to the left of the zero difference origin. On the second vertical dimension (dimension 5 of the CDDC analysis) the missing response category (*?*) is no longer so prominent. This dimension contrasts working full-time (*W*) with working part-time (*w*), showing that Bulgarian, Israeli and Swedish males prefer the former compared to females, while Canadian, USA and Polish males prefer the latter compared to females.

Analysing all effects in one analysis

The data sets are split two ways by gender (male–female) and two ways by survey time (1994–2012). A further separation of effects can be made by setting up the ABBA matrix for 1994 and the CDDC matrix for 2012 in a super-matrix as follows (the ABBA matrix now includes the 19 countries that are common between the two surveys):

$$\begin{bmatrix} C & D & A & B \\ D & C & B & A \\ A & B & C & D \\ B & A & D & C \end{bmatrix} \tag{23.2}$$

The two block circulant matrices ABBA and CDDC for the gender differences are nested within another "ABBA" style block circulant matrix for the two time periods. This super-matrix has 16 columns (4 times the 4 response categories), but has maximum CA dimensionality of only 12 because of the constant row sums of each of the four tables. The coordinates of the row or column points on these 12 dimensions now turn out in four sets of repeated numerical values, with changes of sign depending on which effect is isolated on the respective dimension. The pattern of signs is given in Exhibit 23.4.

Exhibit 23.4:
Pattern of signs in the four sets of coordinates emanating from the CA of the super-matrix (23.2). Each pattern corresponds to a different effect.

					Dimensions						
1	2	3	4	5	6	7	8	9	10	11	12
+	+	+	+	+	+	+	+	+	+	+	+
+	+	+	+	+	+	−	−	−	−	−	−
−	+	+	+	−	−	+	+	+	−	−	−
−	+	+	+	−	−	−	−	−	+	+	+
T	A	A	A	T	T	G	G	G	TG	TG	TG

A = average effect when collapsing over time and gender,
T = time effect, G = gender effect, TG = time×gender effect.

There are inertias on each of these 12 dimensions, 3 dimensions for each effect, and their aggregated sums for the respective effects are as follows:

	A	T	G	TG	*Total*
Effect sums	0.13467	0.08434	0.00758	0.00136	0.22795
Percentages	59.1%	37.0%	3.3%	0.6%	100%

The orders of magnitude of the inertias of the average effect and gender effect are similar to what we have seen before. The time effect is very strong, testifying to large changes in the countries' attitudes over this 18-year time period — the first dimension of the analysis is, in fact, a time effect dimension. A large amount of additional variation has been introduced due to this time effect. The time×gender effect is tiny, which means that the gender-differences have changed very slightly over time.

Each effect can be visualized separately, using the first two dimensions of each set of three classified in Exhibit 23.4. Exhibit 23.5 shows the four maps: the best two dimensions for visualizing the average matrix, equivalent in CA to analysing the sum $\mathbf{A}+\mathbf{B}+\mathbf{C}+\mathbf{D}$, are dimensions 2 and 3; for the time effect $\mathbf{C}+\mathbf{D}-\mathbf{A}-\mathbf{B}$ they are dimensions 1 and 5 (notice the set-up in (23.2) so that the years 2012 minus 1994 is visualized); for the gender effect $\mathbf{C}+\mathbf{A}-\mathbf{D}-\mathbf{B}$

Visualizing the effects

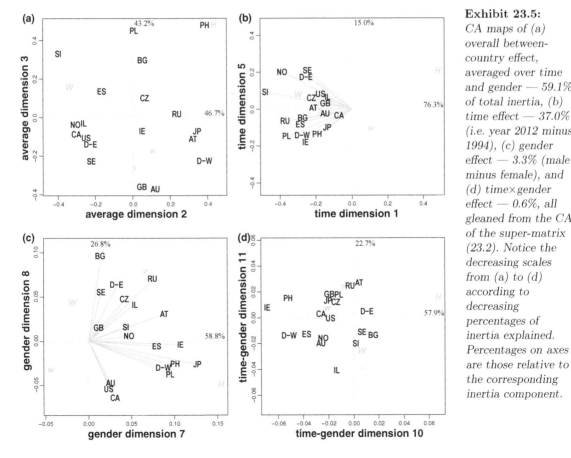

Exhibit 23.5:
CA maps of (a) overall between-country effect, averaged over time and gender — 59.1% of total inertia, (b) time effect — 37.0% (i.e. year 2012 minus 1994), (c) gender effect — 3.3% (male minus female), and (d) time×gender effect — 0.6%, all gleaned from the CA of the super-matrix (23.2). Notice the decreasing scales from (a) to (d) according to decreasing percentages of inertia explained. Percentages on axes are those relative to the corresponding inertia component.

Exhibit 23.6:
Selected plots for Israel and Ireland, showing original percentages for males and females at the two time points.

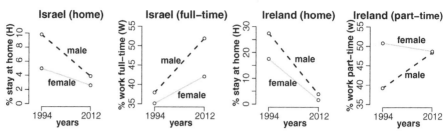

the best dimensions are 6 and 7; and for the time×gender effect $\mathbf{C}+\mathbf{B}-\mathbf{D}-\mathbf{A}$ they are dimensions 10 and 11. The average map in Exhibit 23.5(a), based on the average table over the two time periods and two genders, shows the general positions of the countries with respect to the response categories. The countries' positions have changed somewhat compared to Exhibit 16.4 since attitudes have changed significantly over the 18-year period. Exhibit 23.5(b) shows the three categories *W*, *w* and *H* from left to right (remember this is the first dimension of the whole analysis), indicating the clear movement of all countries away from the "stay at home" attitude towards "work full-time", especially Slovenia (SI). The gender effect, visualized in Exhibit 23.5(c), shows male–female differences consistently towards "stay at home", but less so for countries such as Bulgaria, Sweden and former East Germany, which line up more on the second axis towards "work full-time".

Even though the time×gender effect in Exhibit 23.5(d) explains only a tiny percentage of inertia (0.6%), the way to understand the nature of this effect is to consider examples such as those given in Exhibit 23.6. These plots show male–female differences that change over the years (if the lines shown were more or less parallel, then this combined effect would be nearly zero) — in fact males are changing more towards a liberal attitude than females, which explains why Israel and especially Ireland are away from the "stay at home" point in Exhibit 23.5(d). This could be because the females already showed a more liberal attitude, so their changes are smaller.

SUMMARY:
Analysis of Matched Matrices

1. Matched matrices have the same numbers of rows and columns referring to the same entities in each case.

2. Often these are analysed by CA as matrices concatenated either row- or columnwise. This analysis does not show the differences between the matched matrices optimally.

3. In the case of two matched matrices, they can be arranged in a special circulant block format, in which case the average matrix and the difference between the matrices are separated on different CA dimensions.

4. When there are two binary variables that interactively define four matched matrices, this idea can be generalized to nesting the block matrices for the first variable within a super block matrix for the second variable: then the average effect, individual effects for each binary variable and the combined effect are separated neatly on their respective sets of CA dimensions.

Analysis of Square Tables

## 24

In this chapter we consider the special case when the table of frequencies is square and the rows and the columns refer to the same set of objects in two different states. Such data are found in many situations, for example social mobility tables, confusion matrices in psychology, brand switching tables in marketing research, cross-citations between journals, transition matrices between behavioural states and migration tables. These tables are often characterized by relatively high values down the diagonal, which is such a strong source of association that the more subtle patterns off the diagonal are not seen in the major principal axes. One approach to applying correspondence analysis (CA) to square tables is to split the analysis into two parts: (i) an analysis of the *symmetric* part of the table, which absorbs the main component of inertia, including the diagonal, and (ii) an analysis of the remaining part of the table called the *skew-symmetric* part, which contains the information related to the non-symmetric "flow" between the rows and the columns.

Contents

To give an immediate context to this approach, consider a classic data set on social mobility. This is a historical data set published by Karl Pearson more than 100 years ago on the occupations of fathers and their sons — see Exhibit 24.1. Each father–son pair is counted in one of the cells of the table according to the father's and son's respective occupations. Square tables such as these usually have strong diagonals, since many sons follow their fathers' occupations, but there are some notable asymmetries in the table: for example, in the first line of the table there are 50 fathers in the army, while in the first column there are 84 sons in the army. The flow to the army from other occupations has mostly been from landownership (row 7) and commerce (row 10). Commerce, on the other hand, has had a large outflow to other occupations, with 106 fathers in commerce but only 24 sons, the outflow being mainly to art, divinity, literature and scholarship & science.

Data set 14: Social mobility — occupations of fathers and sons

185

Exhibit 24.1:
Contingency table between the occupations of fathers and sons. For example, of the 50 fathers employed in the army, 28 of their sons were also in the army, 4 went into teaching/clerical work/civil service, and so on.

FATHER'S OCCUPATION	\multicolumn SON'S OCCUPATION														
	ARM	ART	TCC	CRA	DIV	AGR	LAN	LAW	LIT	COM	MED	NAV	POL	SCH	Sums
Army	28	0	4	0	0	0	1	3	3	0	3	1	5	2	50
Art	2	51	1	1	2	0	0	1	2	0	0	0	1	1	62
Teaching...*	6	5	7	0	9	1	3	6	4	2	1	1	2	7	54
Crafts	0	12	0	6	5	0	0	1	7	1	2	0	0	10	44
Divinity	5	5	2	1	54	0	0	6	9	4	12	3	1	13	115
Agriculture	0	2	3	0	3	0	0	1	4	1	4	2	1	5	26
Landowner	17	1	4	0	14	0	6	11	4	1	3	3	17	7	88
Law	3	5	6	0	6	0	2	18	13	1	1	1	8	5	69
Literature	0	1	1	0	4	0	0	1	4	0	2	1	1	4	19
Commerce	12	16	4	1	15	0	0	5	13	11	6	1	7	15	106
Medicine	0	4	2	0	1	0	0	0	3	0	20	0	5	6	41
Navy	1	3	1	0	0	0	1	0	1	1	1	6	2	1	18
Politics...†	5	0	2	0	3	0	1	8	1	2	2	3	23	1	51
Scholarship...°	5	3	0	2	6	0	1	3	1	0	0	1	1	9	32
Sums	84	108	37	11	122	1	15	64	69	24	57	23	74	86	775

*Teaching, Clerical Work & Civil Service   †Politics & Court   °Scholarship & Science

CA of square table

Because this is a contingency table, CA is once more an appropriate method to visualize it (Exhibit 24.2). The table has a high inertia (1.297) because of the strong association between rows and columns, so the asymmetric map is used, with father points in principal coordinates and son points in standard coordinates. If the profile of a father's occupation has all zeros except for the value on the diagonal, then that occupation will lie at the vertex of the respective son's occupation. The second row for the occupation Art is almost like that, with the highest relative value (51 out of 62, or 82%) of fathers having sons in the same occupation, and this fact is reflected by the separating out of Art in Exhibit 22.2, with the father-occupation Art almost reaching the son-occupation vertex point *ART*. The row point Cra(fts) is between the vertex points *ART* and *SCH* (*Scholarship & Science*) because high proportions of sons of fathers in crafts end up in these two occupations (see row 4 of Exhibit 22.1).

Diagonal of table dominates the CA

The problem with trying to visualize a square matrix such as this one is the presence of the strong diagonal which tends to dominate the analysis. Since CA is trying to explain as much inertia as possible, it is not surprising that the focus is on the high source of inertia on the diagonal, to the detriment of the rest of the table which contains the interesting flows between the occupations. To back up this assertion with some figures, the 14 diagonal values account for 70.9% of the total inertia, while the 182 off-diagonal values account for 29.1% — i.e. the total inertia is decomposed as follows:

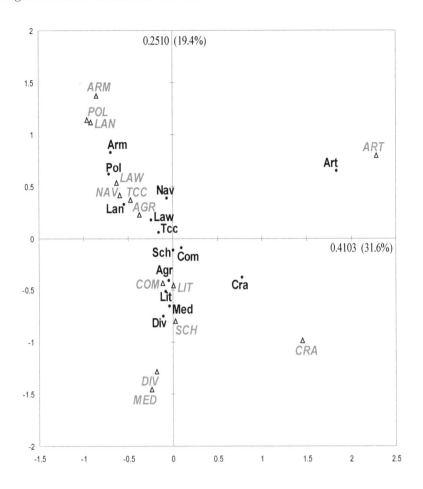

Exhibit 24.2:
Asymmetric CA map of mobility data table in Exhibit 24.1, row points (abbreviations with some lower-case letters) in principal coordinates, column points (in upper-case italic) in standard coordinates. Row points are at weighted averages of the column points, so row points are attracted to their respective column points because of high values on the table's diagonal. Percentage of explained inertia: 51.0%.

total inertia = inertia on diagonal + inertia off-diagonal

$$1.2974 = 0.9200 + 0.3774 \qquad (24.1)$$

$$100\% = 70.9\% + 29.1\%$$

In the two-dimensional display of Exhibit 24.2, 0.6613 (51.0%) of the total inertia is explained, i.e. 0.6361 (49.0%) is error. This error is spread between the diagonal and off-diagonal elements as follows:

error in 2-d = error on diagonal + error off diagonal

$$0.6361 \quad = \quad 0.3717 \quad + \quad 0.2644$$

The errors on and off the diagonal cannot be expressed as percentages of their respective totals in (24.1), but it is a fact that the high inertia on the diagonal is favouring its explanation to the detriment of the smaller component of inertia in the off-diagonal part of the table.

Symmetry and skew-symmetry in a square table

To be able to explain the off-diagonal elements better, as well as quantify their explained inertia and error correctly, the table can be separated into two parts, one part that contains the *symmetric* component of the table, i.e. the average flow between rows and columns, and another part that contains the so-called *skew-symmetric* component quantifying the differential flow. The original table, denoted by $\mathbf{N}$, can be written as follows:

$$\mathbf{N} = \frac{1}{2}(\mathbf{N} + \mathbf{N}^{\mathsf{T}}) + \frac{1}{2}(\mathbf{N} - \mathbf{N}^{\mathsf{T}}) \qquad (24.2)$$

$$= \mathbf{S} + \mathbf{T}$$

where $\mathbf{S}$ is the symmetric part, containing the averages of elements on opposite sides of the diagonal, and $\mathbf{T}$ the skew-symmetric part, containing half of the differences:

$$s_{ij} = \frac{1}{2}(n_{ij} + n_{ji}) \qquad t_{ij} = \frac{1}{2}(n_{ij} - n_{ji}) \qquad (24.3)$$

The following illustrates this decomposition for the top left-hand corner of Exhibit 24.1:

$$\begin{bmatrix} 28 & 0 & 4 & 0 & \cdots \\ 2 & 51 & 1 & 1 & \cdots \\ 6 & 5 & 7 & 0 & \cdots \\ 0 & 12 & 0 & 6 & \cdots \\ \vdots & \vdots & \vdots & \vdots & \ddots \end{bmatrix} = \begin{bmatrix} 28 & 1 & 5 & 0 & \cdots \\ 1 & 51 & 3 & 6.5 & \cdots \\ 5 & 3 & 7 & 0 & \cdots \\ 0 & 6.5 & 0 & 6 & \cdots \\ \vdots & \vdots & \vdots & \vdots & \ddots \end{bmatrix} + \begin{bmatrix} 0 & -1 & -1 & 0 & \cdots \\ 1 & 0 & -2 & -5.5 & \cdots \\ 1 & 2 & 0 & 0 & \cdots \\ 0 & 5.5 & 0 & 0 & \cdots \\ \vdots & \vdots & \vdots & \vdots & \ddots \end{bmatrix}$$

For example, the count of 1 in the second row (father–art) and fourth column (son–crafts) and the count of 12 in the fourth row (father–crafts), second column (son–art), are averaged in $\mathbf{S}$ as 6.5 in both cells, while the deviations (± 5.5) from the average appear in $\mathbf{T}$. The symmetric matrix has the same diagonal as the original table and the property of symmetry: $s_{ij} = s_{ji}$, while the skew-symmetric matrix has zeros on the diagonal and the property of skew-symmetry, namely that elements on opposite sides of the diagonal have the same absolute value but different sign: $t_{ij} = -t_{ji}$.

CA of the symmetric part

CA is now applied to the symmetric and skew-symmetric parts separately. Exhibit 24.3 shows the analysis of the symmetric matrix, showing just one set of profile positions because rows and column coordinates are identical. Apart from the single point for each occupation, this map looks very similar to the configuration of the row points in principal coordinates in Exhibit 24.2, and shows the average association between the occupations. The first percentage on the axes refers to inertia explained relative to the original asymmetric table, while the percentage in italics refers to inertia explained relative to the total inertia of the symmetric part $\mathbf{S}$ that is visualized here. Notice that the row and column margins of $\mathbf{S}$ are the averages of the row and column margins of the asymmetric matrix $\mathbf{N}$: if the latter's row and column masses are $\mathbf{r}$ and $\mathbf{c}$ respectively, then the masses for the rows and columns of $\mathbf{S}$ are $\mathbf{w} = \frac{1}{2}(\mathbf{r} + \mathbf{c})$.

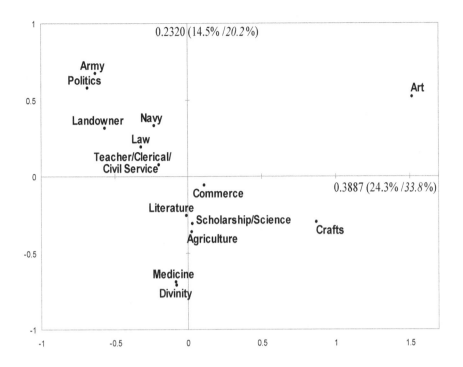

Exhibit 24.3:
CA of symmetric part of Exhibit 22.1. The first percentages are calculated with respect to the total inertia of 1.5991, while percentages in italics are with respect to the inertia of the symmetric part, 1.1485. (The values 1.5991 and 1.1485 are explained on the next page.)

There are two problems to overcome before CA can be applied to the skew-symmetric matrix **T**. First, **T** has positive and negative values; in fact the sum of the elements of the matrix is zero, and it makes no sense to centre it with respect to its margins, which is the first step in the CA algorithm. The algorithm must be changed so that CA analyses the data without the centring step, just the normalization step which leads to the χ^2-distances. This leads to the second problem: the sums of the rows and columns make no sense as masses. The obvious solution is to adopt the same masses as **S**, i.e. the masses in **w** defined above. This looks like we need a special modified algorithm to analyse **T**, but fortunately the results for matched matrices in Chapter 23 solve the problem by a simple set-up of the matrix input to the CA.

CA of the skew-symmetric part

The decomposition (24.2) is identical to what we did in Chapter 23 when the inertia in two matched matrices **A** and **B** was split into an average (or sum) part and a difference part, achived in one step by analysing the matrices in the "ABBA" block circulant matrix format of (23.1). In this case **A** = **N** and **B** = **N**ᵀ, the transpose of **N**, so the block matrix to analyse is:

CA of symmetric and skew-symmetric parts in one step

$$\begin{bmatrix} \mathbf{N} & \mathbf{N}^\mathsf{T} \\ \mathbf{N}^\mathsf{T} & \mathbf{N} \end{bmatrix} \tag{24.4}$$

If **N** is an $I \times I$ matrix, then the block matrix (24.4) is $2I \times 2I$ and yields $2I-1$ dimensions, $I-1$ of which correspond to the dimensions of the symmetric matrix **S** and the remainder to the skew-symmetric matrix **T**. To diagnose which dimensions correspond to which of these two matrices is even easier than before in this special case of square matrices, because the dimensions of the skew-symmetric matrix always occur in pairs of equal principal inertias. In the present social mobility example, where $I = 14$, the 27 principal inertias (eigenvalues) are given in Exhibit 24.4. The seven pairs of dimensions with

Exhibit 24.4:
Principal inertias of all 27 dimensions in the analysis of the 28×28 block matrix (24.4) formed from the social mobility data. The principal inertias that occur in pairs (in boldface) correspond to the skew-symmetric matrix.

Dim.	Princ. inertia	Dim.	Princ. inertia	Dim.	Princ. inertia
1	0.38868	**10**	**0.04184**	**19**	**0.00309**
2	0.23204	**11**	**0.04184**	**20**	**0.00309**
3	**0.15836**	12	0.02287	21	0.00166
4	**0.15836**	13	0.02205	**22**	**0.00115**
5	0.14391	**14**	**0.01287**	**23**	**0.00115**
6	0.12376	**15**	**0.01287**	24	0.00062
7	0.08184	16	0.01036	**25**	**0.00038**
8	0.07074	**17**	**0.00759**	**26**	**0.00038**
9	0.04984	**18**	**0.00759**	27	0.00015

equal principal inertias (shown in boldface), 3 & 4, 10 & 11, 14 & 15, 17 & 18, 19 & 20, 22 & 23, and 25 & 26, correspond to the skew-symmetric analysis, and the other 13 dimensions correspond to the symmetric analysis. The total inertia of the symmetric matrix is the sum of the 13 respective principal inertas: $0.3887 + 0.2320 + 0.1439 + \cdots = 1.1485$, which is 71.8% of the total 1.5991, and the total inertia of the skew-symmetric matrix is the sum of the seven pairs: $2 \times 0.1584 + 2 \times 0.0418 + \cdots = 0.4506$, which is 28.2% of the total. Notice that because $\mathbf{N}$ and $\mathbf{N}^\mathsf{T}$ are placed next to each other, both row- and columnwise, the row and column masses in the analysis of (24.4) will be proportional to the average masses $\mathbf{w}$ (specifically, the masses will be $\frac{1}{4}(\mathbf{r} + \mathbf{c})$ repeated twice to form a $2I \times 1$ vector, where $\mathbf{r}$ and $\mathbf{c}$ are the row and column masses of $\mathbf{N}$). Furthermore, the total inertia of (24.4), equal to 1.5991, is higher than that of the original matrix, given as 1.2974 in (24.1), because the subtables in (24.4) are centred at $\frac{1}{4}(\mathbf{r} + \mathbf{c})$. In the CA of $\mathbf{N}$ both row and column margins, i.e. $\mathbf{r}$ and $\mathbf{c}$, are "closer" to the respective profiles, hence the inertia of $\mathbf{N}$ is lower. Thus, in the analysis of the block matrix (24.4), there is an additional part of inertia due to the differences between $\mathbf{r}$ and $\mathbf{c}$.

Visualization of the symmetric and skew-symmetric parts

Dimensions 1 and 2 are thus the best two for visualizing the symmetric matrix: they explain 0.6217 of the inertia of 1.1485, or 54.0%, and the results are identical to those of Exhibit 24.3. As for matched matrices, CA of the block matrix (24.4) yields twice the sets of results for rows and columns, simple repeats of each other, so it is necessary to use only one set of principal coordinates to obtain the map (see part of the coordinate matrix in Exhibit 24.6, also the Computational Appendix, pages 277–278). Dimensions 3 and 4 are the best for visualizing the skew-symmetric matrix: they explain 0.3167 out of 0.4506, or 70.3% of the inertia of the skew-symmetric part. Notice that two analyses are being performed in one, and the two sets of inertias are judged separately when selecting the ones for visualizing each part. The map of the skew-symmetric part, shown in Exhibit 24.5, has some unusual properties. Firstly, because of the equality of the principal inertias, the coordinates are free to rotate in the two-dimensional map and are not identified with respect to principal axes — hence no axes are drawn in the map. Secondly, the skew-symmetry of the matrix gives a map where again only one set of points is plotted (these are repeated in the CA solution of the block matrix, with a change of sign — see Exhibit 24.6, where the second set of coordinates for dimensions 3 and 4 were used for Exhibit 24.5). In this case the interpretation is not in terms of interpoint distances but rather by looking at triangular areas in the map. For example, Commerce and Scholarship & Science subtend a large triangle with the origin, which is interpreted as a strong differential flow between these two occupations. The clockwise arrow indicates the direction of the flow from fathers to sons: fathers in Commerce have sons who are going to Scholarship & Science relatively frequently (in Exhibit 24.1, the frequency is 15, whereas there is zero flow in the other direction). Thus, the ocupations Landownership, Agriculture, Commerce and Crafts are experiencing outflows to Literature and Scholarship & Science. Some pairs of occupations make very small triangular areas with the origin, for example Army, Politics and Navy, which means that there are no differential flows between these occupations, but they would be experiencing inflows from Agriculture, Crafts, etc.

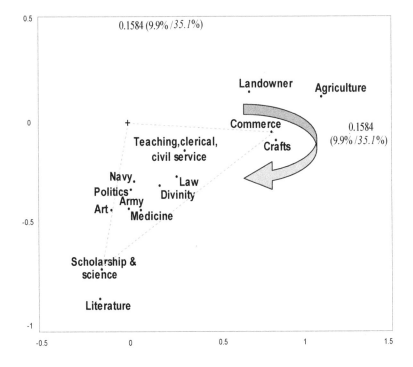

Exhibit 24.5: *CA of skew-symmetric part of Exhibit 24.1. The first percentages are calculated with respect to the total inertia of 1.5991, while percentages in italics are with respect to the inertia of the skew-symmetric part, 0.4506.*

Exhibit 24.6:
*Some of the row
principal
coordinates of the
28 × 28 block matrix
(24.4) using the
social mobility data.
The dimensions for
the symmetric part
(first two in this
case) are simple
repeats, while for
the skew-symmetric
part (dimensions 3
and 4 here) they are
repeated with a
change of sign.*

OCCUPATION	Dim. 1	Dim. 2	Dim. 3	Dim. 4	$\cdots$
Army	−0.632	0.671	−0.011	0.416	$\cdots$
Art	1.521	0.520	0.089	0.423	$\cdots$
Teaching...	−0.195	0.073	−0.331	0.141	$\cdots$
Crafts	0.867	−0.298	−0.847	0.092	$\cdots$
Divinity	−0.077	−0.709	−0.189	0.305	$\cdots$
$\vdots$	$\vdots$	$\vdots$	$\vdots$	$\vdots$	
Army	−0.632	0.671	0.011	−0.416	$\cdots$
Art	1.521	0.520	−0.089	−0.423	$\cdots$
Teaching...	−0.195	0.073	0.331	−0.141	$\cdots$
Crafts	0.867	−0.298	0.847	−0.092	$\cdots$
Divinity	−0.077	−0.709	0.189	−0.305	$\cdots$
$\vdots$	$\vdots$	$\vdots$	$\vdots$	$\vdots$	$\cdots$

SUMMARY:
Analysis of Square
Tables

1. For square tables with the same row and column entities and with large values down the diagonal, the diagonal usually plays a strong role in the analysis, dominating the information off the diagonal.

2. An alternative to a regular CA of such a table is to split the table into two parts: a *symmetric* table and a *skew-symmetric* table, where the latter table — usually of lower inertia than the symmetric part — encapsulates the asymmetries in the table.

3. The symmetric table is analysed in the usual way, while the skew-symmetric table needs a modified CA algorithm which suppresses the centring and normalization of the table with respect to its margins, which have no sense as masses in this case.

4. The masses used for weighting and χ^2-distances in both analyses are the averages of the row and column masses from the original table.

5. Both analyses can be obtained in one single CA of the table and its transpose set up as matched matrices in a block circulant ("ABBA") form. The dimensions corresponding to the symmetric table have unique principal inertias and those for the skew-symmetric table occur in equal pairs.

6. The map of the symmetric table is interpreted in the usual way, showing the overall association between the entities.

7. The map of the skew-symmetric table has a special geometry where the asymmetries in pairs of entities are visualized approximately as the areas of the triangles that they make with the origin, and the direction of the asymmetry is the same for all pairs.

Correspondence Analysis of Networks

## 25

The social mobility application of the previous chapter can be considered as an example of a network that links the set of occupations. Each element of the table linked two occupations with a certain strength of association according to the number of father–son counts for that occupation pair. In network theory a set of items is linked to items of the same set or to items of a different set by a measure of their relationship. Relationships between two sets of items, or *two-mode networks*, fit into the general scheme of correspondence analysis (CA) of a rectangular matrix of associations, usually counts. CA is thus a natural methodology for analysing and interpreting two-mode networks. In this chapter we will concentrate on *one-mode networks*, where the relationships are coded into a square matrix, which presents unique features reminiscent of multidimensional scaling. This square matrix can be either symmetric, for an *undirected network*, or non-symmetric, for a *directed* network.

Contents

The data of Exhibit 25.1 is a classic example of a network, or *graph*. These are the marriages between 15 key Florentine families in the 15th century. This is an example of an *undirected* and *unweighted* network, with a single mode (the 15 families). No direction is implied in the marriage links between pairs of families and all marriages are considered equally important.

Data set 15: The Florentine marriage network

Exhibit 25.2 is one of many possible representations of the marriage network, using package `igraph` in R. The families form the *vertices* of the network and the links between the vertices are called *edges*: thus, in Exhibit 25.2 there are 15 vertices (families) and 20 edges (marriages). Both the vertices and the edges could have weights: for example, the vertices could be weighted by the wealth of the families, and the edges could be weighted by the amount of commercial exchange between them. The network could be called *directed*

Network concepts and terminology

Exhibit 25.1:
Marriages between Florentine families (left-hand list) and the corresponding symmetric matrix coding the marriages, with a 1 in each corresponding row and column. The column (or row) sums of the table are the number of families each one is connected to (called their "degrees" — see next section).

Marriages

Acc — Med
Alb — Gin
Alb — Gua
Alb — Med
Bar — Cas
Bar — Med
Bis — Gua
Bis — Per
Bis — Str
Cas — Per
Cas — Str
Gua — Lam
Gua — Tor
Med — Rid
Med — Sal
Med — Tor
Paz — Sal
Per — Str
Rid — Str
Rid — Tor

	A c c	A l b	B a r	B i s	C a s	G i n	G u a	L a m	M e d	P a z	P e r	R i d	S a l	S t r	T o r
Acc	0	0	0	0	0	0	0	0	1	0	0	0	0	0	0
Alb	0	0	0	0	0	1	1	0	1	0	0	0	0	0	0
Bar	0	0	0	0	1	0	0	0	1	0	0	0	0	0	0
Bis	0	0	0	0	0	0	1	0	0	0	1	0	0	1	0
Cas	0	0	1	0	0	0	0	0	0	0	1	0	0	1	0
Gin	0	1	0	0	0	0	0	0	0	0	0	0	0	0	0
Gua	0	1	0	1	0	0	0	1	0	0	0	0	0	0	1
Lam	0	0	0	0	0	0	1	0	0	0	0	0	0	0	0
Med	1	1	1	0	0	0	0	0	0	0	0	1	1	0	1
Paz	0	0	0	0	0	0	0	0	0	0	0	0	1	0	0
Per	0	0	0	1	1	0	0	0	0	0	0	0	0	1	0
Rid	0	0	0	0	0	0	0	0	1	0	0	0	0	1	1
Sal	0	0	0	0	0	0	0	0	1	1	0	0	0	0	0
Str	0	0	0	1	1	0	0	0	0	0	1	1	0	0	0
Tor	0	0	0	0	0	0	1	0	1	0	0	1	0	0	0
Sums	1	3	2	3	3	1	4	1	6	1	3	3	2	4	3

Acc=Acciaiuoli, Alb=Albizzi, Bar=Barbadori, Bis=Bischeri, Cas=Castellani, Gin=Ginori, Gua=Guadagni, Lam=Lamberteschi, Med=Medici, Paz=Pazzi, Per=Peruzzi, Rid=Ridolfi, Sal=Salviati, Str=Strozzi, Tor=Tornabuoni.

Exhibit 25.2:
Representation of the Florentine marriages network, using R package igraph. *Each marriage between two families (i.e. each 1 in the upper or lower triangle of the matrix in Exhibit 25.1) is indicated by an edge linking two families. The numbers of marriages per family (column sums of the matrix in Exhibit 25.1, i.e. the degrees) are given below the respective labels.*

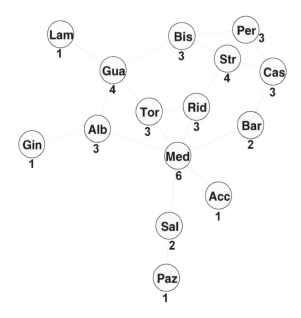

if non-symmetric information were available: for example, from which of the two families the bride came from, or the direction of the trade between two families (i.e. who buys how much from whom). The matrix in Exhibit 25.1 is called an *adjacency matrix* — it is symmetric, with row and column sums equal to the number of edges linked to each vertex, called the *degrees* of the vertices. Notice that the *degree vector*, the vector of marginal sums of the adjacency matrix, is proportional to what would be the masses in the CA of the matrix. In the case of an *edge-weighted* network, the ones in the adjacency matrix would be replaced by their weights, and the marginal sums would then form the *weighted degree* vector, again proportional to CA masses. Directed networks lead to asymmetric square matrices, the subject of Chapter 24 — our attention here is focused on symmetric matrices, which could arise from undirected networks or be the symmetric part of a directed network.

The marriage network is coded in the form of a square symmetric matrix, so let us first examine more closely the properties of such matrices in the context of CA. In the mobility table of Exhibit 24.1, the symmetric part (matrix **S**) was visualized by CA in Exhibit 24.3 (a small part of this matrix is shown on page 188). Of the total inertia (1.1485) of this symmetric matrix, 54.0% was explained in the two-dimensional solution. Most of this total inertia is due to the diagonal of the matrix, which contains high counts of the number of fathers whose sons have continued in the same occupation — in fact, the diagonal constitutes 70.4% of the total inertia. It is not surprising, then, that the display concentrates on explaining the diagonal values more than on the off-diagonal ones, a similar situation to the multiple correspondence analysis (MCA) of the Burt matrix in Chapter 18. In order to reduce the effect of the diagonal, we could divide all the diagonal values by 4, for example, in which case the total inertia of the matrix falls to 0.4509, of which only 15.6% is due to the diagonal, and the off-diagonal values are now better explained. But a strange phenomenon is now discovered on the second dimension: although this is a symmetric matrix, the column coordinates are the negatives of the row coordinates — see column *Dim. 2* of Exhibit 25.3.

At the heart of CA is a singular value decomposition (see the Theoretical Appendix), which is the rectangular generalization of the eigenvalue decomposition of a square matrix. For square symmetric matrices CA can be computed using an eigenvalue decomposition, but the eigenvalues can be negative. Such negative eigenvalues induce so-called *inverse* dimensions, for example the second dimension of Exhibit 25.3. We are usually not interested in these dimensions, since they imply double points for some of the objects being displayed. So we prefer to concentrate on the *direct* dimensions that give exactly the same coordinates for rows and columns, thus just one set of points. For a result such as Exhibit 25.3, dimensions 1 and 3 would be chosen to display the occupations, and then we obtain a result not too much different from Exhibit 24.3, but with improved variance explained on the off-diagonal cells. Rather than this *ad hoc* manipulation of the diagonal elements, a specific procedure is required to shift attention to the off-diagonal elements of the table.

This sounds very much like the case of joint correspondence analysis (JCA, Chapter 19), where we avoided fitting the diagonal subtables of the Burt matrix. So we now implement a similar algorithm that regards the diagonal as

Square symmetric tables revisited: direct and inverse axes

Fitting off-diagonal elements

Exhibit 25.3:
*Principal
coordinates of first
three dimensions of
CA of mobility table
of Exhibit 24.1,
where the diagonal
has been multiplied
by 0.25. The second
dimension is inverse:
the row and column
coordinates have
different signs.*

OCCUPATION	*Dim. 1*	*Dim. 2*	*Dim. 3*	$\cdots$
Row points:				
Army	-0.519	0.097	-0.202	$\cdots$
Arts	0.839	0.266	-0.481	$\cdots$
Teaching...	-0.151	0.002	-0.001	$\cdots$
Crafts	0.920	-0.639	-0.116	$\cdots$
$\vdots$	$\vdots$	$\vdots$	$\vdots$	$\vdots$
Column points:				
Army	-0.519	-0.097	-0.202	$\cdots$
Arts	0.839	-0.266	-0.481	$\cdots$
Teaching...	-0.151	-0.002	-0.001	$\cdots$
Crafts	0.920	0.639	-0.116	$\cdots$
$\vdots$	$\vdots$	$\vdots$	$\vdots$	$\vdots$

missing and attempts to find the best fit to the off-diagonal elements, avoiding the dominating effect of the diagonal. This algorithm imputes new diagonal elements in each iteration based on the CA performed in the previous iteration until convergence at the optimum fit to the off-diagonal elements is achieved. A big difference, however, is that we cannot maintain the same margins as the original matrix in the process of diagonal replacement, as we could do in JCA. Also, we have to be careful to avoid using inverse factors in the iterative substitution of the diagonal elements.

CA of an adjacency matrix   Returning now to the Florentine marriage network example, we notice that the adjacency matrix in Exhibit 25.1 is special in that it has a diagonal of zeros: the edge between a vertex and itself is not observed (unless a marriage took place within a family, which is unlikely). If CA is applied to the adjacency matrix of marriages, the first and third dimensions, amongst several others, turn out to be inverse, so a two-dimensional map would be constructed using dimensions 2 and 4. The question arises once again whether the fit to the off-diagonal values can be improved by replacing the zeros on the diagonal with other values. A reasonable option, which has some spin-offs, is to insert the degree vector down the diagonal, which ensures that there will be no inverse dimensions, as well as preserve the CA masses. The result of the CA is given in Exhibit 25.4, showing quite well the basic structure of the network. Although distances between families are defined by χ^2-distances between profiles, the CA solution turns out to almost equivalent to the one classically obtained in network theory, based on a matrix called the "Laplacian".

The Laplacian matrix   A standard result in network theory is the following. Suppose that the symmetric adjacency matrix $\mathbf{W}$ defines an edge-weighted undirected network on n vertices, with elements w_{ij} that are all positive or zero, with zeros

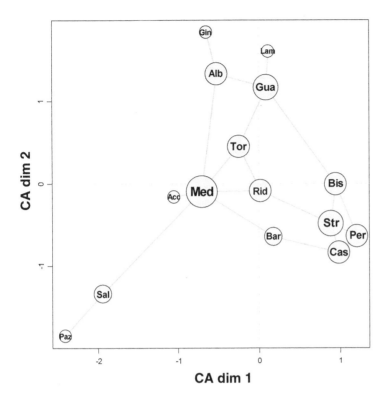

Exhibit 25.4:
CA of adjacency matrix of Florentine marriage network in Exhibit 25.2, with diagonal replaced by the vertex degrees (i.e. column sums). Edges (marriages) have been added and the circle areas are proportional to the respective degrees.

down the diagonal: $w_{ij} \geq 0$, $w_{ii} = 0$. Let the row (or column) sums of $\mathbf{W}$, i.e. the (weighted) degrees, be denoted by $\mathbf{d}$ and denote the diagonal matrix $\mathbf{D}_d = \mathrm{diag}(\mathbf{d})$, called the (weighted) degree matrix. One way to represent the network in a space of given dimensionality K^* is to search for vectors $\mathbf{x}_1, \mathbf{x}_2, \ldots, \mathbf{x}_n$ that minimize the following objective function:

$$f(\mathbf{x}_1, \mathbf{x}_2, \ldots, \mathbf{x}_n) = \sum_{i<j} w_{ij} \|\mathbf{x}_i - \mathbf{x}_j\|^2 \qquad (25.1)$$

Thus, the higher the edge weight w_{ij}, the closer $\mathbf{x}_i$ and $\mathbf{x}_j$ should be*. An identification condition is required to find the minimum of (25.1), and the standard one is that $\mathbf{X}^\mathsf{T}\mathbf{X} = \mathbf{I}$, where the unknown vectors $\mathbf{x}_i$ are rows of the $n \times K^*$ matrix $\mathbf{X}$. The solution is given by the eigenvectors of the *Laplacian*:

$$\text{Laplacian}: \quad \mathbf{L} = \mathbf{D}_d - \mathbf{W} \qquad (25.2)$$

There are no negative eigenvalues of the Laplacian, so no inverse dimensions, but it is not the eigenvectors corresponding to the highest eigenvalues that solve the problem, but rather the smallest ones.

This theory can be transferred into the CA context by introducing the CA masses $\mathbf{c}$ for the vertices (i.e. the degrees in $\mathbf{d}$ divided by their sum), implying the weighted normalization $\mathbf{X}^\mathsf{T}\mathbf{D}_c\mathbf{X} = \mathbf{I}$. It turns out that the (weighted)

---

* For zero/one weights this objective is identical to that of a definition of MCA for what can be considered a two-mode network — see the definition of star plots on page 159.

Laplacian has exactly the same standard coordinates as the CA of the matrix $\mathbf{D}_d + \mathbf{W}$ visualized in Exhibit 25.4, but in the reverse order, so the solutions of the problems are identical. The results of Chapter 23 can be used to verify this, because the analysis of both the sum $\mathbf{D}_d + \mathbf{W}$ and the difference $\mathbf{L} = \mathbf{D}_d - \mathbf{W}$ can be obtained neatly from a single CA of the matrices in the block circulant ("ABBA") format:

$$\begin{bmatrix} \mathbf{D}_d & \mathbf{W} \\ \mathbf{W} & \mathbf{D}_d \end{bmatrix} \qquad (25.3)$$

Separating the sum and difference components by looking at the signs of the repeated eigenvectors, we get the CA performed on $\mathbf{D}_d + \mathbf{W}$ and the Laplacian analysis on the difference matrix $\mathbf{L}$.

A family of
analyses of a
symmetric matrix
In the previous section the diagonal of zeros was replaced by the vertex degrees. Hence, the margins of the table are just multiplied by 2, and their relative values are preserved, i.e. the CA masses remain the same. In addition, the standard coordinates of the CA are identical to those of the original analysis of the adjacency matrix; it is just the principal inertias (eigenvalues) that change — in fact, one can insert any multiple of the degrees on the diagonal and this only affects the eigenvalues. A family of transformations of the adjacency matrix can be defined with this property, in terms of the correspondence matrix $\mathbf{P}$ (the original matrix, denoted by $\mathbf{W}$ above, divided by its grand total) and the marginal sums $\mathbf{c}$ of $\mathbf{P}$ (i.e. the row and column masses, which are identical and equal to the vector $\mathbf{d}$ divided by its sum). This family is defined by the parameters α and β as follows:

$$\mathbf{P}(\alpha, \beta) = \alpha\,\mathbf{P} + \beta\,\mathbf{D}_c + (1 - \alpha - \beta)\,\mathbf{c}\mathbf{c}^\mathsf{T} \qquad (25.4)$$

In the present case, we are just creating mixtures of $\mathbf{P}$ and $\mathbf{D}_c$, so $\alpha + \beta = 1$ (the last term of (25.4) falls out) and the matrix $\mathbf{W} + \mathbf{D}_d$ corresponds to (25.4) with $\alpha = \beta = \frac{1}{2}$. The principal inertias (eigenvalues) of $\mathbf{P}(\alpha, \beta)$ are:

$$\lambda_k(\alpha, \beta) = (\alpha\,\sqrt{\lambda_k}\epsilon_k + \beta)^2 \qquad (25.5)$$

where λ_k are the eigenvalues of the original adjacency matrix $\mathbf{W}$ and $\epsilon_k = 1$ or -1 for direct and inverse dimensions respectively. Thus, the eigenvalues of any such combination (25.4) can be computed, and the corresponding standard coordinates are simply re-ordered in terms of the new descending order of the eigenvalues. As β increases relative to α (the diagonal part increases), all the inverse dimensions disappear — in fact, we are assured of no inverse dimensions for $\mathbf{W} + \mathbf{D}_d$, which is positive semi-definite, as is the Laplacian $\mathbf{D}_d - \mathbf{W}$. An alternative strategy to reduce the effect of the diagonal is to replace it by fractions $k\mathbf{d}$ of the degree vector, from $k = 1$ down to 0 in steps of -0.01 and checking the fit to off-diagonal elements using some measure such as root mean-squared error (RMSE). For the marriage network matrix, the value $k = 0.30$ is found to be optimal (RMSE = 0.1474), corresponding to $\alpha = 1/1.30 = 0.77$, $\beta = 0.23$. The inertia explained by the two CA dimensions, for $k = 0.30$, is 45.4%, compared to the figure of 38.1% when the degree vector replaces the diagonal ($k = 1$) — thus, the effect of the diagonal has been reduced and the percentage of inertia explained is a more accurate measure of display quality.

In the CA map of Exhibit 25.4 there was no direct fitting of the edge weights to the interpoint distances in the network representation. Multidimensional scaling (MDS) specifically aims to optimally visualize distances between objects in a spatial map of low dimensionality. CA is itself an MDS, but visualizes the χ^2-distances between profiles of the data matrix, not actual inter-vertex network distances. One way to measure distances between two vertices is the sum of the weights on the shortest paths linking them. In the present example, the edge weights are all equal to 1 so the shortest path is simply the minimum number of connections between vertices. For example, in Exhibit 25.3, the shortest path between Gin and Bis is 3, and from Sal to Bis it is 4. These shortest-path distances fill a square matrix, with diagonal values zero, which is suitable input to MDS. Using the classical MDS algorithm, which is also based on an eigendecomposition, the map of Exhibit 25.5 is obtained.

Multidimensional scaling of a network

The square symmetric matrix of shortest-path distances has high values for vertices that should be far apart in the map, and low values for those that should be closer. To apply CA to this matrix would not make sense, since CA visualizes a matrix of associations, where high values between objects imply their closeness in the map. It turns out that CA can mimic an MDS by simply reversing the values in the distance matrix, by subtracting the whole

CA can perform MDS

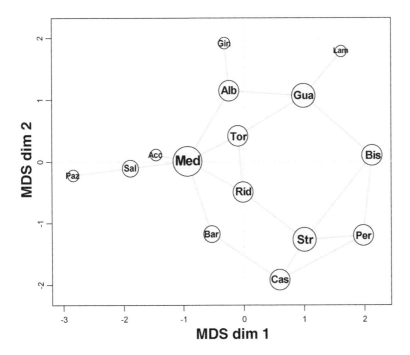

Exhibit 25.5: *Multidimensional scaling of the shortest-path distances between vertices.*

matrix of squared distances from a very large number, at least as large as the largest squared distance. The transformation is:

$$a_{ij} = D - d_{ij}^2 \quad \text{where } D \gg \max\{d_{ij}^2\} \tag{25.6}$$

The higher D is, the closer the CA solution comes to the MDS solution. An adjustment for scale is necessary, since the inertia reduces considerably as D increases. We used the value $D = 1000$ and obtained a result that was almost identical to Exhibit 25.4. The adjustment of the CA solution is as follows:

- Replace the CA inertias λ_k by $\sqrt{\lambda_k}(D/2)$.

- The absolute values of the eigenvalues μ_k of the classical MDS are approximately the transformed inertias above multiplied by the number I of vertices: $\mu_k \approx I\sqrt{\lambda_k}(D/2)$.

To show how close the solutions are, the eigenvalues of the MDS are listed below as well as the inertias from the CA after being adjusted as specified above. Notice that the negative eigenvalues in the MDS, shown in italics, turn up as positive inertias in the CA, also shown in italics.

MDS eigenvalues:
28.41 17.83 6.89 3.98 1.50 0.75 0.37 0.26 0.03 0.00 *-0.83 -1.75 -2.79 -3.78*
CA inertias:
28.66 17.97 6.94 4.01 *3.80 2.82 1.79* 1.51 *0.84* 0.76 0.37 0.26 0.03 0.00

In the MDS the negative eigenvalues are interpreted as non-Euclidean dimensions, whereas in CA they correspond to inverse dimensions.

SUMMARY:
Correspondence
Analysis of
Networks

1. A network is characterised by a *graph* consisting of a set of *vertices* and a set of inter-vertex *edges*.

2. *Undirected* networks can be coded as a square symmetric *adjacency matrix*, with nonnegative values, and zeros down the diagonal. The positive entries of the matrix are *edge weights*, that can be either all equal to 1, for an unweighted network, or different weights quantifying the strengths of connection of the respective vertex pairs.

3. The vertex *degrees* are the sums of the weights of the edges connected to each vertex, i.e. the row or column sums of the adjacency matrix.

4. CA of the adjacency matrix visualizes the network, but the inverse dimensions should be ignored. Replacing the diagonal of zeros with the vertex degrees gives a matrix with no inverse dimensions, and with the same analytical solution as that of the *Laplacian* matrix, which is the diagonal degree matrix minus the adjacency matrix.

5. A distance matrix corresponding to the network can be defined in terms of shortest-path distances between vertices. Classical multidimensional scaling of this matrix gives a distance-based representation of the network. An almost identical solution can be obtained by applying CA to a matrix of the squared distances that have been subtracted from a very large number.

Data Recoding

## 26

In all the chapters up to now we have dealt exclusively with categorical data and frequency tables, either a single table or in sets. In this chapter we will look at other types of data and how they can be recoded, or transformed, in such a way that correspondence analysis (CA) can still be applied as a method of visualization. This strategy is particularly well developed in Benzécri's approach to data analysis, where CA is the central algorithm and different data types are preprocessed before being analysed. The types of data treated here are ratings, preferences, paired comparisons and data on continuous scales. In all of these cases the original CA paradigm should be remembered: CA analyses count data, so if we can transform other types of data to counts of some kind, then it is likely that CA will be appropriate. A standard checklist to perform on the recoded data will be to see if the basic concepts of profile, mass and χ^2-distance make sense in the context of the data.

Contents

We have already met a typical rating scale in Chapter 20, the five-point scale of agreement/disagreement used in the example of science and the environment: *Rating scales*

□	□	□	□	□
strongly agree	*somewhat agree*	*neither agree nor disagree*	*somewhat disagree*	*strongly disagree*

Previously we treated data on a scale such as this as observations on a nominal categorical variable, creating a dummy variable for each category. This was already an example of data recoding, because CA could not be applied to the original data using values 1 to 5, for example — the notion of a profile would make no sense since a set of responses to the four questions [1 1 1 1] (strongly agree to all four statements) and another set [5 5 5 5] (strongly disagree to all four) would have the same profile. Other types of rating scales often found in social surveys and marketing research are:

— 9-point scale (one extra category between points on 5-point scale):

☐ ☐ ☐ ☐ ☐ ☐ ☐ ☐ ☐
*strongly   somewhat   neither agree   somewhat   strongly
agree   agree   nor disagree   disagree   disagree*

— 4-point scale of importance:

☐ ☐ ☐ ☐
*not   fairly   very   extremely
important   important   important   important*

— 7-point semantic differential scale in a customer satisfaction survey

Service ☐ ☐ ☐ ☐ ☐ ☐ ☐ *Service
unfriendly*   *friendly*

— continuous rating scale (e.g. 0 to 10 scale)

Very 0 —————————————— 10 *Very
dissatisfied*   *satisfied*

In this last example the respondent can choose any value between 0 and 10, even with decimal points if desired, but we still think of the data as a rating scale and the recoding will be similar for all the above examples. Notice that when the number of scale points is large, it becomes unwieldy to use the dummy variable coding of multiple correspondence analysis (MCA).

Doubling of ratings   A recoding scheme often used in CA for ratings data is called *doubling*. The idea behind doubling is to redefine each rating scale as a pair of complementary scales, one labelled the "positive", or "high", pole of the scale and the other the "negative", or "low", pole. Before performing the doubling, it is preferable to have rating scales with a lower endpoint of zero, so 1-to-5 and 1-to-7 scales, for example, should first be converted to 0 to 4 and 0 to 6 respectively, simply by subtracting 1. These values define the data assigned to the positive pole of each scale, assuming a high value refers to the substantively positive end of the scale (e.g. high satisfaction, high importance, high agreement). The negative pole of the scale is then defined as M minus the positive pole, where M is the maximum value of the positive pole (4 or 6 in the above examples, or 10 for the 0-to-10 scale). Actually, in the agreement–disagreement scale on the previous page, the high value refers to high disagreement, so the labels "+" and "−" would be reversed to avoid confusion — or we could just reverse this scale beforehand. The idea is illustrated for the agreement ratings in the science and environment data set of Chapter 20. Exhibit 26.1 shows the first five rows of data and their doubled counterparts. For example, the first value for respondent 1 is a 2, subtracting 1 gives the value 1 and its doubled value is 3, hence the values 1 and 3 in the doubled columns for question 1. These columns are labelled $A-$ and $A+$ because the first column quantifies how much the respondent disagrees and agrees respectively with the first question. Similarly, the original value of 3 for the second question becomes a 2 and a doubled value of 2, i.e. equal values for the disagreement and agreement poles $B-$ and $B+$, and so on.

Questions				Qu. A		Qu. B		Qu. C		Qu. D	
A	B	C	D	A−	A+	B−	B+	C−	C+	D−	D+
2	3	4	3	1	3	2	2	3	1	2	2
3	4	2	3	2	2	3	1	1	3	2	2
2	3	2	4	1	3	2	2	1	3	3	1
2	2	2	2	1	3	1	3	1	3	1	3
3	3	3	3	2	2	2	2	2	2	2	2
⋮	⋮	⋮	⋮		⋮		⋮		⋮		⋮

... and so on for 871 rows

Exhibit 26.1: *Raw data (left-hand side) for the variables on science and environment, and the doubled coding (right-hand side), for the first five respondents out of N = 871 (former West German sample).*

The doubled values can be thought of as counts in the following sense. The doubled values 1 and 3 are counts of how many scale points are below and above the observed value of 1. The response of 2 ("agree") on the five-point scale has one scale point below it and three above it. Similarly, the "neither agree nor disagree" response of 3 is in the middle of the scale and has two scale points below and above it. In this way the doubled data table substitutes the original data by measuring association between each respondent and the agreement and disagreement poles of the rating scale.

The counting paradigm

CA is applied to the doubled table on the right-hand side of Exhibit 26.1, which has 871 rows and 8 columns. The rows all have the same sums (16 in this example); hence the respondent masses are equal, which makes sense — there is no reason to give respondents different weights. Each of the four pairs of columns has the same row sums of 4, so there are four linear restrictions on the columns and not just one as in regular CA. Hence the dimensionality of the data matrix is $8 - 4 = 4$. Exhibit 26.2 shows the map of the column points, two points for each question. The positive poles are directly opposite their negative counterparts relative to the origin, as shown by the dashed lines joining the pairs of poles. The fact that question D is out of line with the other three, already seen in Chapter 20, is shown clearly here. We would have expected $D-$ on the right and $D+$ on the left but the direction of this question is practically at right angles to the others.

CA map of doubled ratings

Notice that all four rating scale "axes" pass through the origin of the map. The dashed lines between the poles can be subdivided into four equal intervals, and labelled by the five scale points (shown for question C, using the original 1-to-5 scale where 1 corresponded to strong agreement). The average rating for each question can then be read at the origin on the respective calibrated axis. Thus the average ratings on questions A and C are more to the agreement (+) side of the scale (the actual average for question C is 2.58), while the averages for B and D are slightly to the disagreement side. Another way of thinking about this is to imagine the endpoints of each rating scale axis having weights proportional to the average of the values attributed to the respective poles — thus $C+$ is closer to the origin than $C-$ because it is "heavier".

Exhibit 26.2:

Exhibit 26.2:
CA of doubled
ratings of science
and environment
data, showing
doubled ratings
only. Percentage of
explained inertia is
70.6%. The rating
scale can be
imagined at equal
intervals along each
"axis" connecting
the poles (e.g.,
1-to-5 scale shown
for question C) and
the average for each
question is exactly
at the origin.

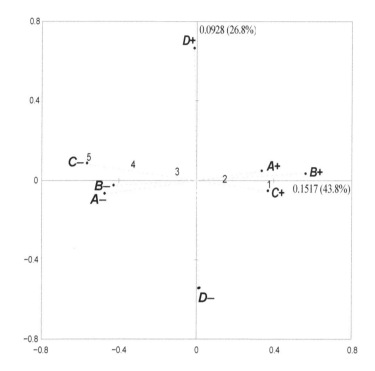

Correlations
interpreted by
alignment of
variables

The alignments of the four rating scale axes in Exhibit 26.2 visually depict the correlations between the variables — specifically the cosines of the angles between them approximate the correlations. Thus we can deduce that variables *A*, *B* and *C* are positively correlated with one another, but uncorrelated with *D*. The four variables have correlation coefficients as follows:

Questions	A	B	C	D
A	1	0.378	0.357	0.036
B	0.378	1	0.436	0.016
C	0.357	0.436	1	−0.062
D	0.036	0.016	−0.062	1

which agrees with our visual deduction. The correlations are not exactly equal to the angle cosines because this map explains only 70% of the inertia. For example, *B* and *C* should make a smaller angle than *A* and *B*, but this would be seen more accurately only in a three-dimensional view of the rating scales.

Positions of
rows and
supplementary
points

Each respondent has a profile and position in the map, as in a regular CA. But, as in MCA of survey data with large samples, the individual positions are not of interest, but rather positions of groups of respondents as supplementary points. For example, to represent males and females in the six different age groups in this data set, the average ratings for these 12 groups are computed and added as supplementary rows (doubled). Their positions are shown in

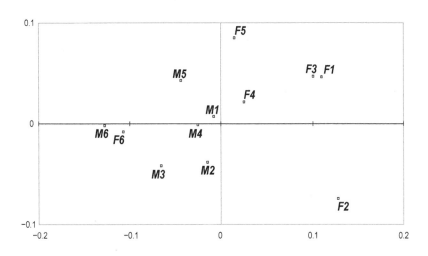

Exhibit 26.3:
*Supplementary
points for males and
females in the six
age groups. The
males are all on the
left-hand side
(disagreement on A,
B and C) while the
females — apart
from the oldest
group F6 — are on
the agreement side.*

Exhibit 26.3. Apart from the oldest group, the female groups are on the right-hand side of the map, that is, the agreement side of questions *A*, *B* and *C*. All the male groups are on the disagreement side of these questions, i.e. less critical of science's role in the environment.

Preference data can be regarded as a special case of ratings data. A typical study in marketing research is to ask respondents to order a set of products from most preferred to least preferred, or a set of product attributes from most important to least important. As an example, suppose that there are six products, *A* to *F*, and that a respondent orders them as follows:

Preference data

most preferred : $B > E > A > C > F > D$: least preferred

This ordering corresponds to the following ranks for the six products:

A	B	C	D	E	F
3	1	4	6	2	5

The six ranks are just like ratings on a 6-point scale, the difference being that the respondent has been forced to use each scale point only once. These data can be doubled in the usual way, with the doubled columns assigned labels where "+" indicates high preference and "−" low preference:

$A-$	$A+$	$B-$	$B+$	$C-$	$C+$	$D-$	$D+$	$E-$	$E+$	$F-$	$F+$
2	3	0	5	3	2	5	0	1	4	4	1

Frequently, respondents are allowed to rank order a smaller set of most-preferred objects (e.g. first three choices), in which case the objects not ranked are considered to be jointly in the last position, which gets the value of a tied rank. For example, if the best three out of six products are rank-ordered, then the three omitted products obtain ranks of 5 each, the average of 4, 5 and 6.

Paired (or *pairwise*) *comparisons* are a freer form of preference rankings. For example, each of the 15 possible pairs of the six products *A* to *F* is presented

*Paired
comparisons*

to the respondent, who selects the more preferred of the pair. The doubled data for each respondent are then established as follows:

  A+: number of times *A* is preferred to the five other products

  A–: number of times the other products are preferred to *A* (= 5 – *A*+)

and so on. Then proceed as before, applying CA to the doubled data.

Data set 16:
European Union
indicators
Continuous data can also be visualized with CA after the data are suitably recoded, and several possibilities exist. As an example, consider the data on the left-hand side of Exhibit 26.4, five economic indicators for the 12 European Union countries in the early 1990s. There are a mixture of measurement scales in these data, with some index values and unemployment rate and change in personal consumption measured in percentages.

Exhibit 26.4:
European Union
economic indicators,
and their ranks from
smallest to largest.

	Original data					*Ranked data*				
COUNTRIES	*Unemp*	*GDP*	*PCH*	*PCP*	*RULC*	*Unemp*	*GDP*	*PCH*	*PCP*	*RULC*
Belgium	8.8	102	104.9	3.3	89.7	7	7	7	7.5	5.5
Denmark	7.6	134.4	117.1	1	92.4	5	12	11	1	8
Germany	5.4	128.1	126	3	90	3	11	12	6	7
Greece	8.5	37.7	40.5	2	105.6	6	2	2	2	12
Spain	16.5	67.1	68.7	4	86.2	12	4	4	11	3
France	9.1	112.4	110.1	2.8	89.7	8	9	9	4.5	5.5
Ireland	16.2	64	60.1	4.5	81.9	11	3	3	12	2
Italy	10.6	105.8	106	3.8	97.4	10	8	8	10	10
Luxemburg	1.7	119.5	110.7	2.8	95.9	1	10	10	4.5	9
Holland	9.6	99.6	96.7	3.3	86.6	9	6	5	7.5	4
Portugal	5.2	32.6	34.8	3.5	78.3	2	1	1	9	1
UK	6.5	95.3	99.7	2.1	98.9	4	5	6	3	11

Une=Unemployment rate (%), *GDP*=Gross Domestic Product/Head (index), *PCH*=Personal Consumption per Head (index), *PCP*=Change in Personal Consumption (%), *RUL*=Real Unit Labour Cost (index).

Recoding
continuous data by
ranks and doubling
A simple recoding scheme is to convert all the observations to ranks, as shown on the right-hand side of Exhibit 26.4. The observations are now ranked within a variable across the countries; for example Luxemburg has the lowest unemployment and so gets rank 1, then Portugal with rank 2 and so on. Tied ranks are given average ranks; for example France and Luxemburg tie for fourth place on the variable *PCP*, so they are given the average 4.5 of ranks 4 and 5. With the transformation of the data to ranks, the doubling can take place as before for each variable: first 1 is subtracted from the ranks to get the positive pole of the scale (the high value) and the negative pole is calculated as 11 minus the positive pole. The CA of the doubled matrix is shown in Exhibit 26.5. Again, the opposite poles of each variable could be connected, but the distances from the origin for each variable are the same in this case because their average ranks are identical — hence, plotting just the positive pole is sufficient. The

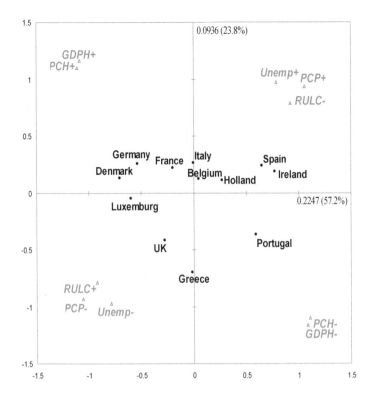

Exhibit 26.5:
*Asymmetric CA
map of European
Union economic
indicators, recoded
as ranks. Inertia
explained is 81.0%.*

map shows two sets of variables, strongly correlated within each set but with low correlation between them. Notice that *RULC* (Real Unit Labour Cost) is negatively correlated with *Unemp* (Unemployment Rate) and *PCP* (Percentage Change in Personal Consumption) (correlations here are nonparametric *Spearman rank correlations* because ranks are used). Each country finds its position in terms of its rank orders on the five variables. Since the ranks are analysed and not the original values, the analysis would be robust with respect to outliers and can be called a *nonparametric* CA of the data.

The transformation of the continuous variables to ranks loses some information, although in our experience the loss is minimal in terms of data visualization, and the robustness of the ranks is an advantage in many situations. However, if all the information in the continuous data is needed, other possibilities exist. For example, a transformation that works well is the following: first, convert all variables to standardized values (so-called z-scores) by subtracting their respective means and dividing by standard deviations; then create two doubled versions of each variable from its standardized z using the recoding positive pole $= (1 + z)/2$ and negative pole $= (1 - z)/2$. Even though it has some negative values, the row and column margins are still positive, and equal for all rows and for all doubled column pairs, so the cases and the variables are weighted equally. CA of this doubled matrix gives a map almost identical

*Other recoding
schemes for
continuous data*

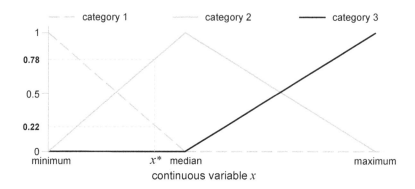

to that of Exhibit 26.5. This is one of the few examples of a data matrix with some negative values that can be validly analysed by CA.

Another solution is to use *fuzzy coding*, which is especially useful when the variables are a mix of categorical and continuous variables. The idea is illustrated in Exhibit 26.6, transforming a continuous variable x (the horizontal axis) to a fuzzy categorical variable with a pre-specified number of categories. For this three-category example above, the transformation is performed using three *membership functions*: for example, straight line *triangular* membership functions. These are defined by three *hinges*, chosen in Exhibit 26.6 as the minimum, median and maximum of the variable's distribution. A value of x such as x^*, shown below the median, is converted to the three values $[\,0.22 \quad 0.78 \quad 0\,]$. The fuzzy values add up to 1, and allow these categories of x to be analysed alongside other categorical variables coded as zero/one dummy variables (called *crisp coding*, as opposed to fuzzy coding).

SUMMARY:
Data Recoding

1. Data on different measurement scales can be recoded to be suitable for CA. As long as the recoded data matrix has meaningful profiles and marginal sums in the context of the application, CA will give valid visualizations of the data.

2. One of the main recoding strategies is to *double* the variables, that is convert each variable to a pair of variables where the sums of the paired values are constant.

3. Doubling can be performed in the case of ratings, preferences and paired comparisons, leading to a map where each variable is displayed by two points directly on opposite sides of the origin. In the case of ratings data, the origin indicates the average value of the variable on the line connecting its extreme poles.

4. Continuous data can be recoded as doubled ranks, leading to a nonparametric form of CA, or can be transformed to a continous pair of doubled variables, using their standardized values.

5. *Fuzzy coding* converts a continuous variable to a set of fuzzy categories, that can be analysed jointly with dummy (0/1) categorical variables..

Canonical Correspondence Analysis

## 27

The objective of correspondence anaysis (CA) is to visualize a table of data in a low-dimensional subspace with optimal explanation of inertia. When additional external information is available for the rows or columns, these can be displayed as supplementary points that do not play any role at all in determining the solution (see Chapter 12). By contrast, we may actually want the CA solution to be directly related to some external variables, in an active rather than a passive way. The context where this often occurs is in environmental research, where information on both biological species composition and environmental parameters are available at the same sampling locations. Here the low-dimensional subspace is required that best explains the biological data but with the condition that the space is forced to be related to the environmental data. This adaptation of CA to the situation where the dimensions are assumed to be responses in a regression-like relationship with external variables is called *canonical correspondence analysis*, or CCA for short.

Contents

To motivate the idea behind CCA, we look again at the marine biological data of Exhibit 10.4, page 77. In addition to the species information at each sampling location on the sea bed, several environmental measurements were made: metal concentrations (lead, cadmium, barium, iron, ...), sedimentary composition (clay, sand, pelite, ...) and other chemical measurements such as hydrocarbon and organic content. Since some of these are highly inter-correlated, we chose three representative variables as examples: barium and iron, measured in parts per million, and pelite as a percentage, shown in Exhibit 27.1 (pelite is sediment composed of fine clay-size or mud-size particles). These variables will be used as explanatory variables within CCA. We prefer to use their values on a logarithmic scale, a typical transformation to convert

Supplementary
continuous
variables

Exhibit 27.1:
*Environmental data
measured at the 13
sampling points (see
Exhibit 10.4); 11
sites in vicinity of
oil-drilling platform
and 2 reference sites
10 km away.*

	STATIONS (SAMPLES)												
VARIABLES	S4	S8	S9	S12	S13	S14	S15	S18	S19	S23	S24	R40	R42
Barium (Ba)	1656	1373	3680	2094	2813	4493	6466	1661	3580	2247	2034	40	85
Iron (Fe)	2022	2398	2985	2535	2612	2515	3421	2381	3452	3457	2311	1804	1815
Pelite (PE)	2.9	14.9	3.8	5.3	4.1	9.1	5.3	4.1	7.4	3.1	6.5	2.5	2.0
$\log(Ba)$	3.219	3.138	3.566	3.321	3.449	3.653	3.811	3.220	3.554	3.352	3.308	1.602	1.929
$\log(Fe)$	3.306	3.380	3.475	3.404	3.417	3.401	3.534	3.377	3.538	3.539	3.364	3.256	3.259
$\log(PE)$	0.462	1.173	0.580	0.724	0.623	0.959	0.724	0.613	0.869	0.491	0.813	0.398	0.301

ratio-scale measurements on a multiplicative scale to an additive scale — their log-transformed values are also given in Exhibit 27.1. This transformation not only has a normalizing effect on the scales of the variables but also reduces the influence of large values.

*Representing
explanatory
variables as
supplementary
variables*
Before entering the world of CCA, let us first display these three variables on the map previously shown in Exhibit 10.5. The way to obtain coordinates for the continuous variables is to perform a weighted least-squares regression of the variable on the two principal axes, using the column standard coordinates γ_1 and γ_2 on the first two dimensions as "predictors" and the column masses as weights, as shown in Chapter 14, page 110. For example, for the regression of $\log(Ba)$, part of the data are as follows:

Stations	$\log(Ba)$	γ_1	γ_2	Weight
S4	3.219	1.113	0.417	0.0601
S8	3.138	−0.226	−1.327	0.0862
S9	3.566	1.267	0.411	0.0686
⋮	⋮	⋮	⋮	⋮
R42	1.929	2.300	0.7862	0.0326

The results of the regression are:

Source	Coefficient	Standardized coefficient
Intercept	3.322	—
γ_1	−0.301	−0.641
γ_2	−0.229	−0.488

$$R^2 = 0.648$$

The usual way of displaying the variable is to use the standardized regression ceofficients as coordinates. As illustrated in Chapter 14, page 111, these are identical to the (weighted) correlation coefficients of $\log(Ba)$ with the two sets of standard coordinates. Repeating the regressions (or, equivalently, calculating the correlation coefficients) the three environmental variables can be

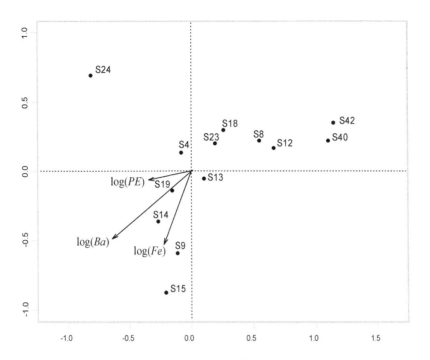

Exhibit 27.2:
Station map of Exhibit 10.5, showing positions of three environmental variables as supplementary variables according to their correlations with the two principal axes.

placed on the map of Exhibit 10.5, which we show in Exhibit 27.2, omitting the species points. The percentage of variance explained (R^2) for each variable is the sum of the squared correlation coefficients, exactly what we called the *quality* of display of a point. For $\log(Ba)$ it is quite high, 0.648 (or 64.8%) as given above, while for $\log(Fe)$ it is 0.326 and for $\log(PE)$ only 0.126.

We now turn the problem around: instead of regressing the continuous explanatory variables on the dimensions, we regress the dimensions on the explanatory variables, always incorporating the masses as weights in the regression. The results of the two regression analyses are given in Exhibit 27.3. Notice that the standardized coefficients are, unfortunately, no longer the correlation coefficients we used to display the variables in Exhibit 27.2, because the explanatory environmental variables are correlated. For example, the cor-

Dimensions as functions of explanatory variables

Response: CA dimension 1			*Response: CA dimension 2*		
Source	*Coeff.*	*Stand. coeff.*	*Source*	*Coeff.*	*Stand. coeff.*
Intercept	−9.316	—	Intercept	14.465	—
$\log(Ba)$	−1.953	−0.918	$\log(Ba)$	−0.696	−0.327
$\log(Fe)$	−4.602	0.398	$\log(Fe)$	−3.672	0.318
$\log(PE)$	0.068	0.014	$\log(PE)$	0.588	0.123
	$R^2 = 0.494$			$R^2 = 0.319$	

Exhibit 27.3:
Regressions of first two CA dimensions on three environmental variables.

relations between $\log(Ba)$ and the two dimensions are -0.641 and -0.488, while in the regression analyses above the standardized regression coefficients are -0.918 and -0.327 respectively.

Constraining the dimensions of CA

Each CA dimension has a certain percentage of variance explained by the environmental variables — 49.4% and 31.9% respectively (see bottom line of Exhibit 27.3). The CA solution, computed on the species data, imposes no restriction on the dimensions, whereas in CCA the condition will be imposed that the dimensions be linear functions of the environmental variables. This will increase the explained variance of the dimensions as a function of the environmental variables to 100%, but at the same time degrade the explanation of the species data. The way the solution is computed is to project the whole data set onto a subspace which is defined linearly by the three environmental variables, and then perform the CA in the usual way in this restricted space. So CCA could just as well stand for "constrained" correspondence analysis — performing CA in a space constrained by the explanatory variables. Having done the CCA (we show the full results later), the regressions of the first two CCA dimensions on the environmental variables are given in Exhibit 27.4. The R^2 for both regressions are now indeed 1, which is what was intended — by construction, the dimensions are now exact linear combinations of the environmental variables.

Exhibit 27.4: Regressions of first two CCA dimensions on three environmental variables.

Response: CCA dimension 1			Response: CCA dimension 2		
Source	Coeff.	Stand. coeff.	Source	Coeff.	Stand. coeff.
Intercept	2.719	—	Intercept	14.465	—
$\log(Ba)$	-2.297	-1.080	$\log(Ba)$	-0.877	-0.412
$\log(Fe)$	1.437	0.124	$\log(Fe)$	12.217	1.058
$\log(PE)$	-0.008	-0.002	$\log(PE)$	-2.378	-0.497
	$R^2 = 1$			$R^2 = 1$	

Constrained and unconstrained spaces in CCA

CCA restricts the search for the optimal principal axes to a part of the total space, called the *constrained space*, while the rest of the space is called the *unconstrained space* (also called *restricted* and *unrestricted*, or *canonical* and *non-canonical* spaces respectively). Within the constrained space the usual CA algorithm proceeds to find the best dimensions to explain the species data. The search for the best dimensions can also take place within the unconstrained space — this space is the one that is linearly unrelated (i.e. uncorrelated) with the environmental variables. So if we are interested in *partialling out* some variables from the analysis, we could do a CCA on these variables and then investigate the dimensions in the unconstrained part of the space.

Decomposition of inertia in CCA

In the present example, the total inertia of the species-by-sites table of Exhibit 10.4 is 0.7826. The inertias in the constrained and unconstrained spaces decompose this inertia into two parts, with values 0.2798 and 0.5028 respectively, i.e. 35.8% and 64.2% of the total inertia. This explains why the original CA dimensions were not strongly correlated with the environmental variables, because CA tries to explain the maximum inertia possible, and there is more inertia in the unconstrained space than in the constrained one. The decomposition of inertia in the CCA is illustrated in Exhibit 27.5, including the decomposition along principal axes. Once the search is restricted to the constrained space (depicted by the shaded area in Exhibit 27.5), the first two dimensions have principal inertias of 0.1895 and 0.0615 respectively, totalling 0.2510 or 89.7% of the constrained inertia of 0.2798. Relative to the original total inertia of 0.7826, these two dimensions are explaining 32.1% (cf. Exhibit 10.5 where the two-dimensional unconstrained CA explained 57.5%). On the other hand, in the unconstrained space (not shaded in Exhibit 27.5), if this space is also of interest, the first two dimensions have principal inertias 0.1909 and 0.1523, totalling 0.3432 which is 68.3% of the unconstrained inertia of 0.5028, or 43.8% of the total inertia. Notice that if the regression of the dimensions of the unconstrained space were made on the environmental variables, there would be no relationship, i.e. regression coefficients of zero and explained variance also zero.

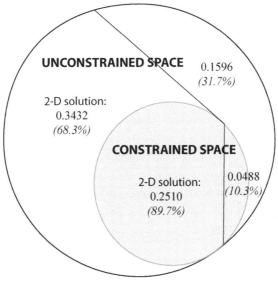

Total inertia = Constrained inertia + Unconstrained inertia
= 0.2510 + 0.0488 + 0.3432 + 0.1596

0.2798 + 0.5028

0.7826

Exhibit 27.5:
Schematic diagram of the decomposition of inertia into parts in the constrained space (shaded) and in the unconstrained space, showing the parts of each explained by respective two-dimensional maps. The parts to the right of the straight lines (inertias of 0.0488 and 0.1596) remain unexplained by the respective two-dimensional solutions.

The CCA triplot The results of CCA in the constrained space involve the usual row and column coordinates, as in CA, with the same scaling options for joint plotting, plus the possibility of adding vectors for the explanatory variables — this is called a *triplot*. The most problematic aspect is how to visualize the explanatory variables: on the one hand, their correlation coefficients with the axes could be used to define their positions, or their standardized regression coefficients in their relationship to the axes. The latter choice, used in Exhibit 27.6, gives direct information to the user how the dimensions are related to the explanatory variables. The sites are now in standard coordinates and the species in principal coordinates, so the basic CA display is a row principal asymmetric map (remember that species are rows in Exhibit 10.4). As far as the sites and species are concerned, the biplot interpretation holds: each site in standard coordinates indicates a biplot axis onto which the species can be projected to estimate their relative abundances at that site (relative to their total abundances across all sites). The site positions along each axis are, by construction, linear combinations of their three standardized values on the environmental variables, using the plotted coefficients. If a site has average

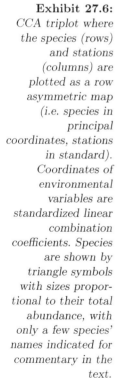

Exhibit 27.6: *CCA triplot where the species (rows) and stations (columns) are plotted as a row asymmetric map (i.e. species in principal coordinates, stations in standard). Coordinates of environmental variables are standardized linear combination coefficients. Species are shown by triangle symbols with sizes proportional to their total abundance, with only a few species' names indicated for commentary in the text.*

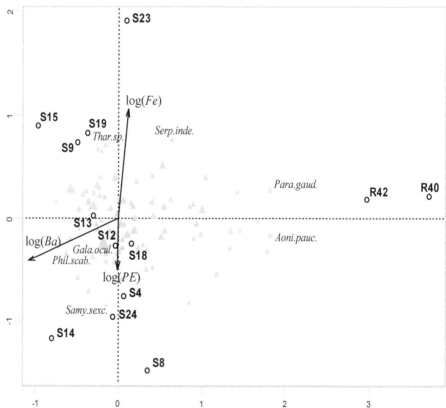

value on an environmental variable the contribution of that variable to its position is zero. So the fact that the reference stations R40 and R42 are so far on the other side of $\log(Ba)$ means that its values must be low in barium, which is certainly true. Likewise, S23, S19, S15 and S9 must be high in iron (especially S23) and S8 and S14 must be high in pelite. This can be confirmed by looking at the actual values in Exhibit 27.1. The relationship between the species and the environmental variables is through the sites that they have in common. Species like *Para.gaud.* and *Aoni.pauc.* are associated with the reference stations, and these reference stations have low barium. Species such as *Thar.sp.* and *Serp.inde.* are associated with stations that have high iron and/or low pelite, while *Samy.sexc.* down below is associated with stations that have high pelite and/or low iron. The reference stations are more or less in the middle of the vertical axis — they are low in both iron and pelite, and this has effectively cancelled out their vertical positionings.

Categorical explanatory variables

If there are categorical variables such as *Region* (e.g. with categories Northeast/Northwest/South) or *Rocky* (e.g. with categories yes/no) as explanatory variables, then these are included as dummy variables in the CCA just as they would be included in a regression analysis. In the CCA solution these dummy variables are not represented by arrows; rather, the sites that are in each category are averaged (and, as always, applying the usual weights in the averaging), so that each category is represented by a point in the CCA map. Continuous variables can also be included as fuzzy categorical variables as a way of investigating possible nonlinearities in their explanation of the dimensions, rather than the straight-line linear effects of Exhibit 27.6.

Weighted averages of explanatory variables for each species

An alternative way of thinking about CCA is as an analysis of the weighted averages of the explanatory variables for each species. Exhibit 27.7 shows a small part of this set of averages, for some of the species that have been referred to before. For example, the frequencies of *Galathowenia oculata* (*Gala.ocul.*) are given in Exhibit 10.4 as 193, 79, 150, etc. for stations S4, S8, S9, etc., and these stations have values for $\log(Ba)$ of 3.219, 3.138, 3.566, etc. So the weighted average for *Gala.ocul.* on that variable is:

$$\frac{193 \times 3.219 + 79 \times 3.138 + 150 \times 3.566 + \cdots}{193 + 79 + 150 + \cdots} = 3.393$$

which is the scalar product of the profile of the species with the values of the variable. The "global (weighted) average" in the last line of Exhibit 27.7 is the same calculation using the totals of all species. Hence we can see that *Gala.ocul.* is quite close to the global average, and so does not play as important a role as it did in the CA of Exhibit 10.5. *Para.gaud.* and *Aoni.pauc.* have low averages on $\log(Ba)$ because of their relatively high frequencies at the reference sites R40 and R42 where barium is very low. *Sami.sexc.* has a high average on $\log(PE)$ and the reason why *Thar.sp.* and *Serp.inde.* lie at the top is more to do with their low pelite averages than their high iron ones.

	Variables		
SPECIES	$\log(Ba)$	$\log(Fe)$	$\log(PE)$
Gala.ocul.	3.393	3.416	0.747
⋮	⋮	⋮	⋮
Serp.inde.	3.053	3.437	0.559
Thar.sp.	3.422	3.477	0.651
Para.gaud.	2.491	3.352	0.534
Aoni.pauc.	2.543	3.331	0.537
Samy.sexc.	3.373	3.409	0.971
⋮	⋮	⋮	⋮
Global average	*3.322*	*3.424*	*0.711*

Partial CCA The idea of partialling out the variation due to some variables can be carried a step further in a *partial CCA*. Suppose that the explanatory variables are divided into two sets, labelled A and B, where the effect of A is not of primary interest, possibly because it is well known, for example spatial or temporal differences. In a first step the effect of the set A of variables is removed, and in the space uncorrelated with these variables a CCA is performed with respect to the set of variables B. There is thus a decomposition of the original total inertia into three parts: the part due to A which is partialled out, and the remainder which decomposes into a part constrained to be linearly related to the B variables (but not to A) and the unconstrained part (unrelated to both A and B).

*SUMMARY:
Canonical
Correspondence
Analysis*

1. In CA the dimensions are found so as to maximize the inertia explained in the solution subspace.

2. In *canonical correspondence analysis* (CCA) the dimensions are found with the same CA objective but with the restriction that the dimensions are linear combinations of a set of additional explanatory variables.

3. CCA necessarily explains less of the total inertia than CA because it looks for a solution in a constrained space, but it may be that this constrained space is of more interest to the researcher.

4. Total inertia can be decomposed into two parts: the part in the constrained space where the CCA solution is sought, and the part in the unconstrained space which is not linearly related to the explanatory variables. In both these spaces principal axes explaining a maximum amount of inertia can be identified: these are the *constrained* and *unconstrained* solutions respectively.

5. In *partial CCA* the effect of one set of variables is first partialled out before a CCA is performed using another set of explanatory variables.

Co-Inertia and Co-Correspondence Analysis

## 28

It is frequently the case that two data sets are available on the same set of individuals and we are interested in relationships between them. The previous chapter dealt with one such situation, where sets of biological and environmental variables were observed at the same sampling points, and the idea was to look for the principal gradients (dimensions) of the biological variables that were directly related to the environmental variables. There was an asymmetry in the way these two sets were considered: the biological variables as responses and the environmental variables as predictors. In this chapter we will look at a general framework for analysing two data matrices, where the set of rows, usually individuals or sampling units, are common to both. Each data set implies a configuration of the individuals in the space of its respective variables. We will be mainly interested in symmetric measures of relationship between the two data sets, answering such questions as: What is the concordance between the two configurations? What are the structures common to both data sets? What are the common dimensions? As specific examples we will concentrate on categorical and count data in the correspondence analysis context.

Contents

Suppose two data matrices, $\mathbf{X}$ $(n \times p)$ and $\mathbf{Y}$ $(n \times q)$, are observed on the same number n of individuals (the rows), with p and q column variables respectively. Euclidean distances between the rows of these matrices can be defined in a very general way, but for our purposes it is sufficient to consider weighted Euclidean distances with weights $g_1, g_2, \ldots, g_p$ for matrix $\mathbf{X}$ and $h_1, h_2, \ldots, h_q$ for matrix $\mathbf{Y}$. Diagonal matrices of these weights are denoted by $\mathbf{D}_g$ and $\mathbf{D}_h$ respectively. We also suppose that the n individuals are weighted by positive weights w_i, $i = 1, 2, \ldots, n$, that sum to 1: $\sum_i w_i = 1$, and that the diagonal matrix of these weights is $\mathbf{D}_w$. If $\mathbf{X}$ and $\mathbf{Y}$ are centred (as they usually are), then the matrix of all covariances between the two sets of variables is $\mathbf{S} = \mathbf{X}^{\mathsf{T}} \mathbf{D}_w \mathbf{Y}$ $(p \times q)$, with general element $s_{jk} = \sum_i w_i (x_{ij} - \bar{x}_j)(y_{ik} - \bar{y}_k)$. Since the term "inertia" is used for weighted variance in the correspondence analysis (CA) context, the term *co-inertia matrix* is similarly used for this covariance matrix that includes the individual weights. But the co-inertias are themselves weighted

Co-inertia analysis

when it comes to combining them into a global measure of covariation between the two sets of variables. The *total co-inertia* is the weighted sum of squared co-inertias:

$$\text{total co-inertia} = \sum_{j=1}^{p} \sum_{k=1}^{q} g_j h_k s_{jk}^2 \tag{28.1}$$

$$= \sum_{j=1}^{p} \sum_{k=1}^{q} \left(\sum_{i=1}^{n} w_i \left[g_j^{\frac{1}{2}} (x_{ij} - \bar{x}_j) \right] \left[h_k^{\frac{1}{2}} (y_{ik} - \bar{y}_k) \right] \right)^2 \tag{28.2}$$

Inside the square brackets in (28.2) one can see the original variables transformed by the square roots of their respective weights, so if the weights were the inverses of the respective variances, this would be a regular standardizing transformation and the total co-inertia would be the sum of squares of the weighted correlations between the two sets of variables. The analysis of the co-inertias, taking into account the different weighting systems for the two sets of variables, is called *co-inertia analysis*. The analysis attempts to reduce the dimensionality of the co-inertia matrix $\mathbf{S} = \mathbf{X}^{\mathsf{T}} \mathbf{D}_w \mathbf{Y}$ in order to optimally visualize the linear relationships between the two sets of data. As special cases there are many methods of multivariate analysis, some of which have already been treated in earlier chapters.

Some special cases of co-inertia analysis   Suppose $\mathbf{X}$ and $\mathbf{Y}$ are two indicator matrices $\mathbf{Z}_1$ and $\mathbf{Z}_2$ of dummy variables coding two categorical variables on n individuals, and suppose that the individuals are equally weighted, i.e. $w_i = 1/n$. Let the vectors of column averages of the two matrices be denoted by $\mathbf{c}_1$ and $\mathbf{c}_2$ respectively — since each row of the matrices consists of zeros and a single 1, the row sums are all 1, as well as the sums of the column averages. Then the co-inertia matrix is $\mathbf{S} = (1/n)(\mathbf{Z}_1 - \mathbf{1}\mathbf{c}_1^{\mathsf{T}})^{\mathsf{T}}(\mathbf{Z}_2 - \mathbf{1}\mathbf{c}_2^{\mathsf{T}}) = (1/n)\mathbf{Z}_1^{\mathsf{T}}\mathbf{Z}_2 - \mathbf{c}_1\mathbf{c}_2^{\mathsf{T}}$. This is just the contingency table $\mathbf{N} = \mathbf{Z}_1^{\mathsf{T}}\mathbf{Z}_2$ cross-tabulating the two variables, divided by the total n of $\mathbf{N}$, which we previously denoted by $\mathbf{P}$, double-centred with respect to its margins. If the distances between the rows of $\mathbf{Z}_1$ and $\mathbf{Z}_2$ are defined by weights equal to the inverses of the elements of $\mathbf{c}_1$ and $\mathbf{c}_2$ respectively, i.e. the chi-square distances, then (28.1) is exactly equal to the total inertia of simple CA: $\sum_j \sum_k (p_{jk} - c_j c_k)^2 / (c_j c_k)$. Thus simple CA is a very special case of co-inertia analysis.

Now extend the two matrices so that

$$\mathbf{X} = [\, \mathbf{Z}_1 \ \mathbf{Z}_2 \ \cdots \ \mathbf{Z}_p \,] \quad \text{and} \quad \mathbf{Y} = [\, \mathbf{Z}_{p+1} \ \mathbf{Z}_{p+2} \ \cdots \ \mathbf{Z}_{p+q} \,]$$

$\mathbf{X}$ and $\mathbf{Y}$ each consists of several indicator matrices coding p and q categorical variables respectively, again on the same n individuals. Using the same argument as above, the co-inertia matrix turns out to be the centred matrix of all two-way cross-tabulations between the p and q variables, divided by its grand total, and the total co-inertia is the average of all the inertias of the pq subtables. Thus we obtain the CA of the stacked matrix in Chapter 17 as a special case as well. If the two sets of categorical variables are the same set, so that $\mathbf{X} = \mathbf{Y}$, then the co-inertia analysis reduces to the analysis of the Burt matrix, i.e. MCA.

Another scenario is where one of the sets of variables is a single categorical variable defining groups of cases. Suppose in general that $\mathbf{X}$ consists of p quantitative variables, possibly on different scales, and that $\mathbf{Y}$ is an indicator matrix $\mathbf{Z}$ based on a single categorical variable that assigns the n rows to q disjoint groups. The natural distance metrics for these two matrices are defined by the inverses of the variances, $1/s_1^2, 1/s_2^2, \ldots, 1/s_p^2$, for $\mathbf{X}$, and the inverses of the column means of $\mathbf{Z}$, which we denote by $1/c_1, 1/c_2, \ldots, 1/c_q$ — in other words, the Euclidean distance on standardized variables for the rows of $\mathbf{X}$ and the chi-square distance for the rows of $\mathbf{Z}$. Then it can be shown that the co-inertia matrix $\mathbf{S}$ between $\mathbf{X}$ and $\mathbf{Z}$ has element $c_k(\bar{x}_{jk} - \bar{x}_j)$, where $\bar{x}_{jk}$ is the mean of the j-th variable of $\mathbf{X}$ in the k-th group, and $\bar{x}_j$ is, as before, the overall mean of the j-th variable. The total co-inertia in this case turns out to be:

$$\sum_{j=1}^{p} \sum_{k=1}^{q} \frac{1}{s_j^2} \frac{1}{c_k} [c_k(\bar{x}_{jk} - \bar{x}_j)]^2 = \sum_{k=1}^{q} c_k \sum_{j=1}^{p} \left(\frac{\bar{x}_{jk} - \bar{x}_j}{s_j} \right)^2 \qquad (28.3)$$

which is the between-group variance of the multivariate (standardized) data in $\mathbf{X}$. This is called *centroid discriminant analysis* because the optimal dimensions for separating the group means will be found in this special case of co-inertia analysis.

As examples of co-inertia analysis in the next two sections, Exhibit 28.1 shows the layout of two matrices collected in a marine ecological survey of 158 sampling stations in the Barents Sea, north of Norway. Sampling took place over a period of four years, from 2006 to 2009. The columns are two sets of species, 142 benthic (sea-bed) species and 41 fish species, that are counted for their abundances in each sample. These two marine communities are interesting to analyse in their own right but also to see how they co-vary. Like most community ecological data sets like these, the data matrices are very sparse, with many zeros and few positive numbers. For several reasons, one of them being some extremely large values in the data set, it was considered better to analyse these data at the presence–absence level. Before considering both matrices in a joint analysis, let us look at just at the benthos data matrix and perform a co-inertia analysis where the second matrix is defined by the categories of a separate variable.

Data set 17: Two ecological abundance matrices at the same locations

The above description of centroid discriminant analysis, based on the co-inertia defined in (28.3), applies to continuous variables in $\mathbf{X}$ for which the usual standardization is appropriate. If the benthos data of Exhibit 28.1 form the $\mathbf{X}$ matrix, then we are back in the CA framework again and the chi-square standardization is appropriate. Let $\mathbf{Y}$ now be the indicator matrix, denoted by $\mathbf{Z}$, consisting of four dummy variables for the four years 2006 to 2009 during which the sampling took place. Interest now is specifically focussed on the

Centroid discriminant analysis for CA

Schematic set-up of two data matrices, with two sets of variables (marine species of benthos and fish respect-ively) observed at the same stations in the Barents Sea. A map is shown indicating the locations of the 158 sampling stations north of the Norwegian northern coastline, with Bear Island at approximately long. 19°, lat. 74.4°.

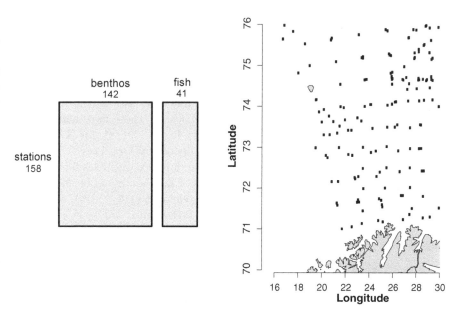

Exhibit 28.2:
Centroid discriminant CA, applied to the benthic species data set. The centroids of the year points 2006 to 2009 for the four respective sub-samples of stations are optimally separated.

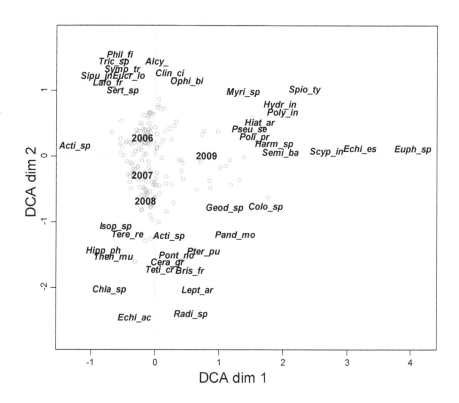

between-year variation and this comparison is forced by looking only at the common inertia between the benthic species and the year dummies. Technically, the rows of **X** are the station profiles and the row weights are the usual row masses, proportional to the number of benthic species observed at the respective stations, known as "species richness". Between-row distances are the usual χ^2-distances for both matrices, i.e. the column weights are the inverses of the relative species counts for **X** and the inverses of the relative frequencies of the years for **Z**. The result is shown in Exhibit 28.2. The contribution biplot scaling is used, where only those species that make more than average contributions to the two-dimensional solution are shown (40 out of the 142 species). The main differences are between the last sampling year 2009 and the first three years, along the first dimension, and the subset of species causing this separation can be identified, for example the main contributor appears to be *Euph_sp* (present at 13 stations in 2009, totally absent in the other years). The first three years are separated on the second dimension and again the species responsible for this separation can be identified: for example, *Echi_ac* at bottom is only present in 6 of the 2008 samples, otherwise absent. To enhance the display, 95% confidence regions in the form of ellipses can be added for each year centroid, shown in Exhibit 28.3 (the way this is computed is explained in the next chapter). The elliptical confidence regions are not overlapping, which is an informal way of concluding that the year averages are significantly separated. This need not be strictly due to temporal differences, however, because

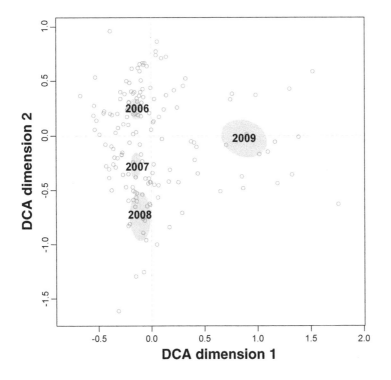

Exhibit 28.3:
The station points of Exhibit 28.2 and year centroids, along with their 95% confidence regions (shaded ellipses).

the annual sampling was not evenly distributed across the region — in 2008, for example, there tended to be more samples taken in the south. One could rather relate the distributions to environmental variables such as depth and temperature. A possible approach in the same spirit is to code these variables into fuzzy categories (see last section of Chapter 26), and perform the same discriminant version of co-inertia analysis.

Co-correspondence analysis As a final example of co-inertia analysis, let us now consider both matrices of Exhibit 28.1, each of which is a candidate for a CA. In general, suppose that we have two data matrices with n common rows, whose elements are nonnegative counts or other nonnegative values where all values within a matrix are on the same scale, but could be different for each matrix: for example, one matrix could be in counts and the other matrix in percentages. Both matrices are suitable for CA, to identify the most important dimensions in the respective space of each data matrix, but now the objective is rather to find common dimensions in the two spaces. The matrices, denoted $\mathbf{N}_1$ and $\mathbf{N}_2$, have different numbers of columns as well as different row sums, so the question is what the row weights will be. An obvious solution is to set the row weights equal to the average of the row masses of the separate matrices, or the masses implied by the matrices together $[\mathbf{N}_1 \ \mathbf{N}_2]$, if it makes sense to make combined row sums of both matrices. The analysis proceeds as before, using the relative values of the respective column margins in the χ^2-distances between row profiles in the two spaces. The total co-inertia between the two matrices is (28.2) with w_i the chosen row weights, g_j the inverse of the j-th column mass of the first matrix, x_{ij} the j-th element of the i-th row profile of the first matrix, $\bar{x}_j$ the j-th column mass (of which g_j is the inverse), and similarly for h_j, y_{ij} abd $\bar{y}_k$ for the second matrix. This total co-inertia is decomposed along the principal dimensions of the analysis (see Appendix A). This form of co-inertia analysis is called *co-correspondence analysis* (CoCA).

Exhibit 28.4 shows the CoCA of the benthic and fish presence–absence data matrices. Row (station) weights were chosen as the average of the row masses of the respective matrices, but could also have been defined as the (relative) species richness at each station, i.e. the sum of each row of the concatenated presence/absence matrices $[\mathbf{N}_1 \ \mathbf{N}_2]$, divided by the grand total. Again, only the species of each set that contribute more than average to the common solution are shown: 53 of the benthic species and 16 of the fish species. The station positions are obtained in each space as weighted averages of the species points using the elements of the station profiles as weights (notice that this is done using the standard coordinates of the species, whereas what are shown in Exhibit 28.4 are their contribution coordinates, which are the standard coordinates "shrunk" by multiplying them by the square roots of the species masses — see Chapter 13). There is a single measure of total co-inertia, equal to 0.2430 in this example, of which 44.4% and 24.4% are explained by the two dimensions common to each matrix. Also shown in Exhibit 28.4 are the direction vectors emanating from the origin of temperature and depth, added

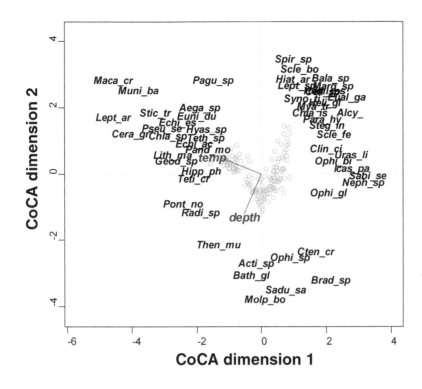

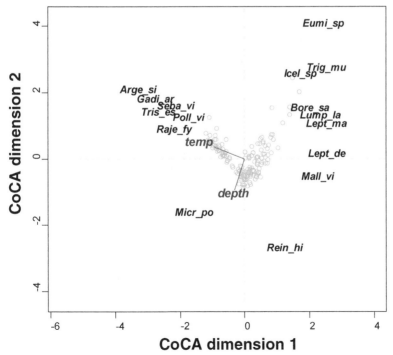

Exhibit 28.4:
Co-correspondence analysis (CoCA) of the benthos and fish species observed at the same stations. The stations are shown as gray circles. Only the species contributing more than average to the dimensions are shown with their abbreviated labels. The contribution biplot scaling is used, so all the omitted species are farther towards the centres of the respective biplots and have lower than average contributions to the dimensions. Contribution coordinates have been scaled up in both displays to enhance legibility. Emanating from the origin are vectors representing the directions of temperature and depth, as supplementary variables. Percentages of co-inertia explained by the two dimensions are 44.4% and 24.4% respectively.

as supplementary variables (see Chapter 27, page 210). This shows that the station points extending to upper left tend to have higher than average temperatures, while the spread of stations diagonally from lower left to upper right correspond to deeper to shallower stations. Species in the corresponding directions indicate higher than average presences at stations with these environmental characteristics.

SUMMARY:
Co-Inertia and Co-Correspondence Analysis

1. The *co-inertia* between two variables, observed on the same individuals, is their weighted covariance, where (optional) weights are assigned to the individuals.

2. The total co-inertia between two sets of variables is a weighted sum of squared co-inertias between all pairs of variables (one from each set). The weights referred to now are associated with the variables, and correspond to the appropriate standardization of the variables when combining them within each of the sets.

3. *Co-inertia analysis* is the study of the total co-inertia between two sets of variables. The total co-inertia is decomposed along principal axes.

4. CA is a special case of co-inertia analysis, when the two sets of variables are indicator matrices of dummy variables coding two categorical variables.

5. The analysis of stacked (or concatenated) contingency tables, each one based on the same individuals, is a co-inertia analysis where the two sets of variables are indicator matrices coding respective categorical variables, as in multiple correspondence analysis (MCA). If the two sets of categorical variables, and thus indicator matrices, are the same, then co-inertia analysis is the same as the MCA of the Burt matrix.

6. *Centroid discriminant analysis*, which optimally separates group centroids, is a co-inertia analysis when one of the sets of variables is an indicator matrix coding the groups.

7. If the two matrices are individually suitable for visualization by CA, a co-inertia analysis can be performed, called *co-correspondence analysis*. The respective weightings by variable (column) masses are respected in each matrix, but a single set of individual (row) masses needs to be defined by, for example, the averages of the row masses of the separate matrices.

Aspects of Stability and Inference

## 29

Apart from the passing mention of the chi-square test, the discussion about significant clustering in Chapter 15 and the visualization of confidence ellipses in Chapter 28, this book has concentrated exclusively on the geometric properties of correspondence analysis (CA) and its interpretation. In these two final chapters we explain some approaches to statistical inference in the context of CA and related methods. In the present chapter we shall be investigating the stability of CA solutions and the sampling properties of statistics such as the total inertia, principal inertias and principal coordinates. The distinction is made between (i) stability of the solution, irrespective of the source of the data, (ii) sampling variability, assuming the data arise out of some form of random sampling from a wider population, and (iii) testing specific statistical hypotheses.

Contents

Throughout this book CA has been described as a method of data description, as a way of re-expressing the data in a more accessible graphical format to facilitate the exploration and interpretation of the observed data. Whether the features in the map are evidence of real phenomena or arise by chance variation is a separate issue. To make statements, or so-called *inferences*, about the population is a different exercise, and is feasible only when the data are validly sampled from a wider population. For the type of categorical data considered in this book, there are many frameworks that allow hypotheses to be tested and inferences to be made concerning the characteristics of the population from which the data are sampled. For example, *log-linear modelling* allows interactions between variables to be formally tested for significance, while *association modelling* is closely connected to CA and enables a wide

Information-transforming versus statistical inference

range of hypotheses to be tested, for example differences between category scale values,. There is, however, a certain amount of statistical inference that can be accomplished within the CA framework, as well as some innovative investigation of variability or stability of the maps, thanks to modern high-speed computing.

Stability of CA By *stability* of the CA solution (the map, the inertias, the coordinates on specific principal axes, etc.), we are referring to the particular data set at hand, without reference to the population from which the data might come. Hence, the issue of stability is relevant in all situations, even for population data or data obtained by convenience sampling. Here we assess how our interpretation is affected by the particular mix of row and column points that are active in determining the map. Would the map change dramatically (and thus our interpretation too) if one of the points were omitted (for example, one of the species in our marine biology example, or one of the authors in the set of texts — see the data sets of Chapter 10)? This aspect of solution stability has already arisen several times when we discussed the concept of influence and how much each point influences the determination of the principal axes. In Chapter 11 the numerical *inertia contributions* were shown to provide indicators of the influence of each point. If a row or column contributes highly to an axis, then it is influential in the solution and the solution would change noticeably if it were omitted. On the other hand, some points contribute very little to the solution, and can be removed without changing the map dramatically — that is, the map is *stable* with respect to including or removing these points.

Sampling variability of the CA solution Now looking outward beyond the data matrix, let us suppose that the data are collected by some sampling scheme from a wider *population*. For example, in the author data set of Exhibit 10.6 we know that the data represent a small part of the complete texts, and if the whole exercise were repeated on a different sample of each text, the counts of each letter would not be the same. It would be perfect, of course, if the sampling exercise could be repeated many times, and each time a CA performed to see if the features observed in the original map remained more or less the same or whether the books' and letters' positions changed. In other words, did what we see in the map arise by chance or was it a real feature of the 12 books being studied?

Bootstrapping the data Since we cannot repeat the study, we have to rely on the actual data themselves to help us understand the sampling variability of the matrix. The usual way to proceed in statistics is to make assumptions about the population and then derive results about the uncertainty in the estimated values, which in the present case are the coordinates of points in the map. A less formal way which avoids making any assumptions is provided by the *bootstrap*[*]. The idea

---

[*] The English expression "pulling yourself up by your own bootstraps" means using your own resources to get yourself out of a difficult situation.

of the bootstrap is to regard the data as the population, since the data are the best representation one has of the population. New data sets are created by resampling from the data in the same way as the data themselves were sampled. In the author data, the sampling has been performed for each text, not for each letter, so this is the way we should resample. For example, for the first book, "Three Daughters", 7144 letters were sampled, so we imagine — notionally, at least — these 7144 letters strung out in a long vector, in which there are 550 *a*s, 116 *b*s, 147 *c*s, ... etc. Then we take a random sample of 7144 letters, *with replacement*, from this vector — the frequencies will not be exactly the same as those in the original table, but will reflect the variability that there is in those frequencies. This exercise is repeated for all the other rows of Exhibit 10.6, until we have a replicated table with the same row totals. This whole procedure can be repeated several times, usually between 100 and 1000 times, to establish many bootstrap replicates of the original data matrix.

An equivalent way to think about (and to execute more efficiently) the resampling is to make use of *multinomial sampling*. Each row profile defines a set of probabilities that can be regarded as the probability of obtaining an *a*, *b*, *c*, etc. in the respective text. Then it is a matter of sampling from a population with these probabilities, which is a computational algorithm, already implemented in R (see the Computational Appendix, page 281). So it is not necessary to create the vector of 7144 letters for example; we need to use only the 26 probabilities of the letters in a multinomial sampling scheme. This is sometimes referred to as *Monte Carlo simulation*.

Multinomial sampling

To illustrate the procedure on the author data, we first computed 100 replicates of the table by the sampling procedure described above. There are two ways to proceed now. The more difficult way is to repeat the CA on each replicate and then somehow compare the results to those obtained originally. The easier way is demonstrated here, called the *partial bootstrap*. Each replicated table can be regarded as a set of row profiles or set of column profiles, so the 100 replicated profiles are simply projected onto the CA map of the original data as supplementary points. Exhibit 29.1 shows the partial bootstrap of the 26 letters — each letter in larger font shows its original position in principal coordinates, with the 100 replicates in a tiny font. Usually we would not show all the replicates, but just show the *convex hull* of each set of points — this is the outer set of points connected by dotted lines in Exhibit 29.1, as if an elastic band has been placed around them.

Partial bootstrap of CA map, with convex hulls

Since the convex hull is sensitive to outlying replicates (for example, see the point for *z* on the right of Exhibit 29.1), it is usually *peeled*; that is, the convex hull of points is removed. The convex hull of the remaining points can be peeled again, and this process repeated until 5% of the outermost points in each subcloud have been removed. The convex hull of the remaining points is thus an estimate of a 95% confidence region for each letter. To make the

Peeling the convex hull

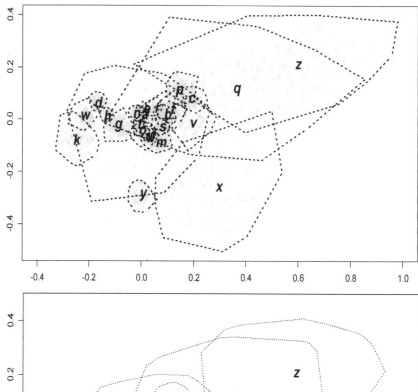

Exhibit 29.1:
*(Partial) bootstrap
of 26 letters, after
100 replications of
the data matrix.
The more frequent
the letter is in the
texts, the more
concentrated (less
variable) are the
replicates. Convex
hulls are shown
around each set of
100 replicated
profiles.*

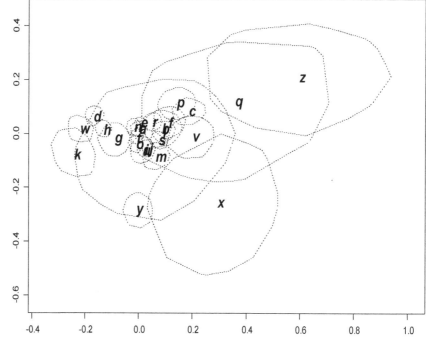

Exhibit 29.2:
*Peeled convex hulls
of points based on
1000 replicates (10
times more than in
Exhibit 29.1),
showing an
approximate 95%
confidence region for
their distribution.*

estimation of these convex regions smoother, we generated 1000 replicates of
each letter and then peeled off as close to 50 of them as possible (Exhibit 29.2),
showing the convex hulls of the remaining subclouds. If two convex hulls do
not overlap then this gives some assurance that the letters are significantly
different in the texts. The actual level of significance is difficult to calculate

because of the lack of formality in the procedure and the issue of multiple comparisons mentioned in Chapter 15. Fortunately, however, the procedure is conservative because of the projections onto the original map. If two convex hulls overlap in the map (for example, *x* and *q*), then it may still be possible that they do not overlap in the full space, but we would not be able to conclude this fact from the map. If they do not overlap in the projection (for example, *k* and *y*), then we know they do not overlap in the full space.

An alternative method for visualizing the confidence regions for each point in a CA map is to use confidence ellipses. These can be based on the replicates in the partial bootstrap above, or can be calculated making some theoretical assumptions. For example, the *delta method* uses the partial derivatives of the eigenvectors with respect to the multinomial proportions to calculate approximate variances and covariances of the coordinates. Then, assuming a bivariate normal distribution in the plane, confidence ellipses can be calculated — these enclose the true coordinates with 95% confidence, just like a confidence interval for single variables. This approach relies on the assumption of independent random sampling, which is not strictly satisfied in the author data because the occurrence of a particular letter is not independent of the occurrence of another (there is a similar problem in ecological sampling, where the same type of species seems to be found in groups in the samples). Nevertheless, the confidence ellipses for the letters in the author data are shown in Exhibit 29.3 and they bear a strong resemblance to the convex hulls in Exhibit 29.2, at least

The delta method

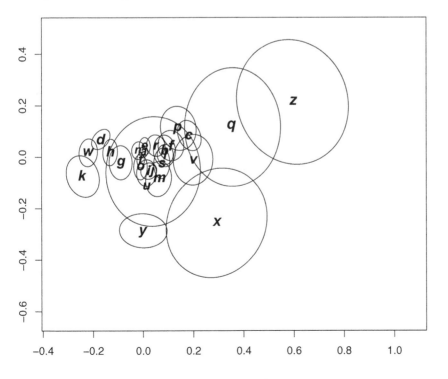

Exhibit 29.3:
Confidence ellipses based on the delta method and the normal approximation to multinomial sampling.

as far as overlapping is concerned. The distribution of bootstrap replicates in Exhibit 29.1 can also be summarized by 95% confidence ellipses, rather than peeled convex hulls.

Testing hypotheses — theoretical approach

The χ^2 test has been mentioned before as a test of independence on a contingency table. For example, the 5×3 table of Exhibit 4.1, which cross-tabulates 312 people on their level of readership and age group, has an inertia of 0.08326 and thus a χ^2 of $312 \times 0.08326 = 25.98$. The p-value for the χ^2 test is computed as 0.0011, a highly significant result. It is also possible to test the first principal inertia of a contingency table using a statistical approximation to the true distribution called an *asymptotic distribution*. The critical points for this test are exactly those that were used in Chapter 15 to test for significant clustering. The first principal inertia has value 0.07037, and its value as a χ^2 component is $312 \times 0.07037 = 21.96$. To test this value, refer to the table in the Theoretical Appendix, page 254, where the critical point at the 0.05 level is shown as 12.68 for a 5×3 table. Since 21.96 is much higher than this value we can conclude that the first dimension of the CA is significant and has not arisen by chance. The second principal inertia is more difficult to test, especially if we assume that the first principal inertia is significant, so we again resort to computer-based methods.

Testing hypotheses — Monte Carlo simulation

Given a hypothesis on the population, and knowing the way the data were sampled, we can set up a Monte Carlo simulation to calculate the null distribution of the test statistic. For example, suppose we want to test both principal inertias of the readership data for significance. The null hypothesis is that there is no association between the rows and columns. The sampling here was not done as in the author data, where the text was sampled within each book — the analogy here would be that we sampled within each education group. In reality, 312 people were sampled and then their education groups and readership categories were ascertained, so that the distribution of the education groups is also random, not fixed. Therefore we need to generate repeated samples of 312 people from the multinomial distribution which corresponds to the whole matrix, not row by row or column by column. The expected probabilities in the 15 cells of the table are equal to the products $r_i c_j$ of the masses. These define a vector of 15 probabilities under the null hypothesis, which will be used to generate simulated multinomial samples of size 312. Two samples are given in Exhibit 29.4 alongside the original contingency table — in total 9999 tables were generated. For each simulated table the CA was performed and the two principal inertias calculated; hence along with the original observed value there are 10000 sets of values in total. Exhibit 29.5 shows the scatterplot of all of these, indicating the pair of values corresponding to the observed contingency table. It turns out there are only 18 values out of 9999 that are larger than the observed first principal inertia; hence, including the actual value, its p-value is estimated as 0.0019. For the second principal inertia there are 580 simulated values larger than the observed one,

EDUCATION GROUPS	Original data			1st Simulation			2nd Simulation			...
	C1	C2	C3	C1	C2	C3	C1	C2	C3	...
E1	5	7	2	2	9	5	4	5	7	...
E2	18	46	20	15	40	38	23	33	37	...
E3	19	29	39	13	36	27	17	34	25	...
E4	12	40	49	11	43	40	14	43	37	...
E5	3	7	16	8	12	13	5	12	16	...

Exhibit 29.4:
The original contingency table of Exhibit 4.1 and two of the 9999 simulated tables under the null hypothesis of no row–column association.

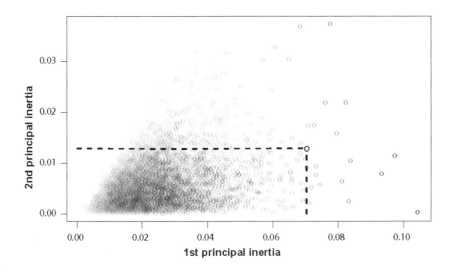

Exhibit 29.5:
Scatterplot of principal inertias from original CA and 9999 simulations of the 5×3 contingency table under the null hypothesis of no row–column association. The actual principal inertias are indicated by the larger circle and dashed lines.

giving a p-value of 0.0581. At the 5% level the first is significant but not the second. At the same time we calculated the total inertia in each simulation — there are 17 simulated values larger than the observed total inertia of 0.08326. Therefore the p-value is 0.0018, which is our Monte Carlo estimate for the χ^2 test, compared to the p-value of 0.0011 based on the χ^2 test.

Permutation tests (or randomization tests, discussed more fully in the next chapter) are slightly different from the bootstrap and Monte Carlo procedures described above. For example, in the "blow-up" of the book points in Exhibit 10.7 we observed that the pairs of books by the same author lay in the same vicinity. It does seem unlikely that this could have occurred by chance, but what is the probability, or p-value, associated with this result? A permutation test can answer this question. First, calculate a measure of proximity between the pairs of books — an obvious measure is the sum of the six distances between pairs, which is equal to 0.4711. Then generate all possible ways of assigning the pairs of authors to the 12 texts; there are exactly $11 \times 9 \times 7 \times 5 \times 3 = 10395$ unique ways to rearrange them into six groups of 2. For each of these re-assignments of the labels to the points in the map,

A permutation test

calculate the sum-of-distance measure. All these values define the *permutation distribution* of the test statistic, which has mean 0.8400 and standard deviation 0.1246. It turns out that there is no other assignment of the labels that gives a sum-of-distances smaller than the value observed in the CA map. Hence the p-value to support the assertion that the pairs of texts are close is $p = 1/10395$, i.e. less than 0.0001, which is a highly significant result. Similar permutation tests were conducted for the subset CAs of the consonants and vowels separately (Exhibits 21.1 and 21.2), yielding $p = 0.0046$ and $p = 0.0065$ respectively. Thus the consonants and the vowels explain almost equally the differences between authors, even though the vowels have less inertia in total. Permutation tests are routinely used in many situations where comparisons are made between groups or where the significance of relationships is being tested. This will be the subject of the next and final chapter.

SUMMARY:
Aspects of
Stability and
Inference

1. *Stability* concerns the data at hand and how much each row or column of data has influenced the display. The level of internal stability can be judged (a) by studying the row and column contributions and (b) by embarking on various re-analyses of the data that involve omitting single points or groups of points and seeing how the map is affected.

2. When the data are regarded as a sample of a wider population, the sampling variability can be investigated through a *bootstrap* resampling procedure to create replicates of the data table. The resampling should respect the row or column margins if these were fixed by the original sampling design.

3. In the *partial bootstrap* the row and/or column profiles of the replicated matrices are projected onto the CA solution as supplementary points. The replicate points can be summarized by drawing convex hulls or confidence ellipses.

4. Various theoretical approaches also exist, which rely on distributional assumptions in the population, for example the delta method and asymptotic theory based on normal approximations of the multinomial distribution.

5. *Monte Carlo* methods and *permutation tests* can be used to test specific hypotheses, relying on generating data under the null hypothesis to simulate (or calculate exactly) the null distribution of chosen test statistics, from which p-values can be deduced.

Permutation Tests

30

When studying single variables in univariate statistics or pairs of variables in bivariate statistics, the testing of group differences and the study of pairwise variable relationships are fairly straightforward and well known, even when the variables might not be normally distributed. In the case of multivariate data when sample differences are measured by distance functions such as the χ^2-distance, possibly involving sample weighting, the theory becomes extremely complex. The concept of permutation testing has been known for a long time, but it is only with the advent of high-speed computing that they have come into their own as viable distribution-free methods for testing hypotheses using multivariate data. In this chapter we will look at several types of permutation tests that make it possible to draw statistical conclusions (inferences) in this more complex setting.

Contents

Let us suppose that an experimenter observed the data in Exhibit 30.1, two sets of values for a test (T) and control (C) group respectively. The question is whether there is a difference in the means of the populations from which these two samples come. The protocol of the experiment might have been to randomize the 25 values between the test and control groups, or to draw two random samples from each population. The difference in the means is 1.81 and the standard two-group t-test that might be applied results in a p-value of 0.063 — thus, according to the convention of the 0.05 significance level the difference would be deemed non-significant. These data suffer from two problems that make the t-test questionable. Firstly, the samples are too small to assume normality in their respective means, so normality of the parent distribution needs to be assumed, but all tests for normality reject this assumption. And secondly, there is a very conspicuous outlier in the test group, as can be seen in the dotplots of Exhibit 30.1. The lack of normality as well

A simple
univariate example

Exhibit 30.1:
Test and control data, plotted as "jittered" dotplots, with overlain boxplots. The outlier for the test group is clearly visible on the right.

as the outlier can be partially rectified by log-transforming the data, in which case all tests accept the assumption of normal distribution and the p-value for the difference is 0.036, a significant result.

Permutation test for difference in means

Permutation testing is a way to avoid the assumptions, keep the data on their original scale and not be too concerned about the outlier. Since the null hypothesis being tested is that the means are the same, i.e. the observations come from the same distribution, whatever that distribution is, the null distribution of the difference in means can be approximated by simulating many versions of the same data, with the 14 T labels and 11 C labels randomly assigned to the 25 observed values. Technically, this is done by fixing the vector of 25 values, then randomly permuting (i.e. shuffling up) the order of the 25 labels, and finally computing the difference in means between the simulated "T" and "C" groups. This operation is repeated very many times to obtain the so-called permutation distribution of the test statistic, i.e. difference in means in this case. There are almost four and a half million ways that we can split the 25 observations into two sets of 14 and 11. But we will only do this for a large random sample, say 9999 permutations of the data, to which we add the original assignment of the observations, with associated difference in mean value of 1.81, totalling 10000 permutations in all. Exhibit 30.2 shows a histogram of this *permutation distribution*, and points out where the observed mean of 1.81 lies. Notice how non-normal this distribution looks, far different

Exhibit 30.2:
Null permutation distribution of the difference between treatment and control means, showing the actual difference obtained.

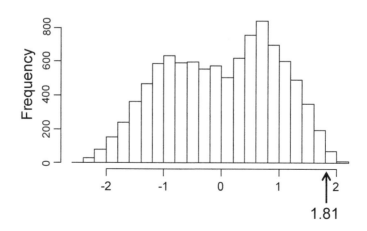

from the assumption of normality that the small sample t-test relies on. In order to estimate the p-value we need to count how many of the simulated values are equal or larger in absolute value to 1.81, i.e. we need to count how many are 1.81 or more or −1.81 or less, since the test is two-sided. It turns out there are 336 out of the 10000 values of the simulated null distribution that are in the tails, giving an estimate of $p = 336/10000 = 0.034$, very close to the p-value of the t-test on the log-transformed data.

The technology of permutation testing transfers quite smoothly to multivariate data, for example the time budget compositional data in Exhibit 22.3, which is visualized using log-ratio analysis in Exhibit 22.4. There are 12 rows of this matrix, composed of combinations of region (USA/Western/Eastern), marital status (married/single) and gender (male/female). Permutation tests can give an indication whether there are signficant differences between regions, between marital status groups and between sexes. For example, the six rows for each sex can be aggregated into two vectors of mean time and then a suitable test statistic would be the log-ratio distance (22.3) between them. Then the 12 labels for gender are randomly permuted, the mean vectors recomputed and the log-ratio distance computed, and this is repeated a large number of times. But in this specific case, there are only 462 unique re-allocations of the male and female labels, so — as for the permutation test at the end of Chapter 29 — it is feasible to run through all of these to get an exact p-value. It turns out that the original log-ratio distance on the unpermuted data, equal to 0.578, is larger than all those based on the permutations, so the p-value is $p = 1/462 = 0.002$, and significant. The permutation distribution is shown in Exhibit 30.3. The same permutation test can be conducted to test the difference between the mean vectors for married and single groups. Since the permutations for the six married and six single groups are identical for the six male and six female groups, the permutation distribution is identical to Exhibit 30.3. But the observed statistic is 0.277 for this comparison, and not significant ($p = 0.15$). Finally, to compare the three regions, where the total

Permutation test in multidimensional space

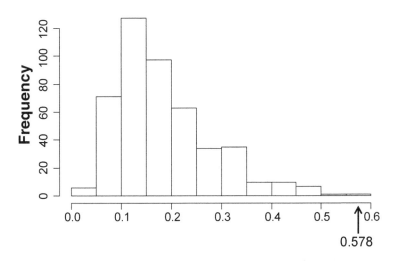

Exhibit 30.3:
Exact null permutation distribution of the log-ratio distance between male and female average time budget vectors, for the data of Exhibit 22.3. The actual observed value for the unpermuted data is indicated.

number of unique permutations is several thousands, we resort to considering 999 random permutations, and also find the regional differences not significant ($p = 0.62$). Because there are three log-ratio distances between pairs of the three regions, the test statistic is the average of these pairwise distances, equal to 0.198.

The way association is measured between two variables depends on the type of data. The most common way for two quantitative variables is the Pearson correlation coefficient, but in the presence of outliers the Spearman rank correlation is often used — each set of observations is converted to ranks and then the Pearson correlation is computed. Suppose there is a second set of 25 observations (contained in vector **y**, say) that we want to correlate with those of Exhibit 30.1 (contained in vector **x**), and because of the outlier in **x** the Spearman correlation is preferred, calculated to be 0.411. To evaluate the significance of this correlation, the null hypothesis is that there is zero correlation between the ranks. Under this hypothesis any value in **y** can be paired with any value in **x**, so we can fix **x**, for example, and then randomly permute the values in **y** and then compute the Spearman correlation. In this case it would be identical to take two vectors of numbers 1 to 25, and permute one of the vectors randomly and compute the Pearson correlation. Doing this a large number of times, say 9999, yields the estimate of the null distribution (Exhibit 30.4) and, if we are interested in testing whether the correlation is significantly positive, the number of simulated correlations greater than or equal to the observed one of 0.411 is counted (there are 211 of them) and expressed as a fraction of 10 000, so $p = 0.021$. The p-value is always an estimate of the true p-value, and its accuracy increases with increasing number of permutations. The margin of error for the above estimate is about 0.003, so the confidence interval would be $[0.018 ; 0.024]$. Generating more permutations gives more accuracy: for example, generating 99999 permutations leads

Exhibit 30.4:
Null permutation distribution of the Spearman rank correlation coefficient for a sample of 25 pairs of observations. There are 211 simulated values greater than or equal to 0.411.

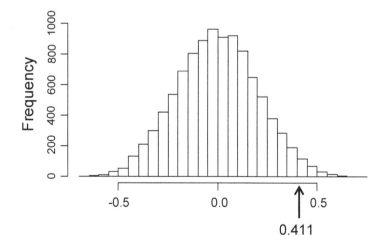

to an estimate of $p = 0.02164$, i.e. $p = 0.022$ to three decimals, a margin of error of 0.001, thus a confidence interval of $[\,0.021\,;0.023\,]$.

Notice that the permutation test for correlation, whether it is the Pearson or Spearman form, maintains the same marginal distributions for the two variables throughout the permutations. So we have to proceed a bit more carefully for bivariate categorical data, since there may or may not be some constraint on the margins due to the sampling protocol. We refer back to the cross-tabulation in Exhibit 4.1 between the two variables education (five categories, *E1* to *E5*) and readership group (three categories, *C1* to *C3*). The original data for the sample of 312 people consist of two vectors of group assignments, with 312 pairs of observations such as (*E2*, *C1*), (*E5*, *C3*), etc., and the sampling was such that neither of the margins was fixed. If the categories of the variables are permuted then their marginal distributions remain the same, which is then not a faithful simulation of the null distribution. In this case one should sample with replacement, i.e. bootstrap, both sets of observations, which will be equivalent to what was done in Exhibit 29.4, but less efficient computationally than using multinomial sampling. In Chapter 29 the inertia measure of association for this table was reported as 0.08326, and the corresponding chi-square test gave a p-value of 0.0011. Using bootstrap instead of permutation (i.e. sampling with replacement instead of sampling without replacement), and generating 9999 bootstrap samples of each variable, the estimated p-value is $p = 12/10000 = 0.0012$.

Permutation tests for bivariate categorical data

Various ways of analysing several categorical variables have been described in previous chapters. The way permutation tests are implemented depends on the way the data have been set up to satisfy the study objective. For example, the stacked tables of Exhibits 17.1 and 17.3 have several demographic variables in blocks of rows and one or more substantive "response" variables in blocks of columns. The null hypothesis is that there is no association between the demographics and the substantive responses, so under this hypothesis the link between the two sets of variables can be broken. The set of demographic characteristics and the set of responses for each respondent should be kept intact, and the permutation (or bootstrapping) should be conducted on the complete demographic sets and response sets — for example, in the data set underlying Exhibit 17.3, 33590 respondents for the rows of the original data table, with five columns for demographic variables and four columns for the variables about working women. The demographic data could be fixed, which obviously preserves the margins and interaction structure of these variables, and the rows of four responses permuted. While breaking the connection between the two sets of variables, this would also preserve the margins and interaction structure of the response variables. This might have some sense for the demographics, assuming there was some sampling protocol that ensured representativity of the population, but seems less defensible for the substan-

Permutation or bootstrap tests for multivariate categorical data

tive responses. Hence, at least for the response variables, bootstrapping seems more appropriate.

Permutation
tests for CCA

In Chapters 27 and 28 we looked at the relationship between two sets of data. In canonical correspondence analysis (CCA, Chapter 27) one set is regarded as multivariate responses that are suitable for CA, while the second set contains explanatory variables of any type, continuous or categorical. In Exhibit 27.5 the total inertia (0.7826) in the response data was decomposed into a part (0.2798) that is directly correlated with the three chosen explanatory variables — barium, iron and pelite (all log-transformed) — and a part (0.5028) that is uncorrelated. The value 0.2798 is 35.8% of the total inertia, and the question is whether this is a significant part. We can proceed as before by generating a null permutation distribution under the null hypothesis that there is no relationship between the two sets of data. This is achieved by randomly permuting the rows of the explanatory variable matrix, say, changing the order randomly of these rows, while keeping the set of three values for each sample together. Even though this sample is small (13 stations), there are over 6 billion possible permutations, hence we again take a large random sample of 9999 permutations, each of which is subjected to a CCA, and each part of inertia accounted for is retained to establish the null distribution. In this case the p-value is $p = 0.073$ (Exhibit 30.5, left-hand side), which would not be judged significant according to convention, but this does not imply that every subset of the explanatory variables is not significant. Going through the same exercise, one variable at a time, barium is very significant (explained inertia = 0.1884, $p = 0.0007$, Exhibit 30.5, right-hand side), but not iron (0.1053, $p = 0.14$) nor pelite (0.0745, $p = 0.36$). A stepwise process of variable selection can proceed as follows. Barium is first introduced into the model and then the other two variables, one at a time, are added, again with permutation. The p-value that results is focused on the additional inertia explained, i.e. how much of the residual inertia is explained after barium is introduced. In this case adding another variable turns out to be not significant (for iron: $p = 0.50$, for pelite: $p = 0.64$), so from a statistical inference standpoint, only barium is worth introducing into the CCA. The CCA could then have barium as a single constraining variable, which would then be identified as the first dimension, while iron and pelite are added as supplementary variables.

Exhibit 30.5:
Permutation
distributions for
explained inertias in
CCA, first for three
explanatory
variables barium,
iron and pelite, and
then for barium
only.

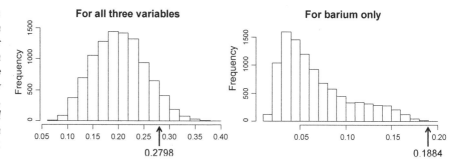

The four parts of inertia in the doubly-matched example, visualized in Exhibit 23.5, can be neatly tested using CCA. Three categorical explanatory variables labelled C (country, 19 categories), T (time period, 2 categories) and G (gender, 2 categories) can be defined, each with $19 \times 2 \times 2 = 76$ combinations, and lined up according to the order of the four matrices **A**, **B**, **C** and **D** stacked on top of one another. Thus C lists the 19 country labels four times, T lists the first year 38 times then the second year 38 times, and G lists male 19 times, female 19 times, male 19 times and female 19 times. The stepwise order of the effects entering the CCA makes no diference to the permutation tests since all the effects of this balanced design are orthogonal. The results are shown in Exhibit 30.6. The first row is the between-country effect, which is equivalent

Effect	DF	Inertia	%	p-value
C	18	0.13467	59.1	< 0.0001
T	1	0.05549	24.3	< 0.0001
$C{:}T$	18	0.02885	12.7	< 0.0001
G	1	0.00337	1.5	< 0.0001
$C{:}G$	18	0.00421	1.8	< 0.0001
$T{:}G$	1	0.00034	0.1	0.004
Residual	18	0.00103	0.5	—
Total	75	0.22795	100.0	

Exhibit 30.6: *The decomposition of total inertia for the example in Exhibit 23.5 and associated p-values from permutation tests for all effects. The colon indicates an interaction. The residual is equivalent to the three-way interaction.*

to the average effect in Exhibit 23.5(a), acounting for 59.1% of the inertia. The next two rows give exactly what we called before the "time effect", which here is the combination of time and country–time interaction (remember that every matched table has the countries as the set of rows), accounting for 37.0% of the inertia (Exhibit 23.5(b)). The next two rows give exactly what we called the "gender effect", which again is the combination of country and country–gender interaction, accounting for 3.3% (Exhibit 23.5(c)). Finally the last two rows sum up to what was called the "time×gender effect", again across countries, so this is what is coded as time–gender interaction in Exhibit 30.6 plus the residual, with a tiny percentage of 0.6% (Exhibit 23.5(d)). The p-value of 0.004 corresponds to this last effect, but the "Residual" from the CCA cannot itself be tested alone. All effects are highly significant according to the permutation tests.

The different variations of co-inertia analysis are multivariate versions of the bivariate correlation. For a permutation test, one of the matrices is fixed, and the other one has its rows permuted, each time the co-inertia analysis is repeated and the total co-inertia is saved to generate the null distribution. For example, in the centroid discriminant analysis of the marine biological presence/absence data of Chapter 28 (see Exhibits 28.2 and 28.3), where stations are compared across the four years, the rows of the indicator matrix for the years are randomly permuted, thereby randomly assigning the year labels to the samples. As could be expected from Exhibit 28.3, the year separation is highly significant, with a p-value from the permutation test of $p < 0.0001$. Similarly, the co-inertia between benthos and fish in the co-correspondence analysis of Exhibit 28.4 is also highly significant, with $p < 0.0001$.

SUMMARY:
Permutation Tests

1. Permutation testing is a distribution-free procedure for performing hypothesis tests on data, especially useful in complex situations where generation of the theoretical null distribution of a statistic is impossible.

2. In general, permutation methods are for computing p-values for statistical tests, whereas bootstrapping is for computing confidence intervals (for single statistics) or confidence regions (for multivariate statistics).

3. Usually, there are two vectors or matrices of data, and one is kept fixed while the observations in the other one are randomly permuted, assuming the null hypothesis of no relationship. This is performed a large number of times (9999 is a reasonable amount to achieve acceptable accuracy), each time computing the statistic in question. These values define the null distribution and the position of the originally observed statistic in this distribution leads to an estimate of the p-value.

4. When the second set of data is a single categorical variable, the permutation test performs an analysis of variance if the first set consists of univariate observations, or a centroid discriminant analysis if it consists of several variables.

5. To test explanatory variables in CCA, a stepwise procedure can be followed by first introducing the variable that is most significant, then fixing it in the analysis and testing each of the remaining ones in turn by permutation of their values. If a second variable explains an additional part of the inertia that is significant, it is also fixed in the analysis and each one of the remaining variables is tested as a third variable. The process stops when no significant additional inertia can be explained.

6. Co-inertia methods investigating relationships between two sets of variables can be tested, like a correlation coefficient, except that whole rows of observations have to be permuted in one of the matrices.

Theory of Correspondence Analysis

Correspondence analysis (CA) is based on fairly straightforward, classical results in matrix theory. The central result is the singular value decomposition (SVD), which is the basis of many multivariate methods such as principal component analysis, canonical correlation analysis, all forms of linear biplots, discriminant analysis and metric multidimensional scaling. In this appendix the theory of CA is summarized, as well as the theory of related methods discussed in the book. Matrix–vector notation is preferred because it is more compact, as well as being closer to the implementation of the method in the R computing language.

Contents

Let $\mathbf{N}$ denote a $I \times J$ data matrix, with positive row and column sums (almost always $\mathbf{N}$ consists of nonnegative numbers, but there are some exceptions such

Correspondence analysis notation

241

as the one described at the end of Chapter 26). For notational simplicity the matrix is first converted to the *correspondence matrix* $\mathbf{P}$ by dividing $\mathbf{N}$ by its grand total $n = \sum_i \sum_j n_{ij} = \mathbf{1}^\mathsf{T}\mathbf{N}\mathbf{1}$ (the notation $\mathbf{1}$ is used for a vector of ones of length that is appropriate to its context; hence the first $\mathbf{1}$ is $I \times 1$ and the second is $J \times 1$ to match the row and column lengths of $\mathbf{N}$).

Correspondence matrix:

$$\mathbf{P} = \frac{1}{n}\mathbf{N} \tag{A.1}$$

The following notation is used (see also the end of Chapter 4):

Row and column masses:

$$r_i = \sum_{j=1}^{J} p_{ij} \qquad c_j = \sum_{i=1}^{I} p_{ij}$$

$$\text{i.e.} \quad \mathbf{r} = \mathbf{P}\mathbf{1} \qquad\qquad \mathbf{c} = \mathbf{P}^\mathsf{T}\mathbf{1} \tag{A.2}$$

Diagonal matrices of row and column masses:

$$\mathbf{D}_r = \mathrm{diag}(\mathbf{r}) \quad \text{and} \quad \mathbf{D}_c = \mathrm{diag}(\mathbf{c}) \tag{A.3}$$

Note that all subsequent definitions and results are given in terms of these relative quantities $\mathbf{P} = [p_{ij}]$, $\mathbf{r} = [r_i]$ and $\mathbf{c} = [c_j]$, whose elements add up to 1 in each case. Multiply these by n to recover the elements of the original matrix $\mathbf{N}$: $np_{ij} = n_{ij}$, $nr_i = i$-th row sum of $\mathbf{N}$, $nc_j = j$-th column sum of $\mathbf{N}$.

Computational algorithm The computational algorithm to obtain coordinates of the row and column profiles with respect to principal axes, using the SVD, is as follows:

CA Step 1 — Calculate the matrix $\mathbf{S}$ of standardized residuals:

$$\mathbf{S} = \mathbf{D}_r^{-\frac{1}{2}}(\mathbf{P} - \mathbf{r}\mathbf{c}^\mathsf{T})\mathbf{D}_c^{-\frac{1}{2}} \tag{A.4}$$

CA Step 2 — Calculate the SVD of $\mathbf{S}$:

$$\mathbf{S} = \mathbf{U}\mathbf{D}_\alpha\mathbf{V}^\mathsf{T} \quad \text{where} \quad \mathbf{U}^\mathsf{T}\mathbf{U} = \mathbf{V}^\mathsf{T}\mathbf{V} = \mathbf{I} \tag{A.5}$$

where $\mathbf{D}_\alpha$ is the diagonal matrix of (positive) singular values in descending order: $\alpha_1 \geq \alpha_2 \geq \cdots$

CA Step 3 — Standard coordinates $\mathbf{\Phi}$ of rows:

$$\mathbf{\Phi} = \mathbf{D}_r^{-\frac{1}{2}}\mathbf{U} \tag{A.6}$$

CA Step 4 — Standard coordinates $\mathbf{\Gamma}$ of columns:

$$\mathbf{\Gamma} = \mathbf{D}_c^{-\frac{1}{2}}\mathbf{V} \tag{A.7}$$

CA Step 5 — Principal coordinates $\mathbf{F}$ of rows:

$$\mathbf{F} = \mathbf{D}_r^{-\frac{1}{2}}\mathbf{U}\mathbf{D}_\alpha = \mathbf{\Phi}\mathbf{D}_\alpha \tag{A.8}$$

CA Step 6 — Principal coordinates $\mathbf{G}$ of columns:

$$\mathbf{G} = \mathbf{D}_c^{-\frac{1}{2}}\mathbf{V}\mathbf{D}_\alpha = \mathbf{\Gamma}\mathbf{D}_\alpha \tag{A.9}$$

CA Step 7 — Principal inertias λ_k:

$$\lambda_k = \alpha_k^2, \quad k = 1, 2, \ldots, K \text{ where } K = \min\{I - 1, J - 1\} \tag{A.10}$$

The rows of the coordinate matrices in (A.6)–(A.9) refer to the rows or columns, as the case may be, of the original table, while the columns of these matrices refer to the principal axes, or dimensions, of which there are $\min\{I - 1, J - 1\}$, i.e. one less than the number of rows or columns, whichever is smaller. Notice how the principal and standard coordinates are scaled:

$$\mathbf{F}\mathbf{D}_r\mathbf{F}^\mathsf{T} = \mathbf{G}\mathbf{D}_c\mathbf{G}^\mathsf{T} = \mathbf{D}_\lambda \tag{A.11}$$

$$\mathbf{\Phi}\mathbf{D}_r\mathbf{\Phi}^\mathsf{T} = \mathbf{\Gamma}\mathbf{D}_c\mathbf{\Gamma}^\mathsf{T} = \mathbf{I} \tag{A.12}$$

In (A.11) the weighted sum of squares of the principal coordinates on the k-th dimension (i.e. their inertia in the direction of this dimension) is equal to the principal inertia (or eigenvalue) $\lambda_k = \alpha_k^2$, the square of the k-th singular value. The standard coordinates in (A.12), however, have weighted sum of squares equal to 1. All coordinate matrices have orthogonal columns, where the masses are always used in the calculation of the (weighted) scalar products.

The *contribution coordinates* are equal to $\mathbf{U}$ for rows and $\mathbf{V}$ for columns, since they are the standard coordinates rescaled by the square roots of the respective masses (see (A.6) and (A.7)), e.g. $\mathbf{D}_c^{\frac{1}{2}}\mathbf{\Phi} = \mathbf{D}_c^{\frac{1}{2}}\mathbf{D}_c^{-\frac{1}{2}}\mathbf{V} = \mathbf{V}$. The contribution coordinates have sums of squares equal to 1 on each dimension: $\mathbf{U}^\mathsf{T}\mathbf{U} = \mathbf{V}^\mathsf{T}\mathbf{V} = \mathbf{I}$, each squared term being the contribution of the row or column to the inertia of the respective dimension. If the columns serve as variables describing the rows, which is usually the case, then it is the column contribution coordinates that are relevant for the plotting.

Notice also that (A.4) can be written in terms of the so-called *contingency ratios* $\mathbf{D}_r^{-1}\mathbf{P}\mathbf{D}_c^{-1} = [p_{ij}/(r_i c_j)]$ as

$$\mathbf{D}_r^{\frac{1}{2}}(\mathbf{D}_r^{-1}\mathbf{P}\mathbf{D}_c^{-1} - \mathbf{1}\mathbf{1}^\mathsf{T})\mathbf{D}_c^{\frac{1}{2}} \tag{A.13}$$

which is a convenient alternative form that is useful for showing relationships with other methods.

The SVD is the fundamental mathematical result for CA, as it is for other dimension-reduction techniques such as principal component analysis, canonical correlation analysis, linear discriminant analysis, and other methods described in this book such as log-ratio analysis (Chapter 22) and co-inertia analysis (Chapter 28). This matrix decomposition expresses any rectangular matrix as a product of three matrices of simple structure, as in (A.5) above: $\mathbf{S} = \mathbf{U}\mathbf{D}_\alpha\mathbf{V}^\mathsf{T}$. The columns of the matrices $\mathbf{U}$ and $\mathbf{V}$ are the *left* and *right singular vectors* respectively, and the positive values α_k, in descending order down the diagonal of $\mathbf{D}_\alpha$, are the *singular values*. The SVD is related to the well-known eigenvalue–eigenvector decomposition (or *eigendecomposition*) of a square symmetric matrix as follows: $\mathbf{S}\mathbf{S}^\mathsf{T}$ and $\mathbf{S}^\mathsf{T}\mathbf{S}$ are square symmetric matrices that have eigendecompositions $\mathbf{S}\mathbf{S}^\mathsf{T} = \mathbf{U}\mathbf{D}_\alpha^2\mathbf{U}^\mathsf{T}$ and $\mathbf{S}^\mathsf{T}\mathbf{S} = \mathbf{V}\mathbf{D}_\alpha^2\mathbf{V}^\mathsf{T}$.

A note on the singular value decomposition (SVD)

Hence the singular vectors are also eigenvectors of these respective matrices and the squared singular values are their eigenvalues, conventionally denoted by λ: $\lambda_k = \alpha_k^2$. The practical utility of the SVD is that if another $I \times J$ matrix $\mathbf{S}_{(m)}$ is constructed using the first m columns of $\mathbf{U}$ and $\mathbf{V}$, $\mathbf{U}_{(m)}$ and $\mathbf{V}_{(m)}$, and the first m singular values in $\mathbf{D}_{\alpha(m)}$: $\mathbf{S}_{(m)} = \mathbf{U}_{(m)}\mathbf{D}_{\alpha(m)}\mathbf{V}_{(m)}^{\mathsf{T}}$, then $\mathbf{S}_{(m)}$ is the least-squares rank m approximation of $\mathbf{S}$ (this result is known as the *Eckart–Young theorem*). Since the objective of finding low-dimensional best-fitting subspaces coincides with the objective of finding low-rank matrix approximations by least-squares, the SVD solves this problem perfectly and in a very compact way. The only adaptation needed is to incorporate the weighting of the rows and columns by the masses into the SVD so that the approximations are by weighted least squares. If a generalized form of the SVD is defined, where the singular vectors are normalized with weighting by the masses, then the CA solution can be obtained in one step. For example, the generalized SVD of the contingency ratios $p_{ij}/(r_i c_j)$, centred at the constant value 1, leads to the standard row and column coordinates directly:

$$\mathbf{D}_r^{-1}\mathbf{P}\mathbf{D}_c^{-1} - \mathbf{1}\mathbf{1}^{\mathsf{T}} = \boldsymbol{\Phi}\mathbf{D}_\alpha\boldsymbol{\Gamma}^{\mathsf{T}} \quad \text{where} \quad \boldsymbol{\Phi}^{\mathsf{T}}\mathbf{D}_r\boldsymbol{\Phi} = \boldsymbol{\Gamma}^{\mathsf{T}}\mathbf{D}_c\boldsymbol{\Gamma} = \mathbf{I} \qquad (A.14)$$

The bilinear CA model

From steps 1 to 4 of the basic algorithm, the data in $\mathbf{P}$ can be written as follows (see also (13.4) on page 101 and (14.9) on page 109):

$$p_{ij} = r_i c_j \left(1 + \sum_{k=1}^{K} \sqrt{\lambda_k}\phi_{ik}\gamma_{jk}\right) \qquad (A.15)$$

(also called the *reconstitution formula*). In matrix notation,

$$\mathbf{P} = \mathbf{D}_r(\mathbf{1}\mathbf{1}^{\mathsf{T}} + \boldsymbol{\Phi}\mathbf{D}_\lambda^{\frac{1}{2}}\boldsymbol{\Gamma}^{\mathsf{T}})\mathbf{D}_c \qquad (A.16)$$

Because of the simple relations (A.8) and (A.9) between the principal and standard coordinates, this bilinear model can be written in several alternative ways — see also (14.10) and (14.11) on pages 109–110.

Transition equations between rows and columns

The left and right singular vectors are related linearly, for example by multiplying the SVD on the right by $\mathbf{V}$: $\mathbf{S}\mathbf{V} = \mathbf{U}\mathbf{D}_\alpha$. Expressing such relations in terms of the principal and standard coordinates gives the following variations of the same theme, called *transition equations* (see Equations (14.1), (14.2), (14.5) and (14.6) on pages 108–109 for the equivalent scalar versions):

Principal as a function of standard (barycentric relationships):

$$\mathbf{F} = \mathbf{D}_r^{-1}\mathbf{P}\boldsymbol{\Gamma} \qquad \mathbf{G} = \mathbf{D}_c^{-1}\mathbf{P}^{\mathsf{T}}\boldsymbol{\Phi} \qquad (A.17)$$

Principal as a function of principal:

$$\mathbf{F} = \mathbf{D}_r^{-1}\mathbf{P}\mathbf{G}\mathbf{D}_\lambda^{-\frac{1}{2}} \qquad \mathbf{G} = \mathbf{D}_c^{-1}\mathbf{P}^{\mathsf{T}}\mathbf{F}\mathbf{D}_\lambda^{-\frac{1}{2}} \qquad (A.18)$$

The equations (A.17) are those that were mentioned as early as Chapter 3, expressing the profile points as weighted averages of the vertex points, where the weights are the profile elements. These are the equations that govern the *asymmetric maps.* The equations (A.18) show that the two sets of principal coordinates, which govern the *symmetric map*, are also related by a barycentric (weighted average) relationship, but with scale factors (the inverse square roots of the principal inertias) that are different on each dimension.

The transition equations are used to situate supplementary points on the map. For example, given a supplementary column point with values in $\mathbf{h}$ ($I \times 1$), it is first divided by its total $\mathbf{1}^\mathsf{T}\mathbf{h}$ to obtain the column profile $\tilde{\mathbf{h}} = (1/\mathbf{1}^\mathsf{T}\mathbf{h})\mathbf{h}$. Then this profile is transposed as a row vector in the second equation of (A.17) to calculate the coordinates $\mathbf{g}$ of the supplementary column:

Supplementary points

$$\mathbf{g} = \tilde{\mathbf{h}}^\mathsf{T}\boldsymbol{\Phi} \qquad (A.19)$$

The total inertia of the data matrix is the sum of squares of the matrix $\mathbf{S}$ in (A.4), or equivalently in (A.13):

Total inertia and χ^2-distances

$$\text{inertia} = \text{trace}(\mathbf{S}\mathbf{S}^\mathsf{T}) = \sum_{i=1}^{I}\sum_{j=1}^{J}\frac{(p_{ij} - r_i c_j)^2}{r_i c_j} \qquad (A.20)$$

The inertia is also the sum of squares of the singular values, i.e. the sum of the eigenvalues:

$$\text{inertia} = \sum_{k=1}^{K}\alpha_k^2 = \sum_{k=1}^{K}\lambda_k \qquad (A.21)$$

The χ^2-distances between row profiles and between column profiles are:

$$\chi^2\text{-distance between rows } i \text{ and } i' : \sum_{j=1}^{J}\left(\frac{p_{ij}}{r_i} - \frac{p_{i'j}}{r_{i'}}\right)^2 \Big/ c_j \qquad (A.22)$$

$$\chi^2\text{-distance between columns } j \text{ and } j' : \sum_{i=1}^{I}\left(\frac{p_{ij}}{c_j} - \frac{p_{ij'}}{c_{j'}}\right)^2 \Big/ r_i \qquad (A.23)$$

To write the full set of χ^2-distances in the form of a square symmetric matrix requires a bit more work. First, calculate the matrix $\mathbf{Q}$ of "χ^2 scalar products" between row profiles, for example, as:

$$\chi^2 \text{ scalar products between rows} : \quad \mathbf{Q} = \mathbf{D}_r^{-1}\mathbf{P}\mathbf{D}_c^{-1}\mathbf{P}^\mathsf{T}\mathbf{D}_r^{-1} \qquad (A.24)$$

Then define the vector $\mathbf{a}$ as the elements on the diagonal of this matrix (i.e. the scalar products of the row profiles with themselves):

$$\mathbf{q} = \text{diag}(\mathbf{Q}) \qquad (A.25)$$

Then the $I \times I$ matrix of squared χ^2-distances is:

$$\text{squared } \chi^2\text{-distance matrix between rows} : \quad \mathbf{q}\mathbf{1}^\mathsf{T} + \mathbf{1}\mathbf{q}^\mathsf{T} - 2\mathbf{Q} \qquad (A.26)$$

To calculate the $J \times J$ matrix of squared χ^2-distances between column profiles, interchange rows with columns in (A.24), defining $\mathbf{Q}$ as $\mathbf{D}_c^{-1}\mathbf{P}^\mathsf{T}\mathbf{D}_r^{-1}\mathbf{P}\mathbf{D}_c^{-1}$ and then following with (A.25) and (A.26).

The χ^2-distance between a row profile $\mathbf{a}_i = (1/r_i)[\ p_{i1}\ p_{i2}\ \cdots\ p_{iJ}\ ]^\mathsf{T}$ and the average row profile $\mathbf{c} = [\,c_1 \; c_2 \; \cdots \; c_j\,]^{\mathsf{T}}$ (written as $\|\mathbf{a}_i - \mathbf{c}\|_c$ on page 31) is $\sum_j (p_{ij}/r_i - c_j)^2/c_j$, which can be expressed in terms of the contingency ratios as $\sum_j c_j [p_{ij}/(r_i c_j) - 1]^2$. Incorporating the mass r_i of each row and summing over the rows, the total inertia in (A.20) is obtained: $\sum_i r_i \|\mathbf{a}_i - \mathbf{c}\|_c = \sum_i \sum_j r_i c_j [p_{ij}/(r_i c_j) - 1]^2$.

Contributions of points to principal inertias

The contributions of the row and columns points to the inertia on the k-th dimension are the inertia components:

$$\text{for row } i: \quad \frac{r_i f_{ik}^2}{\lambda_k} = r_i \phi_{ik}^2 \qquad \text{for column } j: \quad \frac{c_j g_{jk}^2}{\lambda_k} = c_j \gamma_{jk}^2 \qquad (\text{A.27})$$

recalling the relationship between principal and standard coordinates given in (A.8) and (A.9): $f_{ik} = \sqrt{\lambda_k}\phi_{ik}$, $g_{jk} = \sqrt{\lambda_k}\gamma_{jk}$. The square roots of the values in (A.27) are exactly the contribution coordinates proposed for the CA contribution biplot of Chapter 13, i.e. the squared lengths of these coordinates are the contributions to the principal axes.

Contributions of principal axes to point inertias (squared correlations)

The contributions of the dimensions to the inertia of the i-th row and j-th column points (i.e. the squared cosines or squared correlations) are:

$$\text{for row } i: \quad \frac{f_{ik}^2}{\sum_k f_{ik}^2} \qquad \text{for column } j: \quad \frac{g_{jk}^2}{\sum_k g_{jk}^2} \qquad (\text{A.28})$$

As shown in Chapter 11, the denominators in (A.28) are the squared χ^2-distances between the corresponding profile point and the average profile.

Ward clustering of row or column profiles

The clustering of Chapter 15 is described here in terms of the rows; exactly the same applies to the clustering of the columns. The rows are clustered at each step of the algorithm to minimise the decrease in the χ^2 statistic (equivalently, the decrease in the inertia since inertia $= \chi^2/n$, where n is the total of the table). This clustering criterion is equivalent to Ward clustering, where each cluster is weighted by the total mass of its members. The measure of difference between rows can be shown to be the weighted form of the squared chi-squared distance between profiles. Suppose $\mathbf{a}_i$ and r_i, $i = 1, \ldots, I$, denote the I row profiles of the data matrix, and their masses respectively. Then identifying the pair that gives the least decrease in inertia is equivalent to looking for the pair of rows (i, i') that minimizes the following measure:

$$\frac{r_i r_{i'}}{r_i + r_{i'}} \|\mathbf{a}_i - \mathbf{a}_{i'}\|_c^2 \qquad (\text{A.29})$$

The two rows are then merged by summing their frequencies, and the profile and mass are recalculated. The same measure of difference as (A.29) is calculated at each stage of the clustering for the row profiles at that stage (see (15.2) on page 120 for the equivalent formula based on profiles of clusters), and the two profiles with the least difference are merged. Hence (A.29) is the level of clustering in terms of the inertia decrease, or if multiplied by n it is the decrease in χ^2. In the case of a contingency table the level of clustering can be tested for significance using the tables at the end of this Appendix.

Suppose tables $\mathbf{N}_{qs}$, $q = 1, \ldots, Q$, $s = 1, \ldots, S$ are concatenated row- and/or *Stacked tables* columnwise to make a block matrix $\mathbf{N}$. If the marginal frequencies are the same in each row and in each column (as is the case when the same individuals are cross-tabulated separately in several tables), then the inertia of $\mathbf{N}$ is the average of the separate inertias of the tables $\mathbf{N}_{qs}$:

$$\text{inertia}(\mathbf{N}) = \frac{1}{QS} \sum_{q=1}^{Q} \sum_{s=1}^{S} \text{inertia}(\mathbf{N}_{qs}) \qquad (A.30)$$

Suppose the original matrix of categorical data is $N \times Q$, i.e. N cases and Q *Multiple CA* variables. Classical multiple CA (MCA) has two forms. The first form converts the cases-by-variables data into an indicator matrix $\mathbf{Z}$ where the categorical data have been recoded as dummy variables. If the q-th variable has J_q categories, this indicator matrix will have $J = \sum_q J_q$ columns (see Chapter 18, Exhibit 18.1 for an example). Then the indicator version of MCA is the application of the basic CA algorithm to the matrix $\mathbf{Z}$, resulting in coordinates for the N cases and the J categories. The second form of MCA calculates the Burt matrix $\mathbf{B} = \mathbf{Z}^\mathsf{T}\mathbf{Z}$ of all two-way cross-tabulations of the Q variables (see Chapter 18, Exhibit 18.4 for an example). Then the Burt version of MCA is the application of the basic CA algorithm to the matrix $\mathbf{B}$, resulting in coordinates for the J categories ($\mathbf{B}$ is a symmetric matrix). The standard coordinates of the categories are identical in the two versions of MCA, and the principal inertias in the Burt version are the squares of those in the indicator version.

Joint CA (JCA) is the fitting of the off-diagonal cross-tabulations of the Burt *Joint CA* matrix, ignoring the cross-tabulations on the block diagonal. The algorithm we use is an alternating least-squares procedure which successively applies CA to the Burt matrix which has been modified by replacing the values on the block diagonal with estimated values from the CA of the previous iteration, using a chosen dimensionality of the solution. On convergence of the JCA algorithm, the CA is performed on the last modified Burt matrix, $\widetilde{\mathbf{B}}$, which has its diagonal blocks perfectly fitted by construction. In other words, supposing that the solution requested is two-dimensional, then the modified diagonal blocks satisfy (A.15) exactly using just two terms in the bilinear CA model (or reconstitution formula).

Hence the total inertia of $\widetilde{\mathbf{B}}$ includes a part Δ for these "imputed" diagonal *Percentage of* blocks, and so do the first two principal inertias, $\tilde{\lambda}_1$ and $\tilde{\lambda}_2$, which perfectly *inertia explained* explain the part Δ. To obtain the percentage of inertia explained by the two- *in JCA* dimensional solution, the amount Δ has to be discounted both from the total and from the sum of the two principal inertias. The value of Δ can be obtained via the difference between the inertia of the original Burt matrix $\mathbf{B}$ (whose inertias in the diagonal blocks are known) and the modified one $\widetilde{\mathbf{B}}$, as follows (here we use the result (A.30) which applies to the subtables of $\mathbf{B}$, denoted by $\mathbf{B}_{qs}$, and those of $\widetilde{\mathbf{B}}$, whose off-diagonal tables are the same):

$$\text{inertia}(\mathbf{B}) = \frac{1}{Q^2}\left(\sum\sum_{q\neq s}\text{inertia}(\mathbf{B}_{qs}) + \sum_q\text{inertia}(\mathbf{B}_{qq})\right)$$

$$= \frac{1}{Q^2}\left(\sum\sum_{q\neq s}\text{inertia}(\mathbf{B}_{qs}) + (J-Q)\right)$$

$$\text{inertia}(\widetilde{\mathbf{B}}) = \frac{1}{Q^2}\left(\sum\sum_{q\neq s}\text{inertia}(\mathbf{B}_{qs})\right) + \Delta$$

Subtracting the above leads to

$$\text{inertia}(\mathbf{B}) - \text{inertia}(\widetilde{\mathbf{B}}) = \frac{J-Q}{Q^2} - \Delta$$

which gives the value of Δ:

$$\Delta = \frac{J-Q}{Q^2} - \left(\text{inertia}(\mathbf{B}) - \text{inertia}(\widetilde{\mathbf{B}})\right) \qquad (A.31)$$

Discounting this amount from the total and the sum of the principal inertias, assuming a two-dimensional solution, gives the percentage of inertia explained by the JCA solution:

$$100 \times \frac{\tilde{\lambda}_1 + \tilde{\lambda}_2 - \Delta}{\text{inertia}(\widetilde{\mathbf{B}}) - \Delta} \qquad (A.32)$$

Contributions in JCA The previous section showed how to discount the extra inertia as a result of the modified diagonal blocks of the Burt matrix in JCA. There is an identical situation at the level of each point. Each category point j has an additional amount of inertia, δ_j, due to the modified diagonal blocks. In the case of the original Burt matrix $\mathbf{B}$ we know exactly what this extra amount is due to the diagonal matrices in the diagonal blocks: for the j-th point it is $(1-Qc_j)/Q^2$, where c_j is the j-th mass (summing these values for $j = 1,\ldots,J$, we obtain $(J-Q)/Q^2$ which was the total additional amount due to the diagonal blocks of $\mathbf{B}$). Therefore, just as above, we can derive how to obtain contributions of the two-dimensional solution to the point inertias as follows:

$$\text{inertia}(j\text{-th category of }\mathbf{B}) = \text{off-diagonal components} + \frac{1-Qc_j}{Q^2}$$

$$\text{inertia}(j\text{-th category of }\widetilde{\mathbf{B}}) = \text{off-diagonal components} + \delta_j$$

Subtracting the above (the "off-diagonal components" are the same) leads to

$$\text{inertia}(j\text{th category of }\mathbf{B}) - \text{inertia}(j\text{-th category of }\widetilde{\mathbf{B}}) = \frac{1-Qc_j}{Q^2} - \delta_j$$

which gives the value of δ_j:

$$\delta_j = \frac{1 - Qc_j}{Q^2} - \left(\text{inertia}(j\text{-th category of } \mathbf{B}) - \text{inertia}(j\text{-th category of } \widetilde{\mathbf{B}})\right) \tag{A.33}$$

Discounting this amount from the j-th category's inertia and similarly from the sum of the components of inertia in two dimensions gives the relative contributions (qualities) with respect to the two-dimensional JCA solution:

$$\frac{c_j \tilde{g}_{j1}^2 + c_j \tilde{g}_{j2}^2 - \delta_j}{(\sum_k c_j \tilde{g}_{jk}^2) - \delta_j} \tag{A.34}$$

where $\tilde{g}_{jk}$ is the principal coordinate of category j on axis k in the CA of $\widetilde{\mathbf{B}}$ (JCA solution), and the summation in the denominator is for all the dimensions. Notice that $\sum_j \delta_j = \Delta$ (i.e. summing (A.33) gives (A.31)).

The MCA solution can be adjusted to optimize the fit to the off-diagonal tables (this could be called a JCA *conditional on* the MCA solution). The optimal adjustments can be determined by weighted least squares, as described in Chapter 19, but the problem is that the solution is not nested. So we prefer slightly sub-optimal adjustments that retain the nesting property and are very easy to compute from the MCA solution of the Burt matrix. The adjustments are made as follows (see Chapter 19, pages 149–150, for an illustration):   *Adjusted inertias in MCA*

Adjusted total inertia of Burt matrix:

$$\text{adjusted total inertia} = \frac{Q}{Q-1} \left(\text{inertia of } \mathbf{B} - \frac{J-Q}{Q^2}\right) \tag{A.35}$$

Adjusted principal inertias (eigenvalues) of Burt matrix:

$$\lambda_k^{\text{adj}} = \left(\frac{Q}{Q-1}\right)^2 \left(\sqrt{\lambda_k} - \frac{1}{Q}\right)^2, \quad k = 1, 2, \ldots \tag{A.36}$$

Here λ_k refers to the k-th principal inertia of the Burt matrix; hence $\sqrt{\lambda_k}$ is the k-th principal inertia of the indicator matrix. The adjustments are made only to those dimensions for which $\sqrt{\lambda_k} > \frac{1}{Q}$ and no further dimensions are used — hence, the percentages of inertia do not add up to 100%, which is correct since these dimensions cannot fully explain the inertia of $\mathbf{B}$. It can be proved that these percentages are lower bound estimates of those that are obtained in a JCA, and in practice they are close to the JCA percentages.

Subset CA is simply the application of the same CA algorithm to a selected part of the standardized residual matrix $\mathbf{S}$ in (A.4) (*not* to the subset of the original matrix). The masses of the full matrix are thus retained and all subsequent calculations are the same, except they are applied to the subset. Suppose that the columns are subsetted, but not the rows. Then the rows still maintain the centring property of CA; i.e. their weighted averages are at the origin of the map, whereas the columns are no longer centred. Subset MCA   *Subset CA, MCA and JCA* is performed by applying subset CA on a submatrix of the indicator matrix or the Burt matrix. In the case of the Burt matrix, a selection of categories implies that this subset has to be specified for both the rows and columns.

Log-ratio analysis   The definition of the standardized residuals matrix (A.4) in terms of the contingency ratios, given in (A.13), has yet another equivalent form:

$$\mathbf{D}_r^{\frac{1}{2}}(\mathbf{I} - \mathbf{1r^\top})(\mathbf{D}_r^{-1}\mathbf{P}\mathbf{D}_c^{-1})(\mathbf{I} - \mathbf{1c^\top})^\top\mathbf{D}_c^{\frac{1}{2}} \tag{A.37}$$

where, instead of the contingency ratios minus 1 in the centre, there are simply the contingency ratios, with row- and column-centring matrices on either side. The pre-multiplying matrix $\mathbf{I} - \mathbf{1r^\top}$ subtracts the weighted average of each column from the elements of the respective column, and the post-multiplying matrix $(\mathbf{I} - \mathbf{1c^\top})^\top$ subtracts the weighted average of each row from the respective row elements. This is a convenient form to compare with weighted log-ratio analysis (LRA), which is defined on the weighted double-centred matrix of log-transformed data:

$$\mathbf{D}_r^{\frac{1}{2}}(\mathbf{I} - \mathbf{1r^\top})\log(\mathbf{N})(\mathbf{I} - \mathbf{1c^\top})^\top\mathbf{D}_c^{\frac{1}{2}} \tag{A.38}$$

where $\log(\mathbf{N}) = [\log(n_{ij})]$. Now the logarithms of the contingency ratios are $\log(n_{ij}) - \log(n) - \log(r_i) - \log(c_j)$ (since $p_{ij} = n_{ij}/n$), and the row- and column-centrings will remove all the terms except $\log(n_{ij})$. Hence $\log(\mathbf{N})$ in (A.38) can be replaced by $\log(\mathbf{D}_r^{-1}\mathbf{P}\mathbf{D}_c^{-1})$ and the only difference between (A.37) and (A.38) is the log-transformation, showing the close relationship between the two methods. Letting $\mathbf{S}$ in (A.4) be equal to (A.38) and then following with steps (A.5)–(A.10) results in the LRA of the matrix $\mathbf{N}$. The only disadvantage of LRA is that data zeros can not be log-transformed, whereas CA can handle matrices with many zero values, as often found in applications in ecology, linguistics and archaeology.

There is an even more intimate relationship between LRA and CA, thanks to the *Box–Cox transform*:

$$f(x; \alpha) = \frac{1}{\alpha}(x^\alpha - 1) \text{ for } \alpha \neq 0$$
$$= \log(x) \text{ for } \alpha = 0$$

As α tends to 0, $f(x; \alpha)$ tends to $\log(x)$. If the contingency ratios are transformed as $(1/\alpha)[p_{ij}/(r_ic_j)]^\alpha$, then $\alpha = 1$ gives the matrix (A.37) analysed in CA, while letting α descend to 0 (but not equal to 0) gives, in the limit, the matrix (A.38) analysed in LRA.

Analysis of matched matrices   Matched matrices $\mathbf{A}$ and $\mathbf{B}$ are both $I \times J$ matrices of comparable data with the same row and column labels. The CA of the $2I \times 2J$ block matrix

$$\begin{bmatrix} \mathbf{A} & \mathbf{B} \\ \mathbf{B} & \mathbf{A} \end{bmatrix} \tag{A.39}$$

yields the CA of the sum $\mathbf{A} + \mathbf{B}$ and the *uncentred* CA of the difference $\mathbf{A} - \mathbf{B}$. Dimensions corresponding to the sum have coordinate vectors that have repeats of a vector, whereas dimensions corresponding to the difference have coordinate vectors that have repeats of a vector with a sign change:

$$\text{sum dimension: } \begin{bmatrix} \mathbf{x} \\ \mathbf{x} \end{bmatrix} \qquad \text{difference dimension: } \begin{bmatrix} \mathbf{y} \\ -\mathbf{y} \end{bmatrix}$$

The dimensions corresponding to the sum and the difference are interleaved in the analysis of the block matrix, so the major dimensions for each part need to be selected for visualization (see Chapter 23 for an example). The row and column masses of the block matrix (A.39) are the averages of those of the two matched matrices $\mathbf{A}$ and $\mathbf{B}$, which are the same as those of their sum $\mathbf{A} + \mathbf{B}$. The same masses are implicitly used for the analysis of the difference $\mathbf{A} - \mathbf{B}$.

If the data matrix $\mathbf{N}$ is square asymmetric, where both rows and columns refer to the same objects, then $\mathbf{N}$ can be written as the sum of symmetric and skew-symmetric parts:

$$\mathbf{N} = \frac{1}{2}(\mathbf{N} + \mathbf{N}^\top) + \frac{1}{2}(\mathbf{N} - \mathbf{N}^\top) \tag{A.40}$$

$$= \text{symmetric } + \text{ skew-symmetric}$$

CA is applied to each part separately, but with a slight variation for the skew-symmetric part. The analysis of the symmetric part $\frac{1}{2}(\mathbf{N} + \mathbf{N}^\top)$ is the usual CA — this provides one set of coordinates, and the masses are the averages of the row and column masses corresponding to the same object: $w_i = \frac{1}{2}(r_i + c_i)$. The analysis of the skew-symmetric part $\frac{1}{2}(\mathbf{N} - \mathbf{N}^\top)$ is the application of the CA algorithm, again without centring, and using the same masses as in the symmetric analysis; i.e. the "standardized residuals" matrix of (A.4) is rather the "standardized differences" matrix

$$\mathbf{S} = \mathbf{D}_w^{-\frac{1}{2}}[\frac{1}{2}(\mathbf{P} - \mathbf{P}^\top)]\mathbf{D}_w^{-\frac{1}{2}} \tag{A.41}$$

where $\mathbf{P}$ is the correspondence matrix and $\mathbf{D}_w$ is the diagonal matrix of the average masses w_i. This is a special case of matched matrices and both these analyses are subsumed in the regular CA of the block matrix

$$\begin{bmatrix} \mathbf{N} & \mathbf{N}^\top \\ \mathbf{N}^\top & \mathbf{N} \end{bmatrix} \tag{A.42}$$

If $\mathbf{N}$ is an $I \times I$ matrix, then the $2I - 1$ dimensions that emanate from this CA can be easily allocated to the symmetric and skew-symmetric solutions, since the symmetric dimensions have unique principal inertias while the skew-symmetric dimensions occur in pairs with equal principal inertias. Similarly, as for matched matrices, the coordinate vectors for each dimension have two subvectors: for dimensions corresponding to the symmetric analysis these are repeats of each other, while for dimensions corresponding to the skew-symmetric analysis these are repeats with a change of sign (see Chapter 24 for an example).

A square symmetric matrix can be observed as the symmetric part of an asymmetric matrix (see previous section) or, for example, directly observed as an adjacency matrix coding an undirected network (see Chapter 25). The row and column masses are equal in this case, and the square symmetric matrix $\mathbf{S}$ of standardized residuals (A.4) has the form $\mathbf{D}_c^{-\frac{1}{2}}(\mathbf{P} - \mathbf{c}\mathbf{c}^\mathsf{T})\mathbf{D}_c^{-\frac{1}{2}}$. The SVD of $\mathbf{S}$ can coincide with its eigendecomposition (EVD) if $\mathbf{S}$ has no negative eigenvalues, in which case $\mathbf{S} = \mathbf{V}\mathbf{D}_\lambda\mathbf{V}^\mathsf{T}$ and the eigenvalues λ are the same as the singular values. But usually $\mathbf{S}$ will have some negative eigenvalues in which case the corresponding singular values will be the absolute values of these and the corresponding left and right singular vectors will differ in sign. These latter dimensions are called *inverse dimensions*, as opposed to the *direct dimensions* that correspond to the positive eigenvalues (equal to singular values) that have identical left and right sigular vectors. In order to visualize a symmetric matrix using a single set of points, the most important direct dimensions should be selected.

An interesting family of symmetric matrices is defined by parameters α and β, a linear combination of the correspondence matrix $\mathbf{P}$, the diagonal matrix of masses $\mathbf{D}_c$ and the rank one matrix of mass products $\mathbf{c}\mathbf{c}^\mathsf{T}$:

$$\mathbf{P}(\alpha, \beta) = \alpha\mathbf{P} + \beta\mathbf{D}_c + (1 - \alpha - \beta)\mathbf{c}\mathbf{c}^\mathsf{T} \qquad (A.43)$$

for which the dimensions have the same coordinates and only the principal inertias vary (and thus induce different orderings of the dimensions):

$$\lambda(\alpha, \beta) = (\alpha\sqrt{\lambda_k}\epsilon_k + \beta)^2 \qquad (A.44)$$

where λ_k is the k-th eigenvalue of $\mathbf{P}$ and $\epsilon_k = 1$ for direct dimensions, and $= -1$ for inverse dimensions. The set of (α, β) that gives matrices $\mathbf{P}(\alpha, \beta)$ in (A.43) with nonnegative elements forms a convex polygon in the α-β plane.

The Burt matrix $\mathbf{B}$ is a square symmetric matrix central to MCA and its variants. $\mathbf{B}$ has the property that its diagonal is proportional to its margins, so that if $\alpha + \beta = 1$ (i.e. the third term in (A.43) is zero), then varying α (and $\beta = 1 - \alpha$) is simply varying the role of the diagonal values (see also the case in Chapter 25, page 198, where the vertex degrees are inserted on the diagonal of the adjacency matrix). An interesting special case is when the diagonal is set to 0, called the *nullified Burt matrix*, which occurs for $\alpha = Q/(Q-1)$ and $\beta = -1/(Q-1)$ (where Q is the number of categorical variables), for which

$$\lambda(\alpha, \beta) = \left(\frac{Q}{Q-1}\sqrt{\lambda_k}\epsilon_k - \frac{1}{Q-1}\right)^2 = \left(\frac{Q}{Q-1}\right)^2\left(\sqrt{\lambda_k}\epsilon_k - \frac{1}{Q}\right)^2 \quad (A.45)$$

These are exactly the adjusted inertias in (A.36), corresponding to direct dimensions for which $\sqrt{\lambda_k} > 1/Q$ ($\epsilon_k = 1$). The other dimensions of the nullified Burt matrix for which $\sqrt{\lambda_k} < 1/Q$ are inverse ($\epsilon_k = -1$), including several dimensions corresponding to zero inertias of the Burt matrix ($\lambda_k = 0$) for which $\lambda(\alpha, \beta) = 1/(Q-1)^2$.

Canonical correspondence analysis (CCA)   In canonical correspondence analysis (CCA) an additional matrix $\mathbf{X}$ of explanatory variables is available, and the requirement is that the dimensions of the analysis of the correspondence matrix $\mathbf{P}$ be linearly related to $\mathbf{X}$. The total inertia is split into two parts: a part that is linearly related to the explanatory variables, called the inertia in the *constrained space*, and a part that is not, the inertia in the *unconstrained space*. The usual data structure is that the rows of $\mathbf{P}$ are sampling units and $\mathbf{X}$ is an additional set of M variables in columns, i.e. $I \times M$. The first step in CCA is a weighted regression step that calculates the $I \times J$ constrained matrix, by projection of $\mathbf{P}$ (in its form of standardized residuals) onto $\mathbf{X}$, giving a matrix whose columns are linearly related to $\mathbf{X}$. The residual part from this regression step is the unconstrained matrix, whose columns are not linearly related (i.e. uncorrelated) to $\mathbf{X}$. CCA thus consists of applying CA to the constrained matrix and (optionally) to the unconstrained residual matrix. In each application the original row and column masses are maintained for all computations, and the various results such as coordinates, principal inertias, contributions, reconstruction formula, etc. are the same as in a regular application of CA. We assume that the columns of $\mathbf{X}$ are standardized, using the row masses as weights in the calculation of means and variances. If there are some categorical explanatory variables, these are coded as dummy variables, dropping one category of each variable as in a conventional regression analysis, or — even better — including all dummy variables and then using a generalized inverse in Step 2 below. The dummy variables are also standardized in the same way as the columns of $\mathbf{X}$, but in the eventual display are shown as row centroids.

The steps in CCA are as follows:

CCA Step 1 — Calculate the standardized residuals matrix $\mathbf{S}$ *as in CA:*

$$\mathbf{S} = \mathbf{D}_r^{-\frac{1}{2}}(\mathbf{P} - \mathbf{rc}^{\mathsf{T}})\mathbf{D}_c^{-\frac{1}{2}} \tag{A.46}$$

CCA Step 2 — Calculate the $I \times I$ *(weighted) projection matrix, of rank* M, *which projects onto the constrained space:*

$$\mathbf{Q} = \mathbf{D}_r^{\frac{1}{2}}\mathbf{X}(\mathbf{X}^{\mathsf{T}}\mathbf{D}_r\mathbf{X})^{-1}\mathbf{X}^{\mathsf{T}}\mathbf{D}_r^{\frac{1}{2}} \tag{A.47}$$

CCA Step 3 — Project the standardized residuals to obtain the constrained matrix:

$$\mathbf{S}^{\star} = \mathbf{QS} \tag{A.48}$$

CCA Step 4 — Apply CA Steps 1–6 (page 242) to $\mathbf{S}^{\star}$*:*

CCA Step 5 — Principal inertias $\lambda_k^{\star}$ *in constrained space:*

$$\lambda_k^{\star} = \alpha_k^2, \quad k = 1, 2, \ldots, K \text{ where } K = \min\{I - 1, J - 1, M\} \tag{A.49}$$

CCA Step 6 (optional) — Project the standardized residuals onto the unconstrained space:

$$\mathbf{S}^{\perp} = (\mathbf{I} - \mathbf{Q})\mathbf{S} = \mathbf{S} - \mathbf{S}^{\star} \tag{A.50}$$

CCA Step 7 (optional) — Apply CA Steps 1–6 to $\mathbf{S}^{\perp}$*.*

As described in Chapter 24, the principal inertias in (A.49) can be expressed as percentages of the total inertia, or as percentages of the constrained inertia, which is the sum of squares of the elements in $\mathbf{S}^{\star}$, equal to $\sum_{k=1}^{K} \lambda_k^{\star}$.

Co-inertia analysis and co-correspondence analysis

The most general form of *co-inertia analysis* that we need involves the covariance matrix between two data matrices $\mathbf{X}$ $(n \times p)$ and $\mathbf{Y}$ $(n \times q)$, usually centred and optionally normalized, where the n common rows are weighted by the diagonal elements of $\mathbf{D}_w$ and the columns of $\mathbf{X}$ and $\mathbf{Y}$ are weighted by the diagonal elements of $\mathbf{D}_g$ and $\mathbf{D}_h$ respectively. The analysis involves the SVD of $\mathbf{D}_g^{\frac{1}{2}} \mathbf{X}^{\mathsf{T}} \mathbf{D}_w \mathbf{Y} \mathbf{D}_h^{\frac{1}{2}} = \mathbf{U} \mathbf{D}_\alpha \mathbf{V}^{\mathsf{T}}$, with standard coordinates for the two sets of variables computed as $\mathbf{\Phi} = \mathbf{D}_g^{-\frac{1}{2}} \mathbf{U}$ and $\mathbf{\Gamma} = \mathbf{D}_h^{-\frac{1}{2}} \mathbf{V}$, and the two sets of row points $\mathbf{X}\mathbf{\Phi}$ and $\mathbf{Y}\mathbf{\Gamma}$. Various special cases are described in Chapter 28, for example *co-correspondence analysis*, which looks at the relationship between two tables with common rows, each of which is suitable for a regular CA. The contingency ratios minus 1 of the separate matrices (see left-hand side of (A.14)) are conveniently taken as $\mathbf{X}$ and $\mathbf{Y}$, the column weights (masses) are those of the respective tables. The row weights need to be pre-specified: for example, two possible options are the averages of the row masses of the separate matrices, or the masses of the rows of the concatenated matrices.

Table for testing for significant clustering or significant dimensions

In the case of a contingency table based on a random sample, the first principal inertia can be tested for statistical significance. This is the same test as was used in the case of the Ward clustering of Chapter 15. In that case a critical level for clustering, on the chi-square scale (i.e. inertia times the grand total of the table), can be determined from Exhibit A.1, according to the size of the contingency table (see page 118 for the food store example, a 5×4 table for which the critical point according to Exhibit A.1 is 15.24). These critical points are the same for testing the first principal inertia for significance. For example, in the same example of the food stores, given in Exhibit 15.3, the first principal inertia was 0.02635, which if expressed as a chi-square component is $0.02635 \times 700 = 18.45$. Since 18.45 is greater than the critical point 15.24, the first principal inertia is statistically significant at the 5% level.

Exhibit A.1:
Critical values for multiple comparisons test on a $I \times J$ (or $J \times I$) contingency table. The same critical points apply to testing the significance of a principal inertia. The significance level is 0.05 (5%).

					J				
I	3	4	5	6	7	8	9	10	11
3	8.59								
4	10.74	13.11							
5	12.68	15.24	17.52						
6	14.49	17.21	19.63	21.85					
7	16.21	19.09	21.62	23.95	26.14				
8	17.88	20.88	23.53	25.96	28.23	30.40			
9	19.49	22.62	25.37	27.88	30.24	32.48	34.63		
10	21.06	24.31	27.15	29.75	32.18	34.50	36.70	38.84	
11	22.61	25.96	28.90	31.57	34.08	36.45	38.72	40.91	43.04
12	24.12	27.58	30.60	33.35	35.93	38.36	40.69	42.93	45.10
13	25.61	29.17	32.27	35.09	37.73	40.22	42.60	44.90	47.12

Source: Pearson, E.S. & Hartley, H.O. (1972). *Biometrika Tables for Statisticians, Volume 2: Table 51.* Cambridge, UK: Cambridge University Press.

Computation of Correspondence Analysis

In this appendix the computation of CA is illustrated using the object-oriented computing language R, which can be freely downloaded from the website:

http://www.r-project.org

We assume here that the reader has some basic knowledge of this language, which has become the *de facto* standard for statistical computing. If not, the above website gives many resources for learning it. In this chapter various computational issues are discussed and commented on, illustrated with R code. The R scripts for creating the analyses and graphics in this book are given at the website of the CARME network:

http://www.carme-n.org

Contents

The R system provides all the tools necessary to produce CA maps. These tools are encapsulated in R *functions*, and several functions and related material can be gathered together to form an R *package*. A package called **ca** (by Nenadić and Greenacre, see Bibliographical Appendix, page 288) is already available for doing the various types of CA described in this book, to be demonstrated later in this appendix. But before that, we show step-by-step how to perform various computations using R. The three-dimensional graphics package **rgl** will also be demonstrated in the process. In the following a `Courier font` is used for all R commands, which are *slanted*, and R output, which is `upright`. For example, here we create the matrix (13.2) on page 99, calculate its singular value decomposition (SVD) and store it in an R object, and then ask for the part of the object labelled d, which contains the singular values:

The R *program*

```
> table.T   <- matrix(c(8,5,-2,2,4,2,0,-3,3,6,2,3,3,3,-3,-6,
+                       -6,-4,1,-1,-2), nrow=5)
> table.SVD <- svd(table.T)
> table.SVD$d
[1] 1.412505e+01 9.822577e+00 1.376116e-15 7.435554e-32
```

The > is a "prompt" that indicates the start of an R command but is not typed manually. Similarly, a + at the start of the line indicates the continuation of a command and is also not manually entered.

Entering data into R has its peculiarities, but once you have managed to do it, the rest is easy! The `read.table()` function is one of the most useful ways to input data matrices, and the easiest data sources are a text file or an Excel file. For example, suppose we want to input the 5×3 data table on readership given in Exhibit 3.1. Here are three options for reading it in.

1. Suppose the data are in a text file as follows:

	C1	C2	C3
E1	5	7	2
E2	18	46	20
E3	19	29	39
E4	12	40	49
E5	3	7	16

   and suppose the file is called `reader.txt` and stored in the present R working directory. Then, executing the following command in R,

   ```
   > tab <- read.table("reader.txt")
   ```

   results in the table being stored as an R "data frame" object with the name `tab`.

2. Data can be read from an Excel file, as well as other data formats, e.g. Stata, Minitab, SPSS, SAS, Systat and DBF, using the R package **foreign**.

3. A convenient alternative is to copy the file into the clipboard (assuming a Windows platform) by selecting the contents in the text or word processor and copying using the pull-down Edit menu by or right-clicking the mouse, or the Ctrl+C combination. Then use function `read.table()` to read directly from the clipboard:

   ```
   > tab <- read.table("clipboard")
   ```

   This works just as well if the data are in an Excel file, as displayed below. The cells of this table are selected and then copied and the same command `read.table("clipboard")` is performed. On a Macintosh, replace `"clipboard"` by `pipe("pbpaste")`. Notice that the success of this `read.table()` command relies on the fact that the first line of the copied table contains one less entity than the other lines — this is why there is an empty cell in the top left-hand corner of the Excel table, similarly in the text file. If the `read.table()` function finds one less entity in the first line, it assumes that the first line consists of column labels and the subsequent lines have the row labels in the first column. The contents of `table` can be seen by entering

```
> tab
   C1 C2 C3
E1  5  7  2
E2 18 46 20
E3 19 29 39
E4 12 40 49
E5  3  7 16
```

The object includes the row and column names, which can be accessed by typing `rownames(tab)` and `colnames(tab)`, for example:

```
> rownames(tab)
[1] "E1" "E2" "E3" "E4" "E5"
```

Let's start ambitiously with the most spectacular use of R graphics: three-dimensional graphics that you can spin around on your computer screen. Suppose that the profile data of Exhibit 2.1 are input as described before and stored in the data frame `profs` in R; i.e. after entering the data, the object looks like this:

Some examples of R *code*

```
> profs
              Holidays HalfDays FullDays
Norway           0.333    0.056    0.611
Canada           0.067    0.200    0.733
Greece           0.138    0.862    0.000
France/Germany   0.083    0.083    0.833
```

(notice that the row and column names have to be written without blanks, otherwise the data will not be read correctly). A three-dimensional view of these profiles can be achieved using the contributed **rgl** package, but this package first needs to be installed and loaded into the R session, either using the pull-down menu in R (*Packages* → *Install packages*) or with the following R commands:

```
> install.packages("rgl")
> library(rgl)
```

Then the three-dimensional view is obtained using the following commands, producing the views in Exhibit B.1:

Three-dimensional graphics

```
> open3d()
> lines3d(c(0,1.2), c(0,0), c(0,0))
> lines3d(c(0,0), c(0,1.2), c(0,0))
```

```
> lines3d(c(0,0), c(0,0), c(0,1.2))
> lines3d(c(0,0), c(0,1), c(1,0), lwd=2)
> lines3d(c(0,1), c(1,0), c(0,0), lwd=2)
> lines3d(c(0,1), c(0,0), c(1,0), lwd=2)
> points3d(profs, size=6)
> texts3d(profs, text=rownames(profs), adj=c(0,-0.3), font=2)
> texts3d(rbind(c(1.25,0,0), c(0,1.25,0), c(0,0,1.25)),
+          text=colnames(profs))
```

Exhibit B.1:
Three-dimensional views of the country row profiles of the travel data set, using the R package **rgl**: *(a) a particular view after rotating with the mouse; (b) another view looking flat onto the triangular profile space.*

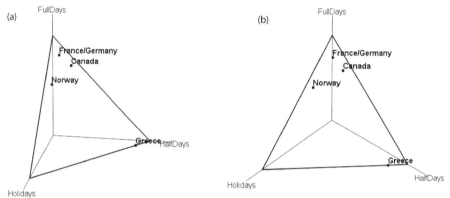

The three-dimensional scatterplot can be rotated at will by pressing the left-hand mouse button and moving the mouse around — the view in Exhibit B.1(b) shows the triangle linking the unit points "flat on", and the country points arranged in the triangle. The mouse wheel allows zooming into the display.

Chi-square statistic, inertia and distances

To have some basic exercise with useful R functions, let's compute a few basic CA statistics: first, the chi-square (χ^2) statistic and inertia, and then the χ^2-distance between profiles, using the readership data in `tab` (see pages 256–257). The χ^2 statistic is computed as follows, where `tab.exp` is the matrix of expected frequencies using the marginal sums of the table:

```
> tab.rowsum <- apply(tab, 1, sum)
> tab.colsum <- apply(tab, 2, sum)
> tab.sum    <- sum(tab)
> tab.exp    <- tab.rowsum %*% t(tab.colsum) / tab.sum
> tab.chi2   <- sum((tab - tab.exp)^2 / tab.exp)
> tab.chi2
[1] 25.97724
> tab.chi2 / tab.sum
[1] 0.08326039
```

In the above code, the matrix multiplication operator `%*%` and the transpose function `t()` are used. The χ^2 statistic (25.98) divided by the sum of the table (312) is the total inertia (0.0833).

To compute the χ^2-distances we first use a `for` loop and then more compactly the function `dist()`. For example, to compute the χ^2-distance between the first two rows of `tab`, we need the profiles and average profiles (column masses), and then a sum is made over the elements of the profiles:

```
> tab           <- as.matrix(tab)
> tab.pro       <- tab / apply(tab, 1, sum)
> tab.colmass <- apply(tab, 2, sum) / sum(tab)
> chidist       <- 0
> for(j in 1:ncol(tab))
+  chidist <- chidist +
+  (tab.pro[1,j] - tab.pro[2,j])^2 / tab.colmass[j]
> sqrt(chidist)
       C1
0.3737004
```

The label `C1` is given to the value of `chidist` because this is the first column of the loop. Notice the use of the `apply()` function, which computes all the sums over the first or second indices (i.e. rows or columns) of `tab`.

A more elegant way is to compute all inter-row distances in one step. The columns of the profile matrix `tab.pro` need to be normalized by dividing by the square roots of the respective column masses, then Euclidean distances can be computed between rows. All this can be achieved in a single command:

```
> dist(tab.pro %*% diag(1/sqrt(tab.colmass)))
          E1        E2        E3        E4
E2 0.3737004
E3 0.6352512 0.4696153
E4 0.7919425 0.5065568 0.2591401
E5 1.0008054 0.7703644 0.3703568 0.2845283
```

Computing
χ^2-*distances*
between all
profiles, using
`dist`

The function `dist()` computes by default the Euclidean distances between the rows of the matrix argument. The columns of `tab.pro` are first normalized by post-multiplying by the diagonal matrix, using function `diag()`, of the inverse square roots of the column masses, leading to χ^2-distances. Notice the triangular structure of the distance object resulting from `dist()`.

Let's now compute a complete CA "by hand", using the algorithm described in Appendix A, specifically the basic objects (A.1) to (A.3) and then the computational steps (A.4) to (A.10).

All the computing
steps of CA

```
### Correspondence matrix
> tab.P <- as.matrix(tab) / sum(tab)

### Row and column masses
> tab.r <- apply(tab.P, 1, sum)
> tab.c <- apply(tab.P, 2, sum)

### CA Step 1: the matrix S
> tab.S <- diag(1/sqrt(tab.r)) %*% (tab.P - tab.r %*% t(tab.c))
+                          %*% diag(1/sqrt(tab.c))
```

```
### CA Step 2: the SVD of S
> tab.svd <- svd(tab.S)
### CA Steps 3 & 4: standard row and column coordinates
> tab.rsc <- diag(1/sqrt(tab.r)) %*% tab.svd$u
> tab.csc <- diag(1/sqrt(tab.c)) %*% tab.svd$v
### CA Steps 5 & 6: principal row and column coordinates
> tab.rpc <- tab.rsc %*% diag(tab.svd$d)
> tab.cpc <- tab.csc %*% diag(tab.svd$d)
### CA Step 7: principal inertias (eigenvalues) and %s
> tab.svd$d^2; round(100 * tab.svd$d^2 / sum(tab.svd$d^2), 2)
[1] 7.036859e-02 1.289180e-02 6.222235e-34
[1] 84.52 15.48  0.00
```

Three eigenvalues are given but only the first two (0.07037 and 0.01289) are nonzero (the dimensionality of CA is one less than the number of rows or columns, whichever is smallest). The contribution coordinates are simply the left and right singular vectors of the "S" matrix `tab.S`, i.e. `tab.svd$u` and `tab.svd$v`. The above R commands constitute the whole basic CA algorithm — simply replace `tab` with any other data matrix to compute the solution.

Plotting the computed CA coordinates

It only remains to plot the results. We use the principal row and column coordinates to make the CA map of Exhibit B.2, and leave it up to the reader to study the various options used in this sequence of commands:

```
### Plot the row and column principal coordinates
> par(mar=c(4.2,4,1,1), mgp=c(2,0.5,0), cex.axis=0.8, font.lab=2)
> plot(rbind(tab.rpc, tab.cpc), type="n", asp=1,
+       xlab="CA dimension 1", ylab="CA dimension 2")
> abline(h=0, v=0, lty=2, col="gray")
> text(tab.rpc, labels=rownames(tab), font=2)
> text(tab.cpc, labels=colnames(tab), font=4, col="gray")
```

Exhibit B.2:
Symmetric CA map of the readership data. Since the data matrix has only three columns, 100% of the inertia is displayed: 84.5% and 15.5% respectively on the two dimensions.

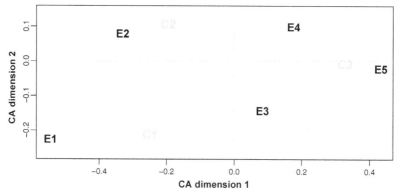

The most important option in the `plot()` function above is `asp=1`, which ensures that the scales on the two dimensions are the same, i.e. *aspect ratio* = 1.

The **ca** package in R is a comprehensive package to compute the various forms of CA, to be illustrated in the following. All the work of the previous section can be executed with one simple statement: `ca(tab)`, resulting in a `ca` object, with a list of components. The plotting is just as simple, wrapping the generic `plot()` function around the object: `plot(ca(tab))` will give Exhibit B.2, only differing in the format. Like **rgl**, this package needs to be installed and loaded into the session. The "smoking" data of Chapter 9 are included in the package, as well as the "author" data of Chapter 10. To illustrate simple CA, the first data set will be used, which can be loaded simply by issuing the command

```
> data(smoke)
```

giving the data frame smoke:

```
> smoke
   none light medium heavy
SM    4     2      3     2
JM    4     3      7     4
SE   25    10     12     4
JE   18    24     33    13
SC   10     6      7     2
```

The CA of the smoking data is obtained easily by saying `ca(smoke)`, giving some minimal default output:

```
> ca(smoke)

Principal inertias (eigenvalues):
             1        2        3
Value     0.074759 0.010017 0.000414
Percentage 87.76%   11.76%   0.49%

Rows:
               SM        JM        SE        JE        SC
Mass      0.056995  0.093264  0.264249  0.455959  0.129534
ChiDist   0.216559  0.356921  0.380779  0.240025  0.216169
Inertia   0.002673  0.011881  0.038314  0.026269  0.006053
Dim. 1   -0.240539  0.947105 -1.391973  0.851989 -0.735456
Dim. 2   -1.935708 -2.430958 -0.106508  0.576944  0.788435

Columns:
              none     light    medium     heavy
Mass      0.316062 0.233161 0.321244  0.129534
ChiDist   0.394490 0.173996 0.198127  0.355109
Inertia   0.049186 0.007059 0.012610  0.016335
Dim. 1   -1.438471 0.363746 0.718017  1.074445
Dim. 2   -0.304659 1.409433 0.073528 -1.975960
```

Several numerical results are listed which should be familiar: the principal inertias and their percentages, and then for each row and column the mass, χ^2-distance to the centroid, the inertia, and the standard coordinates on the first two dimensions. The features of this package will be described in much more detail later on, but just to show the graphical output, simply wrap the plot() function around ca(smoke) to get the default symmetric CA map in Exhibit B.3:

> *plot(ca(smoke))*

Exhibit B.3:
*Symmetric map of
the data set* smoke,
using the ca
package.

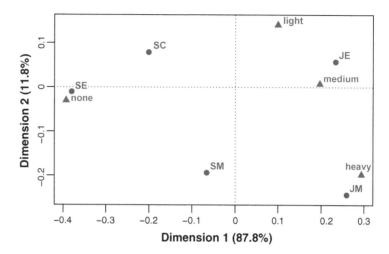

Notice that both principal axes have been inverted compared to the map of Exhibit 9.5 — this easily happens with different software, and is of no consequence (below we show how one can access the coordinates if a different orientation of an axis is preferred). To obtain the asymmetric maps, add the option map="rowprincipal" or map="colprincipal" to the plot() function; for example, Exhibit 9.2 (also with axes inverted) is obtained with the following command:

> *plot(ca(smoke), map="rowprincipal")*

*Numerical results
of CA: inertias
and contributions*
The complete set of the numerical results can be obtained using the summary() function around ca(smoke): summary(ca(smoke)). Here the summary results are given of the funding example of Chapter 11, where the CA numerical diagnostics were explained, assuming that the table has been read into fund:

> *summary(ca(fund))*

```
Principal inertias (eigenvalues):

dim    value      %     cum%  scree plot
1      0.039117  47.2   47.2  ************************
2      0.030381  36.7   83.9  ******************
3      0.010869  13.1   97.0  ******
4      0.002512   3.0  100.0
       -------- -----
Total: 0.082879 100.0

Rows:
        name   mass  qlt  inr |  k=1 cor ctr |  k=2 cor ctr |
1  |  Gel  |   107  916  137 |   76  55  16 |  303 861 322 |
2  |  Bic  |    36  881  119 |  180 119  30 | -455 762 248 |
3  |  Chm  |   163  644   21 |   38 134   6 |   73 510  29 |
```

```
 4 |  Zol |  151  929  230 |  -327 846 413 |   102  83  52 |
 5 |  Phy |  143  886  196 |   316 880 365 |    27   6   3 |
 6 |  Eng |  111  870  152 |  -117 121  39 |  -292 749 310 |
 7 |  Mcr |   46  680   10 |    13   9   0 |  -110 671  18 |
 8 |  Bot |  108  654   67 |  -179 625  88 |   -39  29   5 |
 9 |  Stt |   36  561   12 |   125 554  14 |    14   7   0 |
10 |  Mth |   98  319   56 |   107 240  29 |   -61  79  12 |
Columns:
      name   mass  qlt  inr |  k=1  cor  ctr |   k=2  cor  ctr |
 1 |     A |   39  587  187 |  478  574 228 |    72  13   7 |
 2 |     B |  161  816  110 |  127  286  67 |   173 531 159 |
 3 |     C |  389  465   94 |   83  341  68 |    50 124  32 |
 4 |     D |  162  968  347 | -390  859 632 |   139 109 103 |
 5 |     E |  249  990  262 |  -32   12   6 |  -292 978 699 |
```

In the first part of the summary output, the principal inertias are given, as well as their percentages of the total, cumulative percentages and a bar chart called a *scree plot* (see Exhibit 11.3). In the second part there are two tables for the rows and columns respectively, in identical formats. By default, the results for two dimensions are given; if more are required, for example 4, the option nd=4 should be added to the ca() function: ca(fund, nd=4). All quantities are multiplied by 1000 to make the output more readable:

- mass: masses ($\times$1000) of the respective row and column points;
- qlt: quality of representation (out of 1000) of the point in the solution of chosen dimensionality, in this case two-dimensional (see Exhibit 11.8);
- inr: part of total inertia (out of 1000) of the point in the full space of the rows or columns (see permill values in Exhibit 11.1);
- k=1 and k=2: principal coordinates on first two dimensions, multiplied by 1000 (once again, the axes have been reversed compared to those in Exhibits 10.2 and 10.3 — do a plot(ca(fund)) to see for yourself);
- cor: relative contributions (out of 1000) of each dimension to the inertia of individual points (see Exhibit 11.6) — these are also interpreted as squared correlations ($\times$1000);
- ctr: contributions (out of 1000) of each point to the principal inertia of a dimension (see Exhibit 11.4 for contributions to the first dimension).

Adding a supplementary row or column to an existing plot "by hand" is achieved using the barycentric relationship between standard coordinates of the column points, say, and the principal coordinates of the row points; in other words, profiles lie at weighted averages of vertices. The example at the top of page 94 shows how to situate the supplementary point *Museums*, which has data vector [4 12 11 19 7], summing up to 53. Calculating its profile and then the profile's scalar products with the standard column coordinates gives the supplementary point's coordinates in the map (the standard coordinates are in the colcoord component of the ca object):

Supplementary profiles

```
> fund.pro <- c(4,12,11,19,7)/53
> fund.csc <- ca(fund)$colcoord
```

```
> t(fund.pro) %*% fund.csc
            [,1]      [,2]
[1,] -0.3143203 0.3809511
```

(the sign of the second axis is again reversed in this solution compared to Exhibit 12.2).

Of course, the `ca` function has an option for supplementary points: either `suprow` for rows or `supcol` for columns. For example, if the matrix `fund` included an 11th row with the *Museums* data, then the analysis of the active part of the table (rows 1 to 10), along with row 11 as a supplementary point, would be achieved using `ca(fund, suprow=11)`.

Supplementary continuous variables

Adding a continuous supplementary variable involves performing a weighted linear regression of the variable on the standard coordinates of the existing solution. As an example, consider the row sums of the `fund` data set as a possible variable to relate to the dimensions. Notice this is not a supplementary point, which would define a profile that is exactly at the centre of the display. The objective is to see if the number of researchers applying for funding is related to the CA dimensions. This is performed as follows, using the logarithm of the count as the supplementary variable:

```
> fund.s   <- apply(fund, 1, sum)
> fund.rsc <- ca(fund)$rowcoord[,1:2]
> fund.lm  <- lm(log(fund.s) ~ fund.rsc, weights=fund.s)
> fund.lm$coefficients[2:3]
fund.rscDim1  fund.rscDim2
 -0.07654946    0.18589467
```

Notice that the regression includes the row masses as weights — here the row sums were equivalently used. The two coefficients would then be used to plot the supplementary variable, usually as an arrow from the centre. The second coefficient of the regression indicates a tendency of the disciplines with more applicants to receive funding more than average, but this effect is not significant, as can be seen if one looks at a summary of the regression, using `summary(fund.lm)`. At the time of writing the **ca** package has no supplementary variable option, but this will be introduced in a future version.

Options in ca package

The **ca** package comprises functions for simple, multiple and joint CA with support for subset analyses and the inclusion of supplementary points. Furthermore, it offers functions for the graphical display of the results in two and three dimensions. The package is comprised of the following components:

- Simple CA:
  - Main computational function: `ca()`
  - Printing and summaries: `print.ca()` and `summary.ca()` (and `print.summary.ca()`)
  - Plotting: `plot.ca()` and `plot3d.ca()`
- Multiple CA and joint CA:
  - Main computational function: `mjca()`

— Printing and summaries: `print.mjca()` and `summary.mjca()`
(and `print.summary.mjca()`)
— Plotting: `plot.mjca()` and `plot3d.mjca()`
• Data sets:
— `smoke`, `author` and `wg93`

Note that the `print`, `summary`, `plot` and `plot3d` functions are generic and the suffixes `.ca` and `.mjca`, indicating the corresponding method, can be omitted. The package contains further functions, such as `iterate.mjca()` for the updating of the Burt matrix in JCA.

A list of all available objects that are returned by `ca` (i.e. the *values* of the `ca` object), is obtained with `names`, for example using the `smoke` data: *Output of `ca` function*

```
> names(ca(smoke))
 [1] "sv"         "nd"         "rownames"   "rowmass"    "rowdist"
 [6] "rowinertia" "rowcoord"   "rowsup"     "colnames"   "colmass"
[11] "coldist"    "colinertia" "colcoord"   "colsup"     "N"
[16] "call"
```

The output of `ca()` is structured as a list-object; for example, the row masses are obtained with

```
> ca(smoke)$rowmass
```

(for more details of the returned objects, see the help file obtained by `?ca` or equivalently `help(ca)`).

Optional arguments for the `ca()` function include an option for setting the dimensionality of the solution (`nd`), options for marking selected rows and/or columns as supplementary ones (`suprow` and `supcol` respectively — see page 264) and options for setting subset rows and/or columns (`subsetrow` and `subsetcol` respectively) for subset CA (see below).

In the case of supplementary points, an asterisk is appended to the variable names in the output; for example, the summary for the CA of the `smoke` data, where the `none` category (the first column) is treated as supplementary, is:

```
> summary(ca(smoke, supcol=1))
```

In the corresponding section of the output the following is given:

```
  ...
  Columns:
        name   mass  qlt  inr    k=1 cor  ctr    k=2 cor  ctr
  1 | (*)non | <NA>   55 <NA> |  292  39 <NA> | -187  16 <NA> |
  ...
```

showing that masses, inertias and contributions are "not applicable", but the `qlt` and `cor` values are still valid, since they do not depend on the masses.

A subset analysis is achieved using the option `subsetrow` or `subsetcol`. The following reproduces the subset analyses in Chapter 21, i.e. subsets of consonants and vowels in the `author` dataset (which is part of the **ca** package): *Subset analysis*

```
> data(author)
> vowels    <- c(1,5,9,15,21)
```

```
> consonants <- (1:26)[-vowels]
> summary(ca(author, subsetcol=consonants))


Principal inertias (eigenvalues):

  dim    value     %     cum%  scree plot
   1    0.007607  46.5   46.5  ************************
   2    0.003253  19.9   66.4  ***********
   3    0.001499   9.2   75.6  *****
   4    0.001234   7.5   83.1  ****
   :       :       :      :    :
        -------  -----
  Total: 0.01637 100.0


Rows:
        name  mass  qlt  inr |   k=1 cor ctr  |  k=2 cor ctr
  1  |  td( |   85   59   29 |    7   8   1  | -17  50   7 |
  2  |  d() |   80  360   37 |  -39 196  16  | -35 164  31 |
  3  |  lw( |   85  641   81 | -100 637 111  |   8   4   2 |
  4  |  ew( |   89  328   61 |   17  27   4  |  58 300  92 |
  :  |   :  |    :    :    : |    :   :   :  |   :   :   : |


Columns:
        name  mass  qlt  inr |   k=1 cor ctr  |  k=2 cor ctr
  1  |    b |   16  342   21 |  -86 341  15  |  -6   2   0 |
  2  |    c |   23  888   69 | -186 699 104  | -97 189  66 |
  3  |    d |   46  892  101 |  168 783 171  | -63 110  56 |
  4  |    f |   19  558   33 | -113 467  33  | -50  91  15 |
  :  |    : |    :    :    : |    :   :   :  |   :   :   : |
```

Similarly, the summary of vowels analysis is obtained using `subsetcol=vowels`.

Visualization options in the **ca** *package*   The graphical representation of CA and MCA solutions is often done using *symmetric* maps, and this is the default option in the `plot()` function (`map="symmetric"`). The complete set of `map` options is as follows:

—   `"symmetric"`   Rows and columns in principal coordinates (default), i.e. inertia of points equal to principal inertia (eigenvalue, or square of singular value)

—   `"rowprincipal"`   Rows in principal, columns in standard coordinates

—   `"colprincipal"`   Columns in principal, rows in standard coordinates

—   `"symbiplot"`   Row and column coordinates scaled to have inertias equal to the singular values

—   `"rowgab"`   Rows in principal coordinates, columns in standard coordinates times masses (according to a proposal by Gabriel)

—   `"colgab"`   Columns in principal coordinates, rows in standard coordinates times masses

— "rowgreen"   Rows in principal coordinates, columns in standard
coordinates times square root of masses
(i.e. contribution coordinates, according to a
proposal by Greenacre; see Chapter 13,
Contribution Biplots)

— "colgreen"   Columns in principal coordinates, rows in standard
coordinates times square root of masses
(i.e. contribution coordinates)

For true biplots, it might be desired to use arrows from the origin to the points regarded as variables. For example, the following does a contribution biplot of the **author** data, and draws an arrow to each letter (column) point:

```
> data(author)
> plot(ca(author), map="rowgreen", arrows=c(FALSE, TRUE))
```

By default, supplementary points are added to the plot with a different symbol. The symbols can be defined with the **pch** option in **plot.ca()**. This option takes four values in the following order: plotting point character or symbol for (i) active rows, (ii) supplementary rows, (iii) active columns and (iv) supplementary columns. The set of **pch** characters can be obtained using the **ca** package function **pchlist()**, with no arguments. As a general rule, options that contain entries for rows and for columns contain the entries for the rows first and then those for the columns. For example, the colour of the symbols and labels are specified with the **col** and **col.lab** options; by default they are **col=c("blue", "red")** and **col.lab=c("blue", "red")**, blue for rows and red for columns, but can be reset by the user.

The option **what** controls the content of the plot. It can be set to "all", "active", "passive" or "none" for the rows and for the columns. For example, a plot of only the active (i.e. excluding supplementary) points is created by using **what=c("active", "active")**.

In addition to the **map** scaling options, various options allow certain values to be added to the plot as graphical attributes. The option **mass** selects if the masses of rows or columns should be indicated by the size of the point. Similarly, relative or absolute contributions can be indicated by the colour intensity in the plot by using the **contrib** option.

The **plot.ca()** function (generically **plot()** of a **ca** object) conveniently has a value with components **row** and **col** where the coordinates of the chosen joint plot can be obtained. For example,

```
> plot.ca.author <- plot(ca(author), map="rowgreen",
+                        arrows=c(FALSE,TRUE))
```

will make the plot and then the row and column coordinates of the biplot are in the components **plot.ca.author$row** and **plot.ca.author$col**.

The option **dim** selects which dimensions to plot, the default being **dim=c(1,2)**, i.e. the first two dimensions are plotted, first dimension horizontally, second vertically. A plot of the second and third dimensions, for example, is obtained by setting **dim=c(2,3)**. Another possibility for adding the third dimension to the plot is given with the functions `plot3d.ca()` and `plot3d.mjca()`. These two functions rely on the **rgl** package for three-dimensional graphics in R (see pages 257–258). Their structure is kept similar to their counterparts for two dimensions; for example,

```
> plot3d(ca(smoke, nd=3))
```

creates a three-dimensional display, shown in Exhibit B.4.

Exhibit B.4:
Three-dimensional display of a simple CA (compare with two-dimensional map in Exhibit B.3).

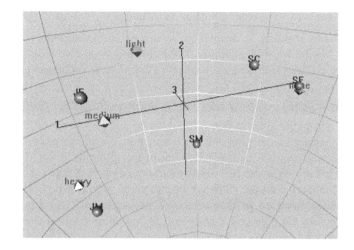

This display can be rotated using the left mouse button. The mouse wheel or the right button allows zooming in and out of the display, and holding the wheel while moving the mouse adds more or less perspective to the view.

MCA in **ca** *package*

MCA and JCA are performed with the function `mjca()`. The structure of the function is kept similar to its counterpart from simple CA. The two most striking differences are the format of the input data, which is a response pattern matrix where the rows are the individual cases and the columns the categorical variables. Within the function, the response pattern matrix is analysed as an indicator matrix or a Burt matrix, depending on the type of analysis. Options for the columns only are given, for example supplementary columns (`supcol`) or subsets of categories (`subsetcat`). The "approach" to MCA is specified by the `lambda` option in `mjca()`:

- `lambda="indicator"`: Analysis based on a simple CA of the indicator matrix
- `lambda="Burt"`: Analysis based on a simple CA of the Burt matrix
- `lambda="adjusted"`: Analysis based on the Burt analysis with an adjustment of inertias (default option)
- `lambda="JCA"`: Joint correspondence analysis, iterating to optimize fit to the off-diagonal tables of the Burt matrix.

By default, function `mjca()` performs an adjusted analysis, i.e. the option `lambda="adjusted"`, which is the solution of the Burt analysis where the dimensions are rescaled to optimize fit to the off-diagonal tables. For JCA (`lambda="JCA"`), the diagonal tables of the Burt matrix are updated iteratively by weighted least squares, using the internal function `iterate.mjca()`. This updating function has two convergence criteria, namely `epsilon` and `maxit`. Option `epsilon` sets a convergence criterion by means of maximum absolute difference of the Burt matrix in an iteration step compared to the Burt matrix of the previous step. The maximum number of iterations is given by the option `maxit`. The program iterates until any one of the two conditions is satisfied. Setting one option to `NA` results in ignoring that criterion; for example, exactly 50 iterations without considering convergence are performed with `maxit=50` and `epsilon=NA`.

As with simple CA, the solution is restricted by the **nd** option to two dimensions. However, eigenvalues are given for all possible dimensions, which number $(J - Q)$ for the "indicator" and "Burt" versions of MCA. In the case of an adjusted analysis or a JCA, the eigenvalues are given only for those dimensions k, where the singular values from the Burt matrix λ_k (i.e. the principal inertias, or eigenvalues, of the indicator matrix) satisfy the condition $\lambda_k > 1/Q$.

Now suppose the raw data from the data set on working women is in an Excel file as shown below: four questions from `Q1` to `Q4`, country (`C`), gender (`G`), age (`A`), marital status (`M`) and education (`E`). To input the data into R, copy the

Preparation of multivariate categorical data

columns to the clipboard as before, using function `read.table()` (see page 256). But now the table does not have row names, and so no blank in the top left-hand cell; hence the option `header=TRUE` needs to be specified:

```
> women <- read.table("clipboard", header=TRUE)
```

The column names of data frame `women` are obtained using function `colnames`:

```
> colnames(women)
 [1] "Q1" "Q2" "Q3" "Q4" "C"  "G"  "A"  "M"  "E"
```

To obtain the table in Exhibit 16.4:

```
> table(women[,"C"], women[,"Q3"])
```

Notice how the columns of women can be accessed by name or by number. In order to facilitate referencing the columns, it is convenient to use the attach() function, which allows all the column names to be available as if they were regular object names (to make these names unavailable the inverse operation detach() should be used). So the following is equivalent to obtain Exhibit 16.4:

```
> attach(women)
> table(C, Q3)
     Q3
C       1    2    3    4
   1  256 1156  176  191
   2  101 1394  581  248
   3  278  691   62   66
   4  161  646   70  107
   .    .    .    .    .

   .    .    .    .    .
  21  243  448  484   25
  22  468  664   92   63
  23  203  671  313  120
  24  738 1012  514  230
```

(cf. Exhibit 16.4).

There are a few missing values for gender, which has missing value code of 9, so these should be removed, or missing values in the gender column number 6 can be assigned R's missing value code NA, as follows:

```
> women[,6][women[,6]==9] <- NA
```

To get the interactively coded country–gender variable and the frequencies on which the percentages in Exhibit 16.6 are computed:

```
> CG <- 2 * (C - 1) + G
> table(CG, Q3)
     Q3
CG     1   2   3   4
   1 117 596 114  82
   2 138 559  60 109
   3  43 675 357 123
   4  58 719 224 125
   .   .   .   .   .

   .   .   .   .   .
  47 347 445 294 111
  48 390 566 218 118
```

The combinations of CG and A are coded as follows in order to construct the variable with 288 categories that interactively codes country, gender and age group, used for the analysis of Exhibit 16.8:

```
> CGA <- 6 * (CG - 1) + A
```

To stack or concatenate tables, the functions `rbind()` and `cbind()` enable   *Stacked tables*
binding rows or columns together. The five cross-tabulations corresponding
to Question 3 depicted in Exhibit 17.1 can be built up as follows in a `for`
loop:

```
> women.stack <- table(C, Q3)
> for(j in 6:9) {
+   women.stack <- rbind(women.stack, table(women[,j], Q3))
+ }
```

The contents of `women.stack` will contain several rows corresponding to missing data codes for the demographic variables marital status and education
(assuming that missing gender values have already been changed to NAs).
These would have to be omitted before the CA is performed, which can be
done in three different ways: (i) by deleting these rows from the stacked matrix; (ii) by changing the missing value codes to NAs as described on the
previous page; or (iii) by declaring the missing rows outside the subset of
interest in a subset CA as described in Chapter 21 (this is possibly the best
option, since it keeps the sample size in each table the same).

To build up the table in Exhibit 17.3, the stacked tables for each of the
four questions (each with five tables) are bound columnwise using `cbind`. An   *Extracting stacked*
alternative way is to use the `mjca()` function to generate the Burt matrix,   *table from Burt*
which is the matrix of all two-way tables, and then select the set of stacked   *matrix*
tables directly from the square Burt matrix. First look at the row names (same
as the column names) of this matrix:

```
> rownames(mjca(women)$Burt)
 [1] "Q1:1" "Q1:2" "Q1:3" "Q1:4" "Q2:1" "Q2:2" "Q2:3" "Q2:4" "Q3:1" "Q3:2"
[11] "Q3:3" "Q3:4" "Q4:1" "Q4:2" "Q4:3" "Q4:4" "C:1"  "C:2"  "C:3"  "C:4"
[21] "C:5"  "C:6"  "C:7"  "C:8"  "C:9"  "C:10" "C:11" "C:12" "C:13" "C:14"
[31] "C:15" "C:16" "C:17" "C:18" "C:19" "C:20" "C:21" "C:22" "C:23" "C:24"
[41] "G:1"  "G:2"  "A:1"  "A:2"  "A:3"  "A:4"  "A:5"  "A:6"  "M:1"  "M:2"
[51] "M:3"  "M:4"  "M:5"  "M:9"  "E:0"  "E:1"  "E:2"  "E:3"  "E:4"  "E:5"
[61] "E:6"  "E:7"  "E:98" "E:99"
```

The required submatrix consists of the first 16 columns, from `Q1:1` to `Q4:4`
and rows 17 onwards, but dropping the missing value categories `M:9`, `E:98`
and `E:99` (rows 54, 63 and 64):

```
> women.stack <- mjca(women)$Burt[-c(1:16, 54,63,64),1:16]
> women.stack
     Q1:1 Q1:2 Q1:3 Q1:4 Q2:1 Q2:2 Q2:3 Q2:4 Q3:1 ...
C:1  1353  215   33  178   63  491 1043  182  256 ...
C:2  1576  382   70  296   27  632 1431  234  101 ...
 :     :    :    :    :    :    :    :    :    :  ...
E:6  2888  543  137  305  420 1483 1619  351  969 ...
E:7  2858  388  101  329  516 1441 1325  394 1103 ...
```

One advantage of this latter way is that row and column labels are given
with the table, although the labels will preferably be changed to better labels
for the plots, using functions `rownames()` and `colnames()`, e.g. changing the
country labels to country codes, the gender codes to "M" and "F", etc.

Data preparation   In Chapter 18 the data on working women, for West and East German sam-
for MCA   ples, were analysed using the indicator and Burt versions of MCA. Assuming
that the `women` data frame (with 33590 rows) read previously is available and
"attached", the two German samples have country codes 2 and 3 repectively.
The part of the `women` corresponding to these two samples can be accessed
using a *logical* vector which we call `germany`:

```
> germany <- C == 2 | C == 3
> womenG  <- women[germany,]
```

The first command creates a vector of length 33590 with values `TRUE` corre-
sponding to the rows of the German samples, otherwise `FALSE` (the symbol |
is the logical "or"). The second command then passes only those rows with
`TRUE` values to the new data frame `womenG`. There are 3421 rows in `womenG`,
whereas the matrix analysed in Chapter 18 has 3418 rows — three cases that
have some missing demographic data have been eliminated (called *listwise
deletion* of cases with missing data). Variables gender and marital status have
missing value codes 9, whereas for education they are 98 and 99. The steps
needed to eliminate the missing rows use the same method as above to flag
the rows and then eliminate them, but be careful to first detach `women` and
then attach `womenG`, since they have the same column names:

```
> detach(women)
> attach(womenG)
```

Listwise deletion
of missing values
```
> missing <- G == 9 | M == 9 | E > 90
> womenG  <- womenG[!missing,]
> dim(womenG)
[1] 3418    9
```

Notice the negation `!missing` which changes the `FALSE` (is not missing) to
`TRUE` for the complete cases.

The indicator version of MCA for the first four columns (the four questions
on women working or staying at home) is obtained simply as follows:

MCA of indicator   `> mjca(womenG[,1:4], lambda="indicator")`
matrix
```
Eigenvalues:
                 1         2         3         4        5         6
Value     0.693361  0.513203  0.364697  0.307406  0.21761  0.181521
Percentage 23.11%    17.11%    12.16%    10.25%    7.25%    6.05%
                 7         8         9        10        11        12
Value     0.164774  0.142999  0.136322  0.113656  0.100483  0.063969
Percentage 5.49%     4.77%     4.54%     3.79%     3.35%     2.13%

Columns:
                  Q1:1      Q1:2      Q1:3      Q1:4      Q2:1      Q2:2
Mass          0.182929  0.034816  0.005778  0.026477  0.013239  0.095012 ...
ChiDist       0.605519  2.486096  6.501217  2.905510  4.228945  1.277206 ...
Inertia       0.067071  0.215184  0.244222  0.223523  0.236761  0.154988 ...
Dim. 1       -0.355941 -0.244454 -0.279167  2.841498 -0.696550 -0.428535 ...
Dim. 2       -0.402501  1.565682  3.971577 -0.144653 -2.116572 -0.800930 ...
```

and the Burt version:

```
> mjca(womenG[,1:4], lambda="Burt")
```

MCA of Burt
matrix

Eigenvalues:

	1	2	3	4	5	6
Value	0.480749	0.263377	0.133004	0.094498	0.047354	0.03295
Percentage	41.98%	23%	11.61%	8.25%	4.13%	2.88%
	7	8	9	10	11	12
Value	0.027151	0.020449	0.018584	0.012918	0.010097	0.004092
Percentage	2.37%	1.79%	1.62%	1.13%	0.88%	0.36%

Columns:

	Q1:1	Q1:2	Q1:3	Q1:4	Q2:1	Q2:2 ...
Mass	0.182929	0.034816	0.005778	0.026477	0.013239	0.095012 ...
ChiDist	0.374189	1.356308	3.632489	2.051660	2.354042	0.721971 ...
Inertia	0.025613	0.064046	0.076244	0.111452	0.073363	0.049524 ...
Dim. 1	0.355941	0.244454	0.279167	-2.841498	0.696550	0.428535 ...
Dim. 2	-0.402501	1.565682	3.971577	-0.144653	-2.116572	-0.800930 ...

Notice that the eigenvalues (principal inertias) of the Burt version are the squares of those of the indicator one. The standard coordinates are identical apart from possible sign changes. The total inertia can be computed in the two cases as the sum of squared singular values, as in the simple CA case:

```
> sum(mjca(womenG[,1:4], lambda="indicator")$sv^2)
[1] 3
```

```
> sum(mjca(womenG[,1:4], lambda="Burt")$sv^2)
[1] 1.145222
```

The contributions of each subtable of the Burt matrix to the total inertia is given in the component called `subinertia` of the `mjca` object, so the sum of these also gives the total inertia:

```
> sum(mjca(womenG[,1:4], lambda="Burt")$subinertia)
[1] 1.145222
```

Since the total inertia is the average of the 16 subtables, the inertias of individual subtables are 16 times the values in $subinertia:

```
> 16*mjca(womenG[,1:4], lambda="Burt")$subinertia
        [,1]      [,2]      [,3]      [,4]
[1,] 3.0000000 0.3657367 0.4261892 0.6457493
[2,] 0.3657367 3.0000000 0.8941517 0.3476508
[3,] 0.4261892 0.8941517 3.0000000 0.4822995
[4,] 0.6457493 0.3476508 0.4822995 3.0000000
```

To obtain the positions of the supplementary columns:

```
> summary(mjca(womenG, lambda="Burt", supcol=5:9))
```

Principal inertias (eigenvalues):

dim	value	%	cum%	scree plot
1	0.480749	42.0	42.0	************************
2	0.263377	23.0	65.0	**************
3	0.133004	11.6	76.6	*******
4	0.094498	8.3	84.8	*****

```
5      0.047354   4.1   89.0  **
:         :        :      :    :
       --------- -----
Total: 1.145222 100.0
```

Columns:

		name	mass	qlt	inr		k=1	cor	ctr		k=2	cor	ctr	
1	|	Q1:1 |	183	740	22 |		247	435	23 |		-207	305	30	|
2	|	Q1:2 |	35	367	56 |		169	16	2 |		804	351	85	|
3	|	Q1:3 |	6	318	67 |		194	3	0 |		2038	315	91	|
:		:	:	:	:		:	:	:		:	:	:	
17	|	(*)C:2 |	<NA>	283	<NA> |		-89	48	<NA> |		195	234	<NA>	|
18	|	(*)C:3 |	<NA>	474	<NA> |		188	81	<NA> |		-413	393	<NA>	|
19	|	(*)G:1 |	<NA>	26	<NA> |		-33	5	<NA> |		67	21	<NA>	|
20	|	(*)G:2 |	<NA>	24	<NA> |		34	5	<NA> |		-68	19	<NA>	|
21	|	(*)A:1 |	<NA>	41	<NA> |		-108	12	<NA> |		-170	29	<NA>	|
22	|	(*)A:2 |	<NA>	52	<NA> |		-14	0	<NA> |		-172	52	<NA>	|
:		:	:	:	:		:	:	:		:	:	:	

The supplementary categories are marked by a * and have no masses, inertia values (inr) nor contributions to the principal axes (ctr).

Adjusted MCA solution   To obtain the adjusted MCA solution, i.e. the same standard coordinates as in MCA but rescaled to be close to the optimal JCA solution, either use the lambda option "adjusted" or leave out this option since it is the default:

```
> summary(mjca(womenG[,1:4]))
```

Principal inertias (eigenvalues):

```
dim    value      %     cum%  scree plot
1      0.349456  66.3  66.3  *************************
2      0.123157  23.4  89.7  *********
3      0.023387   4.4  94.1  *
4      0.005859   1.1  95.2
       -------- -----
Total: 0.526963
```

Columns:

		name	mass	qlt	inr		k=1	cor	ctr		k=2	cor	ctr	
1	|	Q1:1 |	183	996	22 |		-210	687	23 |		-141	309	30	|
2	|	Q1:2 |	35	822	56 |		-145	53	2 |		549	769	85	|
3	|	Q1:3 |	6	562	67 |		-165	8	0 |		1394	554	91	|
4	|	Q1:4 |	26	1000	97 |		1680	1000	214 |		-51	0	1	|
5	|	Q2:1 |	13	505	64 |		-412	119	6 |		-743	387	59	|
6	|	Q2:2 |	95	947	43 |		-253	424	17 |		-281	522	61	|
:		:	:	:	:		:	:	:		:	:	:	

(again, it is of no consequence that the first dimension has now been inverted compared to previous solutions).

The adjusted total inertia, used to calculate the percentages above, is calculated just after (19.5) on page 149. The first two adjusted principal inertias (eigenvalues) are calculated just after (19.6) on page 149 (see also (A.35) and (A.36)).

To obtain the JCA of the same data as in Exhibit 19.3, use the `lambda = "JCA"` option, which gives the optimal fit to the off-diagonal subtables of the Burt matrix. In this case percentages of inertia are not given for individual axes, but only for the solution space as a whole (by default, two-dimensional) since the axes are not nested:

```
> summary(mjca(womenG[,1:4], lambda="JCA"))

Principal inertias (eigenvalues):

1       0.353452
2       0.128616
3       0.015652
4       0.003935
        --------
Total: 0.520617

Diagonal inertia discounted from eigenvalues: 0.125395
Percentage explained by JCA in 2 dimensions: 90.2%
(Eigenvalues are not nested)
[Iterations in JCA: 31 , epsilon = 9.33e-05]
Columns:
      name   mass  inr     k=1   k=2    cor ctr
1 | Q1:1 |   183   22 |    204 -129 |  962  24 |
2 | Q1:2 |    35   56 |    144  503 |  772  22 |
3 | Q1:3 |     6   67 |    163 1260 |  512  22 |
4 | Q1:4 |    26   97 |  -1637  -45 |  990 154 |
5 | Q2:1 |    13   64 |    394 -764 |  534  21 |
6 | Q2:2 |    95   43 |    250 -276 |  943  29 |
:     :     :    :      :     :      :   :
```

Joint correspondence analysis

Also notice the single set of contributions for the two-dimensional solution, not dimension by dimension: one squared correlation for each point with respect to the plane (i.e. quality) and one set of contributions to the inertia in two dimensions.

Furthermore, in the JCA solution, the "total" inertia is the inertia of the modified Burt matrix, which includes a part due to the modified diagonal blocks — this additional part is the "`Diagonal inertia discounted from eigenvalues: 0.125395`" which has to be subtracted from the total to get the total inertia due to the off-diagonal blocks. Since the solution requested is two-dimensional and fits the diagonal blocks exactly by construction, the first two eigenvalues also contain this additional part, which has to be discounted as well. The proportion of (off-diagonal) inertia explained is thus:

$$\frac{0.3534 + 0.1286 - 0.1254}{0.5206 - 0.1254} = 0.9024$$

i.e. the percentage of 90.2% reported above — see Theoretical Appendix, (A.32). The denominator above, the adjusted total $0.5206 - 0.1254 = 0.3952$, can be verified to be the same as:

$$\text{inertia of } \mathbf{B} - \frac{J-Q}{Q} = 1.1452 - \frac{12}{16} = 0.3952$$

Subset MCA does not involve defining a subset of variables (columns of the response pattern data matrix), but rather a subset of categories. A useful application is to the subset of substantive categories, excluding the missing value categories. Again, the default implementation adjusts the inertias to give percentages of explained inertia without the diagonal table subsets, and the percentages do not sum to 100%, as was the case for the adjusted analysis of the complete set of categories.

Subset MCA

```
> summary(mjca(womenG, subsetcol=(1:16)[-seq(4,16,4)]))

Principal inertias (eigenvalues):

   dim    value      %    cum%   scree plot
   1      0.123352  69.6  69.6   ********************
   2      0.023647  13.3  82.9   ****
   3      0.005831   3.3  86.2   *
   4      0.003651   2.1  88.3   *
          --------  -----
   Total: 0.177293

Columns:
        name    mass  qlt  inr    k=1  cor  ctr    k=2  cor  ctr
   1 | Q1:1 |   183   854   37 |   162  850   39 |   11    4    1 |
   2 | Q1:2 |    35   871   94 |  -531  799   79 | -159   72   37 |
   3 | Q1:3 |     6   811  111 | -1363  540   87 |  965  271  227 |
   4 | Q2:1 |    13   681  107 |   781  459   66 |  543  222  165 |
   5 | Q2:2 |    95   933   72 |   308  915   73 |  -43   18    7 |
   6 | Q2:3 |   120   880   65 |  -324  879  103 |  -10    1    1 |
   :     :     :     :     :       :    :    :      :    :    :
```

Alternatively, the Burt matrix can be computed and then the subset CA applied to that square part of the Burt matrix corresponding to the subset (see re-arranged Burt matrix in Exhibit 21.3).

```
> womenG.B <- mjca(womenG)$Burt
> subset   <- c(1:16)[-seq(4,16,4)]
> summary(ca(womenG.B[1:16,1:16], subsetrow=subset,
+                                 subsetcol=subset))
```

(Output not given here, but remember that the inertias will not be adjusted)

Another possibility is to select the demographic rows of the Burt matrix and the subset of substantive categories as columns:

```
> summary(ca(womenG.B[17:38,1:16], subsetcol=subset))
```

Analysis of matched matrices

When two matrices are matched row- and columnwise, an analysis of the sum (equivalently, average) and difference between the matrices can be achieved by setting up the two matrices in a block matrix format. This is illustrated here for the example of a mobility table in Chapter 24, a square asymmetric matrix which is used along with its transpose as the matched matrices. The matrix sum and matrix difference are then the symmetric and skew-symmetric parts. After reading the mobility table into a data frame named mob, the sequence of commands to set up the block matrix and then do the CA is as follows. Notice that mob has to be first converted to a matrix; otherwise we cannot bind the rows and columns together properly to create the block matrix mob2 (see (24.4)).

```
> mob  <- as.matrix(mob)
> mob2 <- rbind(cbind(mob,t(mob)), cbind(t(mob), mob))
> summary(ca(mob2))
```

Principal inertias (eigenvalues):

```
dim    value      %     cum%  scree plot
 1    0.388679  24.3   24.3  *************************
 2    0.232042  14.5   38.8  ***************
 3    0.158364   9.9   48.7  **********
 4    0.158364   9.9   58.6  **********
 5    0.143915   9.0   67.6  *********
 6    0.123757   7.7   75.4  ********
 7    0.081838   5.1   80.5  *****
 8    0.070740   4.4   84.9  *****
 9    0.049838   3.1   88.0  ***
10    0.041841   2.6   90.6  ***
11    0.041841   2.6   93.3  ***
12    0.022867   1.4   94.7  *
 :       :        :      :
26    0.000381   0.0  100.0
27    0.000147   0.0  100.0
     --------  -----
Total: 1.599080 100.0
```

Rows:

	name	mass	qlt	inr	k=1	cor	ctr	k=2	cor	ctr
1 |	Arm |	43	426	54 |	-632	200	44 |	671	226	84 |
2 |	Art |	55	886	100 |	1521	793	327 |	520	93	64 |
3 |	Tcc |	29	83	10 |	-195	73	3 |	73	10	1 |
: |	: |	:	:	: |	:	:	: |	:	:	:
15 |	ARM |	43	426	54 |	-632	200	44 |	671	226	84 |
16 |	ART |	55	886	100 |	1521	793	327 |	520	93	64 |
17 |	TCC |	29	83	10 |	-195	73	3 |	73	10	1 |
: |	: |	:	:	: |	:	:	: |	:	:	:

Columns:

	name	mass	qlt	inr	k=1	cor	ctr	k=2	cor	ctr
1 |	ARM |	43	426	54 |	-632	200	44 |	671	226	84 |
2 |	ART |	55	886	100 |	1521	793	327 |	520	93	64 |
3 |	TCC |	29	83	10 |	-195	73	3 |	73	10	1 |
: |	: |	:	:	: |	:	:	: |	:	:	:
15 |	Arm |	43	426	54 |	-632	200	44 |	671	226	84 |
16 |	Art |	55	886	100 |	1521	793	327 |	520	93	64 |
17 |	Tcc |	29	83	10 |	-195	73	3 |	73	10	1 |
: |	: |	:	:	: |	:	:	: |	:	:	:

The principal inertias coincide with Exhibit 22.4, and since the first two dimensions correspond to the symmetric part of the matrix, each set of coordinates is just a repeat of the same set of values.

Dimensions 3 and 4, with repeated principal inertias (eigenvalues), correspond to the skew-symmetric part and their coordinates turn out as follows (to get more than the default two dimensions in the summary, change the original command to `summary(ca(mob2, nd=4))`):

```
Rows:
        name    k=3 cor ctr    k=4 cor ctr
1  | Arm |   -11   0   0 |  416  87  47 |
2  | Art |    89   3   3 |  423  61  62 |
3  | Tcc |  -331 211  20 |  141  38   4 |
:  |  :  |    :   :   :  |   :   :   :
15 | ARM |    11   0   0 | -416  87  47 |
16 | ART |   -89   3   3 | -423  61  62 |
17 | TCC |   331 211  20 | -141  38   4 |
:  |  :  |    :   :   :  |   :   :   :

Columns:
        name    k=3 cor ctr     k=4 cor ctr
1  | ARM |  -416  87  47 |   -11   0   0 |
2  | ART |  -423  61  62 |    89   3   3 |
3  | TCC |  -141  38   4 |  -331 211  20 |
:  |  :  |    :   :   :  |    :   :   :
15 | Arm |   416  87  47 |    11   0   0 |
16 | Art |   423  61  62 |   -89   3   3 |
17 | Tcc |   141  38   4 |   331 211  20 |
:  |  :  |    :   :   :  |    :   :   :
```

which shows that the skew-symmetric coordinates reverse sign within the row and column blocks, but also swap over, with the third axis row solution equal to the fourth axis column solution and vice versa. Only one set of coordinates is needed to plot the objects in each map, but the interpretation of the maps is different, in terms of areas of triangles, as explained in Chapter 24.

Canonical correspondence analysis (CCA)   The **ca** package does not contain CCA, but it is available in the very comprehensive **vegan** package (see web resources in the Bibliographical Appendix). Not only does **vegan** have CCA but also CA, principal component analysis (PCA) and redundancy analysis (RDA), which is the constrained version of PCA. Some features of the **ca** package, such as supplementary points and contribution biplots, are not available in **vegan**. Since this package is usually used in an ecological context, like the example in Chapter 27, we shall use **vegan**'s terminology: "sites" (samples, the rows), "species" (the columns) and "variables" (additional columns used as "explanatory" variables that constrain the solution). Using **vegan** is just as easy as using **ca**: the main function is called `cca()` and can be used in either of the two following formats:

```
cca(X, Y, Z)
cca(X ~ Y + condition(Z))
```

where X is the sites×species matrix of counts, Y is the sites×variables matrix of explanatory data and Z is the sites×variables matrix of conditioning data for a partial CCA. The second format is in the form of a regression-type model formula. If only X is specified, the analysis is a CA — so try one of the previous analyses, for example summary(cca(author)) to compare the results with the **ca** package version summary(ca(author)). Notice that the books are referred to as "sites" and the letters as "species", and that the default plotting option in **vegan**, for example plot(cca(author)), is the same as the **ca** package plotting option map="colprincipal". If X and Y are specified, the analysis is a CCA. If X, Y and Z are specified, the analysis is a partial CCA.

Assuming now that the biological data of Chapters 10 and 27 are read into the data frame bio as a 13×92 table, and that the three variables Ba, Fe and PE are read into env as a 13 × 3 table whose columns are log-transformed to variables with names logBa, logFe and logPE; then the CCA can be performed simply as follows:

```
> summary(cca(bio, env))

Call:
cca(X = bio, Y = env)

Partitioning of mean squared contingency coefficient:

                Inertia Proportion
Total            0.7826     1.0000
Constrained      0.2798     0.3575
Unconstrained    0.5028     0.6425

Eigenvalues, and their contribution to the
          mean squared contingency coefficient

                       CCA1    CCA2    CCA3     CA1     CA2     CA3
Eigenvalue           0.1895  0.0615  0.0288  0.1909  0.1523  0.0416
Proportion Explained 0.2422  0.0786  0.0368  0.2439  0.1946  0.0531
Cumulative Proportion 0.2422 0.3208  0.3576  0.6014  0.7960  0.8492

                        CA4     CA5     CA6     CA7     CA8     CA9
Eigenvalue           0.0278  0.0254  0.0230  0.0165  0.0146  0.0108
Proportion Explained 0.0356  0.0324  0.0293  0.0211  0.0187  0.0137
Cumulative Proportion 0.8847 0.9171  0.9465  0.9676  0.9863  1.0000

Accumulated constrained eigenvalues
Importance of components:

                       CCA1    CCA2     CCA3
Eigenvalue           0.1895  0.0615  0.02879
Proportion Explained 0.6773  0.2198  0.10288
Cumulative Proportion 0.6773 0.8971  1.00000

Scaling 2 for species and site scores
--- Species are scaled proportional to eigenvalues
--- Sites are unscaled: weighted dispersion equal on all dimensions
```

Species scores

```
                CCA1       CCA2       CCA3       CA1       CA2       CA3
Gala.ocul   0.173239    0.24592   -0.07091   0.635963  -0.06348   0.031990
Chae.seto   0.574797   -0.27082    0.01181  -0.502916  -0.67421   0.093354
Amph.falc   0.295388   -0.11407    0.07598  -0.222414   0.04180  -0.005020
Myse.bide  -0.527109   -0.50526   -0.10398  -0.078991   0.17668  -0.484208
     :          :          :          :          :         :         :
```

Site scores (weighted averages of species scores)

```
         CCA1      CCA2       CCA3      CA1      CA2       CA3
S4    -0.1079    0.39195  -2.02902  -0.2161   0.1892  -0.33692
S8    -0.6697    0.59302   4.10104  -0.9267   1.1969  -0.04428
S9     0.7372   -1.37510  -0.65372  -0.6732  -0.8301   0.50189
S12   -0.7843   -0.39022   2.65081  -0.9392   1.8756  -0.64818
  :       :         :         :        :        :        :
```

Site constraints (linear combinations of constraining variables)

```
         CCA1      CCA2       CCA3      CA1      CA2       CA3
S4    -0.06973   0.75885  -2.29951  -0.2161   0.1892  -0.33692
S8    -0.35758   1.47282   2.27467  -0.9267   1.1969  -0.04428
S9     0.48483  -0.72459  -0.66547  -0.6732  -0.8301   0.50189
S12    0.02536   0.27129  -0.14677  -0.9392   1.8756  -0.64818
  :       :         :         :        :        :        :
```

Biplot scores for constraining variables

```
         CCA1      CCA2     CCA3   CA1 CA2 CA3
logBa  0.9957   -0.08413  0.03452   0   0   0
logFe  0.6044   -0.72088  0.33658   0   0   0
logPE  0.4654    0.55594  0.68710   0   0   0
```

Notice the following:

— the `mean squared contingency coefficient` is another name for the total inertia;

— the **constrained** inertia is the part of the total in the space of the explanatory variables, and the **unconstrained** inertia is the remainder, uncorrelated with the explanatory variables;

— the principal inertias (`eigenvalues`) in the constrained space are headed CCA1, CCA2, etc., and the principal inertias in the unconstrained space CA1, CA2, etc.; each set of values is in descending order and expressed relative to the total inertia;

— the `accumulated constrained eigenvalues` are those in the constrained space and expressed as proportions of the total in the constrained space;

— all other proportions are expressed relative to the total inertia;

— `scaling 2` is the default scaling where rows (sites) are in standard coordinates, and columns (species) in principal coordinates, i.e., the `"colprincipal"` scaling in the `plot.ca()` function; scaling 1 would be the other way round, i.e. the `"rowprincipal"` scaling;

— the `Species scores` are the column principal coordinates;

— the `Site constraints` are the row standard coordinates;

— the `Biplot scores for constraining variables` are the weighted correlation coefficients between the explanatory variables and the site coordinates.

Inference using resampling

For inferential purposes, methods such as multinomial sampling, bootstrapping and permutation testing are easily performed in R using various functions and packages. For example, the 100 replications of the `author` data, shown in the partial bootstrap CA map of Exhibit 29.1, are obtained by multinomial sampling using the function `rmultinom()` for each row of the table. The relative frequencies in each row define the multinomial distribution and each row's total defines the sample size. For example, if the row sums of the table are in object `author.rowsums`, then 100 replicate samples of the j-th row of the table are obtained by

```
> rmultinom(n=100, size=author.rowsum[j], prob=author[j,])
```

Notice that the probabilities do not have to be specifically computed and can be specified simply by giving the corresponding row. Points defining the convex hulls are obtained using the function `chull()`. In Exhibit 29.2, 1000 replications are made and then the points defining the convex hulls are removed to get the convex hulls of the remaining points, and this process of peeling continues until approximately 950 points (i.e. 95%) remain.

The Monte Carlo simulation of Exhibit 29.5 also makes use of function `rmultinom()`, using the expected relative frequencies of the whole table in Exhibit 29.4, under the null hypothesis of independence, as parameters of the multinomial distribution, with option `size` equal to 312 in the function. A total of 9999 tables with the same total of 312 are generated from this distribution and CA is applied to each table to obtain principal inertias that form the null distribution, against which the actual principal inertias are compared — see Exhibit 29.5.

Permutation testing and bootstrapping

For permutation tests and bootstrapping the key R function is `sample()`, which does sampling without replacement, to get sample permutations, or sampling with replacement (using option `replace=TRUE`), to obtain bootstrap samples. For example, the data of Exhibit 30.1 consist of two vectors of 25 elements each, say `values` and `groups` respectively, containing the numerical values and the group labels (T or C). A random permutation of the labels is obtained simply by

```
> group.perm <- sample(group)
```

and so, under the null hypothesis of no difference between the groups, the two group means and their difference are computed according to the labels in `group.perm`. This is repeated 9999 times to obtain the null distribution of the difference between means in Exhibit 30.2, against which the actual difference is judged.

Similarly, to generate the permutation distribution of a correlation between two variables, as in Exhibit 30.4, one variable's values are randomly permuted and the correlation recomputed. The same strategy is applied to obtain null distributions of statistics that quantify the relationship between two sets of variables, as in the CCA of Exhibit 30.5. In this case complete rows of one of the sets are permuted, not single values. For example, suppose the two sets of variables are in matrices `Y` and `X`, each with `n` rows. Then the rows of `X` are randomly permuted using the following:

```
> X.perm <- X[sample(1:nrow(X)),]
```

Here the row numbers are permuted, which induces the permutation of the rows of the matrix.

The **vegan** package incorporates this test, which can be obtained with the `anova()` function around the `cca()` analysis. For example, in Exhibit 30.5 the p-value corresponding to the observed proportion 0.2798 of explained inertia was estimated as 0.073. An estimate can be obtained in **vegan**, where `bio` and `env` are the matrices defined earlier on page 279, as follows:

```
> anova(cca(bio, env))
```

```
Permutation test for cca under reduced model

          Df ChiSquare      F Pr(>F)
Model      3   0.27984 1.6696  0.074 .
Residual   9   0.50281
---
Signif. codes:  0 *** 0.001 ** 0.01 * 0.05 . 0.1   1
```

The observed proportion 0.27984 of explained inertia is reported in the column `ChiSquare`, while the column `F` is a "pseudo-F" statistic, as if an analysis of variance is being performed (for more details, see the **vegan** documentation). Notice that the p-values are estimates and can vary from one permutation test to the next, depending on how many permutations are used. To improve the estimation of the p-value, simply increase the number of permutations — by default, the number of permutations in the above is 999 (see the help documentation for the function `anova.cca`). The same function can be used to test variables in a stepwise fashion, but the CCA has to be specified in the format of a formula, as follows:

```
> anova(cca(bio ~ env[,1] + env[,2] + env[,3]))
```

```
Permutation test for cca under reduced model
Terms added sequentially (first to last)

           Df ChiSquare      F Pr(>F)
env[, 1]    1   0.18842 3.3726  0.004 **
env[, 2]    1   0.05668 1.0145  0.488
env[, 3]    1   0.03474 0.6218  0.741
Residual    9   0.50281
---
Signif. codes:  0 *** 0.001 ** 0.01 * 0.05 . 0.1   1
```

The variables are entered in the order specified in the model formula, so here we see that the first variable (log of barium) enters with high significance, the others not, as was found in Chapter 30 (page 238).

Chapter 15 deals with Ward clustering of the row or column profiles, where the profiles are weighted by their masses. The R function for performing hierarchical clustering is `hclust()`, which does not allow differential weights in the option for Ward clustering (see formula (15.2) on page 120); neither does the function `agnes()` in the package **cluster**. In his book *Correspondence Analysis and Data Coding with Java and R* (see Bibliographical Appendix), Fionn Murtagh gives many R scripts for CA and especially data recoding, all of which are available on the website **www.correspondances.info**. In particular, on pages 21–26 he describes a program for hierarchical clustering by Ward's method, with incorporation of weights, which is exactly what was used in Chapter 15, but which is otherwise unavailable in

R. Assuming you have downloaded Murtagh's code from his website, and have read the table of data of Exhibit 15.3 as the data frame `food`, then the cluster analysis of the row profiles in Exhibit 15.5 can be achieved using his function `hierclust` as follows:

```
> food.rpro    <- food /apply(food,1,sum)         # row profiles
> food.r       <- apply(food,1,sum) / sum(food)    # row masses
> food.rclust <- hierclust(food.rpro, food.r)
> plot(as.dendrogram(food.rclust))
```

Producing a CA map with specific characteristics that is ready for publication is not a trivial task. Here three different technologies are shown that were used in this book to produce the graphical exhibits. These technologies coincide with the three successive editions of the book.

Graphical options

This book was typeset by the author in LaTeX. LaTeX itself and various LaTeX macros can produce maps directly, without passing through another graphics package. Most of the figures produced in the first part of the book were produced using the macro package PicTeX. As an example, the following code, which is embedded in the LaTeX text of the book itself, produced Exhibit 9.2, the asymmetric map of the smoking data:

LaTeX graphics

```
\beginpicture
\setcoordinatesystem units <2.5cm,2.5cm>
\setplotarea x from -2.40 to 1.70, y from -1.6 to 2.25
\accountingoff
\gray
\setdashes <5pt,4pt>
\putrule from 0 0 to  1.7  0
\putrule from 0 0 to -1.4  0
\putrule from 0 0 to   0 2.25
\putrule from 0 0 to   0 -1.6
\put {+} at 0 0
\black
\small
\put {Axis 1} [Br] <-.2cm,.15cm> at 1.70 0
\put {0.0748 (87.8\%)} [tr] <-.2cm,-.15cm> at 1.70 0
\put {Axis 2} [Br] <-.1cm,-.4cm> at 0 2.25
\put {0.0100 (11.8\%)} [Bl] <.1cm,-.4cm> at 0 2.25
\setsolid
\putrule from 1.3  -1.3  to 1.4  -1.3
\putrule from 1.3  -1.32 to 1.3  -1.28
\putrule from 1.4  -1.32 to 1.4  -1.28
\put {\it scale} [b] <0cm,.25cm> at 1.35 -1.3
\put {0.1} [t] <0cm,-.2cm> at  1.35 -1.3
\multiput {$\bullet$} at
 0.06577   0.19373
-0.25896   0.24330
 0.38059   0.01066
-0.23295  -0.05775
 0.20109  -0.07891
 /
```

```
\sf
\put {SM} [l]   <.15cm,0cm>  at   0.06577   0.19373
\put {JM} [r]   <-.15cm,0cm> at  -0.25896   0.24330
\put {SE} [bl]  <.15cm,0cm>  at   0.38059   0.01066
\put {JE} [r]   <-.15cm,0cm> at  -0.23295  -0.05775
\put {SC} [tl]  <.15cm,0cm>  at   0.20109  -0.07891
\gray
\multiput {$\circ$} at
 1.4384   0.3046
-0.3638  -1.4094
-0.7180  -0.0735
-1.0745   1.9760
 /
\sl
\put {none}   [b]   <0cm,.2cm>  at   1.4384   0.3046
\put {light}  [b]   <0cm,.2cm>  at  -0.3638  -1.4094
\put {medium} [T]   <0cm,-.3cm> at  -0.7180  -0.0735
\put {heavy}  [b]   <0cm,.2cm>  at  -1.0745   1.9760
\black
\endpicture
```

Comparing the above code with Exhibit 9.2 itself shows how each line and each set of characters is laboriously placed in the plotting area. One advantage of this approach, however, is that once you have set the units on the horizontal and vertical coordinate axes to be the same (2.5 cm per unit in the example above), then you are assured that the aspect ratio of 1 is perfectly preserved in the eventual result.

Excel graphics Many of the figures produced for the second edition were made in Excel using XLSTAT — see, for example, the maps in Chapters 17–19. XLSTAT is a commercial add-on for Excel, and contains a comprehensive set of statistical analyses, including many discussed in this book, as well as some graphical options not available in Excel. A certain amount of trimming of the maps was needed in the XLSTAT ouput, which is given conveniently in an Excel sheet: for example, redefining the maxima and minima on the axes, and also stretching the graph window vertically or horizontally until the aspect ratio appeared correct. The graphic was then copied as a metafile and pasted into *Adobe Illustrator*, where further trimming and character redefinition were performed. The aspect ratio may become slightly deformed when copying into *Adobe Illustrator*, with a vertical unit appearing slightly longer than a horizontal unit, so some resizing may be necessary at this stage as well. The graphic was then saved as a PDF file and then included in the LaTeX file using the \includegraphics instruction, for example:

```
\begin{figure}[h]
\center{\includegraphics[width=10cm,keepaspectratio]{Ex18_2.pdf}}
\caption{\sl MCA map of four questions on women working:
         total inertia = 3, percentage inertia in map: 40.2\%.}
\end{figure}
```

R graphics Finally, many maps, especially in the new chapters of this third edition, were produced directly in R. These were exported as PDF files and opened in *Adobe Illustrator*, fine-tuned and then saved for including in the LaTeX code, as shown above.

Glossary of Terms

In this appendix an alphabetical list of the most common terms used in this book is given, along with a short definition of each. Words in italics refer to terms which are themselves contained in the glossary.

• *adjacency matrix* — in *network* theory, a matrix that codes the edges among a set of nodes (one-mode network) or between two sets of nodes (two-mode network); the value of an edge is 0 if there is no connection, otherwise 1 for an unweighted network or the weight for a *weighted network*.

• *adjusted principal inertias* — a modification of the results of a *multiple correspondence analysis* that gives a more realistic estimate of the inertia accounted for in the solution.

• *arch effect* — the tendency for points in a CA map to form a curve, owing to the particular geometry of CA where the profiles lie inside a *simplex*; also called the "horseshoe" effect.

• *aspect ratio* — the ratio between a unit length on the horizontal axis and a unit length on the vertical axis in a spatial representation; should be equal to 1 for a CA *map*.

• *asymmetric map* — a joint display of the rows and columns where the two clouds of points have different normalizations (also called scalings), usually one in *principal coordinates* and the other in *standard coordinates*; the asymmetric map is often a *biplot*.

• *biplot* — a joint display of points representing the rows and columns of a table such that a *scalar product* between a row point and a column point approximates optimally the corresponding element in the table.

• *biplot axis* — a line in the direction of a point vector in a *biplot* onto which the other set of points can be projected in order to estimate values in the table being analysed.

• *bootstrapping* — a computer-based method of investigating the variability of a statistic, by generating a large number replicate samples from the observed sample.

• *Box–Cox transformation* — a particular power transformation which has as limiting case the logarithmic transformation as the power parameter tends to zero.

• *Burt matrix* — a particular matrix of *stacked tables*, consisting of all two-way cross-tabulations of a set of Q categorical variables, including the cross-tabulations of each variable with itself.

• *calibration* — in *biplots*, the process of putting a scale on a *biplot axis* with specific tic-marks and values; in CA, where *profiles* are being mapped, this is a scale in units of proportions or percentages.

• *canonical correspondence analysis (CCA)* — extension of CA to include external explanatory variables; the CA solution is constrained to have dimensions which are linearly related to these explanatory variables.

- *centroid* — the weighted average point.

- *centroid discriminant analysis* — an analysis where attention is focused on group means rather than individual observations, where the group means (i.e. centroids) are weighted by their respective group sizes or aggregated weights.

- *chi-square distance* (χ^2-*distance*)— weighted *Euclidean distance* between *profiles*, where each squared difference between profile elements is divided by the corresponding element of the average profile.

- *chi-square statistic* — the statistic used commonly for testing the *independence model* for a *contingency table*; calculated as the sum of squared differences between observed frequencies and frequencies expected according to the model, each squared difference being divided by the corresponding expected frequency.

- *co-correspondence analysis* — a form of *correspondence analysis* applied to two tables with the same rows, aiming to show the relationship between the tables rather than the relationships within each one; a special case of *co-inertia analysis*.

- *co-inertia* — just like the *inertia* measures the variance within a table in terms of weighted average sum of squared distances of the row points to the centroid, so co-inertia measures the covariance between two tables with common rows as the weighted average sum of cross-products of the two sets of row points to their respective averages.

- *co-inertia analysis* — a general approach to analysing the *co-inertia* between two tables that have the same rows, identifying dimensions that are common to both tables.

- *composition* — a set of observed nonnegative values that sum to 1, or to 100% when expressed as percentages (also called a *profile* in *correspondence analysis* when computed on a vector of nonnegative values by dividing by their sum); the components of a composition are called "parts", and the fact that the sum of the parts is a constant is referred to as their being "closed".

- *compositional data analysis* — the analysis of data in the form of observed *compositions*.

- *contingency ratio* — for a *contingency table*, the observed frequency divided by the expected frequency according to the *independence model*.

- *contingency table* — a cross-tabulation of a set of individuals according to two categorical variables; hence the grand total of the table is the number of individuals.

- *contribution to inertia* — component of *inertia* accounted for by a particular point on a particular *principal axis*; these are usually expressed relative to the corresponding *principal inertia* on the axis (giving a diagnostic of how the axis is constructed) or relative to the inertia of the point (giving a measure of how well the point is explained by the axis).

- *correspondence analysis (CA)* — a method of displaying the rows and columns of a table as points in a spatial map, with a specific geometric interpretation of the positions of the points as a means of interpreting the similarities and differences between rows, the similarities and differences between columns and the association between rows and columns.

- *degree vector* — in *network* theory, the vector of the number of edges (i.e. degrees) connected to each node (unweighted network), or the sum of the weights of the edges connected to each node (*weighted network*).

- *dimensionality* — the number of dimensions inherent in a table needed to reproduce its elements exactly in a CA *map*.

- *direct dimension* — in the analysis of a square symmetric matrix, a dimension defined by an *eigenvector* with positive *eigenvalue* (see *indirect dimension*).

- *doubling* — a recoding scheme where a row (or column) is recoded as a pair of rows (or columns) in order to map the extremes, or poles, of a scale; used in CA to analyse ratings, preferences and paired comparisons.

- *dummy variable* — a variable that takes on the values 0 and 1 only; used in one form of *multiple correspondence analysis* to code multivariate categorical data.

- *eigenvalue* — a quantity inherent in a square matrix, part of a decomposition of the matrix into the product of simpler matrices; in general, a square matrix has as many eigenvalues and associated *eigenvectors* as its rank; in the context of CA, eigenvalue is a synonym for *principal inertia* (see *singular value decomposition*).

- *eigenvector* — a vector associated with an *eigenvalue* of a square matrix, defining a dimension inherent in the matrix (see *singular value decomposition*).

- *Euclidean distance* — distance measure between vectors where squared differences between corresponding elements are summed, followed by taking the square root of this sum.

- *fuzzy coding* — a transformation of a continuous variable into a set of categories, where the value in each category is between 0 and 1 (inclusive) and the sum of the values across the categories is equal to 1; this is a more general coding than the *dummy variable* coding in an indicator matrix, where values are strictly 0 or 1.

- *homogeneity* — in the context of multivariate categorical data a measure of how far case scores are from their corresponding category scale values.

- *homogeneity analysis* — theoretically equivalent to multiple correspondence analysis, it quantifies the categories of multivariate categorical data in order to minimize the loss of homogeneity between cases and categories.

- *identification condition* — a condition which needs to be imposed on an optimization problem in order to obtain a unique solution.

- *independence model* — (also called the "homogeneity hypothesis") a model for the counts in a *contingency table*, which assumes that the rows (or columns) are sampled randomly from the same population; i.e. the expected relative frequencies (proportions) in each row, or in each column, are the same.

- *indicator matrix* — the coding of a multivariate categorical data set in the form of *dummy variables*.

- *inertia* — weighted sum of squared distances of a set of points to their *centroid*; in CA the points are *profiles*, weights are the *masses* of the profiles and the distances are *chi-square distances*.

- *interactive coding* — the formation of a single categorical variable from all the category combinations of two categorical variables.

- *inverse dimension* — in the analysis of a square symmetric matrix, a dimension defined by an *eigenvector* with negative *eigenvalue* (see *direct dimension*); the *singular vectors* corresponding to these eigenvectors are reversed in sign.

- *joint correspondence analysis (JCA)* — an adaptation of *multiple correspondence analysis* to analyse all unique two-way cross-tabulations of a set of Q categorical variables while ignoring the cross-tabulations of each variable with itself.

- *Laplacian matrix* — in *network* theory, the matrix equal to the diagonal matrix of the *degree vector* minus the *adjacency matrix*.

- *log-ratio analysis (LRA)* — an approach highly related to *correspondence analysis*, but applied to the logarithms of a table with strictly positive values; the weighted form of LRA, where rows and columns are weighted by the table margins, as in CA, is usually preferred over the unweighted form.

- *map* — a spatial representation of points (row and column profiles in CA) with a distance or scalar product (*biplot*) interpretation.

- *mass* — the marginal total of a row or a column of a table, divided by the grand total of the table; used as weights in CA.

- *matched matrices* — matrices that have exactly the same rows and columns.

- *multiple correspondence analysis (MCA)* — for more than two categorical variables, the CA of the *indicator matrix* or *Burt matrix* formed from the variables.

- *network* — a set of objects, called nodes, some of which are joined by links called edges; when the edges have no sense of direction, for example two people linked by friendship, it is called an undirected network; if the edges have a sense of direction, for example two countries between which migration flows are different, it is called a directed network; the nodes can be of two different types and the edges can then connect nodes between the two types, for example companies and clients, in which case the network is called a two-mode network, as opposed to a single-mode network where edges are internal to a single group (see also *weighted network*).

- *optimal scale* — a set of scale values assigned to the categories of several categorical variables, which optimizes some criterion such as maximum correlation (with another variable) or maximum discrimination (between a set of groups).

- *outlier* — a point on the periphery of a display that is well separated from the general scatter of points.

- *partial bootstrap* — in CA, the display of many replicate samples, obtained by *bootstrapping*, as supplementary points in the map of the original table.

- *permutation test* — a distribution-free strategy of statistical inference achieved through the generation of data permutations, either all possible ones or a large random sample, assuming a null hypothesis, leading to the null distribution of a test statistic and thus an estimate of the p-value associated with the observed value of the statistic.

- *principal axis* — a direction of spread of points in multidimensional space that optimizes the *inertia* displayed; can be thought of equivalently as an axis which best fits the points in a weighted least-squares sense.

- *principal coordinates* — coordinates of a set of points projected onto a *principal axis*, such that their weighted sum of squares along an axis equals the *principal inertia* on that axis.

- *principal inertia* — the *inertia* displayed along a *principal axis*; also referred to as an *eigenvalue*.

- *profile* — a row or a column of a table divided by its total; the profiles are the points visualized in CA.

- *root mean-squared error (RMSE)* — when comparing a set of estimated values with their true values, the square root of the average sum of squared differences.

- *scalar product* — for two point vectors, the product of their lengths multiplied by the cosine of the angle between them; directly proportional to the projection of one point on the vector defined by the other.

- *shortest path distance* — in *network* theory, the minimum number of edges connecting two nodes.

- *simplex* — a triangle in two dimensions, a tetrahedron in three dimensions and generalizations of these geometric figures in higher dimensions; in CA J-dimensional *profiles* lie inside a simplex defined by J *vertices* in $(J-1)$-dimensional space.

- *singular value decomposition (SVD)* — a matrix decomposition similar to that of *eigenvalues* and *eigenvectors*, but applicable to rectangular matrices; the squares of the singular values are eigenvalues of particular square matrices, and the left and right singular vectors are also eigenvectors.

- *skew-symmetric matrix* — a square matrix with zeros on the diagonal and the property that the elements above the diagonal have the same absolute value as those opposite them below the diagonal, but with opposite sign.

- *stacked table* — a set of tables concatenated rowwise or columnwise or both, often based on cross-tabulating the same individuals. See also *matched matrices*, which are often analysed in a stacked format.

- *standard coordinates* — coordinates of a set of points such that their weighted sum of squares along an axis equals 1.

- *symmetric map* — a joint display of the rows and columns where the two clouds of points have the same normalization in *principal coordinates* ; strictly speaking, the symmetric map is not a *biplot*.

- *subcomposition* — a reduced set of parts of a *composition*, where the parts are re-closed to sum to 1.

- *subcompositional coherence* — a desirable property of *compositional data analysis* whereby relationships between the parts of a *subcomposition* remain the same as in an extended *composition*.

- *subset correspondence analysis* — a variant of CA which allows subsets of rows and/or columns to be analysed, while maintaining the same geometry of the full table.

- *supplementary point* — a point which has a position (a *profile* in CA) with mass set equal to 0; in other words a supplementary point is displayed on the map but has not been used in the construction of the map.

- *transition relationship* — the relationship between the row and column co-ordinates in a map.

- *vertex* — a unit profile, i.e. a profile with all elements 0 except one with value 1.

- *Ward clustering* — a specific hierarchical clustering algorithm which minimizes the within-cluster inertia at each clustering step, equivalent to maximizing the between-cluster inertia.

- *weighted Euclidean distance* — similar to *Euclidean distance*, but with a positive weighting factor for each squared difference term.

- *weighted network* — a *network* where the edges have positive values attached to them, quantifying the strength of the edges between pairs of nodes; more specifically called an edge-weighted network.

Bibliography of Correspondence Analysis

Because of the didactic aim of this book, no references have been given in the chapters. This Appendix aims to highlight the main bibliographical sources for learning more about correspondence analysis.

Although the theory of CA dates back to much earlier in the 20th century, CA as presented in this book originates in the work of Jean-Paul Benzécri and his co-workers in France in the 1960s, which was published in the two volumes of *Analyse des Données* (literally, Data Analysis): *Benzécri's school of data analysis*

—Benzécri, J.-P. & collaborateurs (1973) *Analyse des Données. Tôme 1: La Classification. Tôme 2: L'Analyse des Correspondances*. Paris: Dunod.

However, even with a knowledge of French, these books remain inaccessible to most readers who are not initiated into Benzécri's particular notational style, different from the more pragmatic matrix-vector notation. The following book by Le Roux & Rouanet gives an authentic account of Benzécri's approach to analysing large data sets, which the authors have coined as "geometric data analysis". They have also maintained a complex notational style, however, which limits understanding:

—Le Roux, B. & Rouanet, H. (2004) *Geometric Data Analysis: From Correspondence Analysis to Structured Data*. Dordrecht: Kluwer.

One of the best publications in English for understanding Benzécri's work is the book by Fionn Murtagh, who was also a student of Benzécri. Not only does he communicate much more of the Benzecrian philosophy (there is also a foreword by Benzécri himself, with an English translation), but also the book is innovative in its approach and highly oriented to computing, providing many interesting applications and details of R programming.

—Murtagh, F. (2005) *Correspondence Analysis and Data Coding with Java and R*. London: Chapman & Hall/CRC.

Brigitte Escofier, one of the leading and most creative members of Benzécri's original group from the French city of Rennes, has been commemorated posthumously by a collection of her most important articles (in French):

—Escofier, B. (2003) *Analyse des Correspondances: Recherches au Coeur de l'Analyse des Donées*. Rennes, France: Presses Universitaires de Rennes.

In 1984 two English books on CA appeared almost simultaneously, expressing Benzécri's work in a more conventional mathematical notation: *The two English books published in 1984*

—Lebart, L., Morineau, A. & Warwick, K. (1984) *Multivariate Descriptive Statistical Analysis*. Chichester, UK: Wiley.

—Greenacre, M.J. (1984) *Theory and Applications of Correspondence Analysis*. London: Academic Press.

These books are both out of print, but Greenacre's book is now available online for free download at `www.carme-n.org`. Lebart et al.'s book gives a less detailed description of CA itself but a broader view of its use in the context of large-scale social surveys. Greenacre's book attempts to be a complete account of the method's theory and practice at that time. Both these books serve as good literature sources for work up to 1984.

The Gifi system   A group in Holland, originally led by Jan de Leeuw under the *nom-de-plume* of Albert Gifi, was involved with the most important developments of CA outside France, and still remains very active today. This group mostly explored the use of MCA — which they called *homogeneity analysis* — as a quantification technique embedded in classical multivariate analysis to achieve nonlinear generalizations of multivariate methods. The work of the Gifi group is exposed in the book:

—Gifi, A. (1990) *Nonlinear Multivariate Analysis*. Chichester, UK: Wiley.

As an excellent summary of the "Gifi system", see:

—Michalidis, G. & de Leeuw, J. (1998) The Gifi system for descriptive multivariate analysis. *Statistical Science*, 13, 307–336. (Can be googled.)

The Japanese school   Founded by Chikio Hayashi, this group developed, in parallel to the French and Dutch schools, an equivalent system of data analysis called "quantification of qualitative data", imbued with its own cultural aspects. Several books by Shizuhiko Nishisato describe this approach, renamed as "dual scaling", concentrating more on the algebraic properties of the quantified scale values, although the book by Nishisato does contain many graphical displays:

—Nishisato, S. (2006) *Multivariate Nonlinear Descriptive Analysis*. London: Chapman & Hall/CRC.

Nishisato's book contains many historical details and a very comprehensive reference list of CA-related literature, but no details about computing.

CARME books   As mentioned in the Preface, international conferences on correspondence analysis have been organized every four years since 1991. The first three were held in 1991, 1995 and 1999 at the Central Archive for Empirical Social Research, Cologne, then in 2003 at the Universitat Pompeu Fabra, Barcelona, in 2007 at the Erasmus University, Rotterdam, in 2011 at Agrocampus in Rennes, and in 2015 at the University of Naples, these conferences have come to be known as the CARME conferences, standing for "Correspondence Analysis and Related Methods". Several publications have emerged as a direct product of these conferences, in the form of books collectively written by statisticians and social scientists to reflect the development of the theoretical and practical aspects of CA and related methods, all peer-reviewed and extensively edited by Jörg Blasius and Michael Greenacre:

— Greenacre, M. & Blasius, J., eds (1994) *Correspondence Analysis in the Social Sciences.* London: Academic Press.

— Blasius, J. & Greenacre, M., eds (1998) *Visualizing Categorical Data.* San Diego: Academic Press.

— Greenacre, M. & Blasius, J., eds (2006) *Multiple Correspondence Analysis and Related Methods.* Boca Raton, FL: Chapman & Hall/CRC.

— Blasius, J. & Greenacre, M., eds (2015) *Visualization and Verbalization of Data.* Boca Raton, FL: Chapman & Hall/CRC.

These four volumes, to which over 100 authors have contributed, are highly recommended for further reading. The third volume is particularly oriented to computing as well, and many sources of computer software are given by the individual authors. Half of the latest volume is devoted to the history of multivariate methods, written by experts in each area, and is the ultimate reference for the methodological origins of this area of statistics.

In addition, a special journal issue was published in 2009, as a result of the CARME conference in Rotterdam:

— Blasius, J., Greenacre, M., Groenen, P. and van der Velden, M., eds (2009) Special issue on correspondence analysis and related methods. *Computational Statistics and Data Analysis* 53, 3103–3106.

Several other books have appeared in recent years related to CA. The first *Other books* three books below are available at `www.multivariatestatistics.org` for free download, thanks to the support of the BBVA Foundation in Spain:

— Greenacre, M. (2008) *La Práctica del Análisis de Correspondéncias.* Bilbao: BBVA Foundation.
(The Spanish translation of the second edition of *Correspondence Analysis in Practice.*)

— Greenacre, M. (2010) *Biplots in Practice.* Bilbao: BBVA Foundation.
(A comprehensive practical treatment of all types of biplots, including CA, MCA, CCA and the related methods of PCA and LRA.)

— Greenacre, M. & Primicerio, R. (2015) *Multivariate Analysis of Ecological Data.* Bilbao: BBVA Foundation.
(Aimed specifically at ecologists, based on a short course, this book includes the full spectrum of multivariate descriptive methods including cluster analysis, generalized linear models and classification and regression trees.)

— Husson, F., Lê, S. & Pagès, J. (2011) Exploratory Multivariate Analysis by Example Using R. Boca Raton, FL: Chapman & Hall/CRC.
(Covering a wide range of multivariate methods, this book is based on the `FactoMineR` package in R.)

—Beh, E. & Lombardo, R. (2015) *Correspondence Analysis: Theory, Practice and New Strategies.* Chichester, UK: John Wiley.
(A comprehensive treatment of correspondence analysis, with some novelties developed by the authors themselves, for example when categorical variables are ordinal.)

— Friendly, M.. & Meyer, D. (2016) *Discrete Data Analysis with R.* Boca Raton, FL: Chapman & Hall/CRC.
(An applied treatment of modern methods for the analysis of categorical data, both discrete response data and frequency data, including correspondence analysis. Uses R packages vcd, vcdExtra and ca.)

The R connection The aim of *Correspondence Analysis in Practice* is not only to present a didactically structured text about the method, but also the computational aspects, using the R programming environment. One of the many books on R programming which can be recommended to anyone starting off, as well as an excellent introduction to statistical concepts and methods, is:

—Crawley, M. (2005) *Statistics: An Introduction using R.* Chichester, UK: John Wiley.

The ca package for R is discussed in more detail in the following article:

—Nenadić, O. & Greenacre, M. (2007) Correspondence analysis in R, with two- and three-dimensional graphics: The ca package. *Journal of Statistical Software.* Free download from http://www.jstatsoft.org.

Some web resources The following websites can be consulted for further information and software about CA and related methods:

— http://www.carme-n.org
(Correspondence Analysis and Related Methods Network, with R scripts and data from *Correspondence Analysis in Practice, Second and Third Editions*)

— http://gifi.stat.ucla.edu
(Jan de Leeuw's website for the Gifi system and R functions)

— http://www.correspondances.info
(Fionn Murtagh's website for his book, with R scripts and data sets)

— http://www.datavis.ca
(Michael Friendly's website for data visualization, books and software)

— http://www.imperial.ac.uk/bio/research/crawley/statistics
(Michael Crawley's material from his book *Statistics: An Introduction using R*)

— http://www.issp.org
(website of International Social Survey Programme)

Epilogue

Correspondence analysis (CA) has been presented in this book as a versatile method of data visualization, applicable in a wide variety of situations. This epilogue serves to elaborate further on certain aspects of the method that arise frequently in discussions, and to add some personal thoughts.

The interpretation of the symmetric map remains one of the method's most controversial aspects, even though it is the option of choice for CA maps. This map displays both rows and columns in principal coordinates — that is, the projections of the row profiles and the projections of the column profiles are shown in a joint map even though, strictly speaking, they occupy different spaces. We have seen (see, for example, Chapters 9 and 10) that the difference between the symmetric map and the asymmetric map (where the points do lie in the same space) is the rescaling along principal axes by the square roots of the respective principal inertias. Thus the directions indicated by the points in principal coordinates and by their counterparts in standard coordinates are almost the same when the square roots of the principal inertias are not too different — an example can be seen by comparing the directions of the biplot axes in Exhibit 13.4 with those defined by the corresponding profile points in Exhibit 12.2. In such a case, the biplot style of interpreting the display is valid whether the display is symmetric or asymmetric. If the square roots of the principal inertias are very different, however, there can be problems with the biplot style interpretation of the symmetric map — see, for example, the differences in the directions defined by the smoking categories in Exhibits 9.2 and 9.5. Even so, the distortion induced by using the symmetric map as if it were a true biplot is not so great, as discussed in the following paper by Gabriel:

The symmetric map

— Gabriel, R. (2002) Goodness of fit of biplots and correspondence analysis. *Biometrika*, 89, 423–436.

This means that the scaling debate is really an academic issue and, as far as the practice of CA is concerned, hardly worth all the discussion that it has generated. In my opinion, the symmetric map is still one of the best scaling options for CA, and is thus the default option in our **ca** package for R. If the data matrix is to be interpreted asymmetrically, with the rows (say) representing "observational units" (e.g. demographic groups in sociology such as marital status and educational levels, or sampling locations in ecology or archeology, or texts in linguistics, etc.) and the columns representing "variables" (e.g. response categories in sociology, species in ecology, artefacts in archaeology or stylistic indicators in linguistics, etc.), then the contribution biplot is a good alternative. This version of the biplot displays optimally the distances between the sampling units and gives a valid biplot interpretation of the units projected onto the variable directions, as well as giving meaningful lengths to the variable vectors, which is especially useful when there are a lot of variables.

"You can't have your cake and eat it too!" This English saying is unfortunately true in this case, as well as the similar expression *"You can't get everything in life!"* It would be wonderful if we could represent optimally in a single map the following three desirable aspects for interpretation:

(i) the distances between the row profiles,

(ii) the distances between the column profiles and

(iii) the scalar products between row and column points, which reconstruct the original data (i.e. the biplot).

But the reality is that we can see at most only two of these three represented optimally at the same time. In the symmetric map we see optimal representations of the chi-square distances for the row profiles and for the column profiles; hence row-to-row distances and column-to-column distances can be interpreted (i.e. (i) and (ii)). The row–column relationship is not optimally represented, but can still be interpreted with reasonable assurance, taking into account the remarks on the previous page. In the asymmetric map we see the optimal representation of one set of profiles, say the row profiles, while the column vertices give the extreme profiles as reference points and also lie on the biplot axes for interpreting the optimal row–column relationship (i.e. (i) and (iii)). The contribution CA biplot is a variation of the asymmetric map which also shows one set of profiles, say the row profiles, but pulls in the column vertices by the square roots of their masses to improve the joint representation (i.e. (i) and (iii)). In this biplot the lengths of the column vectors on the biplot axes can be related to their contributions on the principal axes (see Chapter 13), but there is no distance interpretation between the column points.

"Symmetrical normalization" in SPSS Apart from R and a brief mention of XLSTAT, I have not discussed other software packages that include CA, among which are Minitab, Stata, Statistica, SPAD, SAS and SPSS. Because SPSS is widely used, a comment about its options is necessary here. In SPSS's CA program in the *Categories* module, an alternative biplot is given that has not been illustrated in this book, called the "symmetrical normalization", which may be confused with the symmetric map described in this book. It is not exactly the same thing, however, since it uses standard coordinates scaled by the square roots of the singular values (i.e. fourth roots of the principal inertias) instead of the singular values themselves: in other words (referring to (A.8) and (A.9) on page 242), $\mathbf{\Phi D}_\alpha^{\frac{1}{2}}$ and $\mathbf{\Gamma D}_\alpha^{\frac{1}{2}}$ instead of the symmetric map's $\mathbf{\Phi D}_\alpha$ and $\mathbf{\Gamma D}_\alpha$. SPSS's "symmetrical normalization" gives optimal representation of scalar products but non-optimal representations of distances since neither rows nor columns are

represented in principal coordinates. Hence this display gives only one of the three desired properties (i.e. the biplot property (iii) but not the distance properties (i) and (ii)). Even though the difference between this display and the symmetric map is also a matter of scale factors along the two axes, and in most cases is hardly distinguishable to an untrained eye, we would not recommend this map in practice, because it represents no benefit (in fact, a loss) over existing options. If the principal inertias on the two axes are fairly close, then, as before, the relative positions of points in the "symmetrical normalization" are practically identical to those in the symmetric map, but the symmetric map is definitely preferable since it shows the chi-square distances to their true scale. For purposes of comparison, this option is provided in our R package **ca**, where it is called the "symmetric biplot", specified as follows in the plotting: `map="symbiplot"` (see page 266). Curiously, the symmetric map, one of the most popular display options of French researchers, has never been available in SPSS (module *Categories*), and it is still not possible in IBM SPSS version 20 to obtain a joint map of the rows and columns in principal coordinates. The best one can do is to select the "principal" normalization, which gives the row and column principal coordinates in numerical form, but the program does not allow a joint map of them, preferring separate maps. Unless the user's raw respondent-level data are in SPSS format, the CA program in SPSS is not the best option available. However, the other optimal scaling programs in *Categories*, for multiple correspondence analysis (called by its synonym, homogeneity analysis, in previous versions) and nonlinear principal component analysis (*CatPCA*), are very useful for social science applications.

The issue of rare categories and their effect on the χ^2-distance and the CA solution is also one that has generated much discussion, especially in ecological circles, almost entirely without justification. For example, C. R. Rao has stated that "since the chi-square distance uses the marginal proportions in the denominator, undue emphasis is given to the categories with low frequencies in measuring affinities between profiles" — see page 42 of the following article: *Rare (low-frequency) categories*

— Rao, C.R. (1995) A review of canonical coordinates and an alternative to correspondence analysis using Hellinger distance. *Qüestiió*, 19, 23–63. Downloadable from the website of the Institute of Statistics of Catalonia at `www.idescat.cat/sort/questiio`.

Similarly, in the ecological context of analysing abundance counts of species, Pierre Legendre states that "a difference between abundance values for a common species contributes less to the distance than the same difference for a rare species, so that rare species may have an unduly large influence on the analysis." — see page 271 of this article:

—Legendre, P. (2001) Ecologically meaningful transformations for ordination of species data. *Oecologia*, 129, 271–280.

But the fact is that in CA each category is weighted in the analysis proportional to its mass, which reduces the role played by low-frequency categories in the analysis. It can be shown very simply by looking at the numerical contributions of each category to the principal axes that rare categories generally have low influence in the solution — i.e. the solution would be almost the same if they were removed from the analysis entirely.

As an illustration, for the species abundance data set of Chapter 10 (see page 77), we calculated the relative abundance for the 10 most abundant and 10 least abundant species and compared this to their respective percentage contributions to the first two axes of the CA map in Exhibit 10.5. The results are as follows:

| | | *Contributions to axes* | |
Species	*Relative abundance*	*Axis 1*	*Axis 2*
10 most abundant	74.6%	77.3%	89.3%
10 least abundant	0.4%	0.8%	0.5%

This illustrates that the rare species do not make an excessive contribution to the two-dimensional solution — the contributions are very much in line with the abundances in each subset of species. Only a few times, in our experience, do low-frequency categories make an excessively large contribution to the major principal axes, in which case they should be removed or combined with another category. A case in point is in sociological applications, when low frequency categories such as missing values coincide in the same subgroup of respondents. These categories can dominate an MCA solution, often defining the first principal axis as in Exhibits 18.2 and 18.5. This situation can be rectified by using a subset analysis, or by combining rare response categories with others in a sensible way. The analogous situation in ecology would be when several rare species co-occur in the same samples, but this is not a common situation — usually rare species occur randomly in different samples.

In a statistical report in the journal *Ecology*, I have refuted the belief that rare objects generally perturb the results of CA:

— Greenacre, M. (2013) The contribution of rare objects in correspondence analysis. *Ecology*, 94, 241–249.

Low-frequency categories are often outliers

Having said that the rows or columns with low frequencies generally have low influence on the solution, because of low mass, it is true that these points are often outliers in the CA map, owing to their strange profiles. Outliers draw attention to themselves and it is probably for this reason that the impression is created that they may be affecting the analysis strongly. As shown in Chapter 13 and mentioned above, the contribution biplot would solve this problem by "pulling in" these points by the square root of their masses, which effectively eliminates the low frequency outliers since these end up closer to the origin. This also demonstrates graphically that their influence on each principal axis is mostly quite low. When points are represented in contribution coordinates the distance property between them is sacrificed (once more supporting the saying *"You can't have your cake and eat it too!"*).

I have performed hundreds of CAs and especially when there are many points and labels the graphing of the results becomes problematic. Gradually I have homed in on a way to make an acceptable plot, illustrated by the following case. I was involved in a research paper on the fatty acid compositions of *amphipods* (family of small marine species). These compositional data suggest the use of log-ratio analysis (see Chapter 22) but there are approximately a hundred zeros in the data set, so the next best thing is to use CA (see Chapter 22). There are four species of amphipods, abbreviated C.g, T.a, T.c and T.l, and each has been sampled several times in the winter (W) and the summer (S): for example, the label C.gS stands for the summer samples of species abbreviated as C.g. The data set has 57 samples distributed among the eight species–season groups, and 40 variables, divided into three compositional data sets of fatty acids (e.g. *20:4(n-3)*), fatty alcohols (with *_Alc* suffix, e.g. *14:0_Alc*) and lipid classes (e.g. *WE* = wax esters).

Trying to represent all 57 samples, their group means, as well as the 40 variables in one ordination plot, is a challenge. The contribution biplot reduces the number of variables to 16, so this is much more manageable. Exhibit E.1 shows these variables, along with the group means linked to the individual samples displayed as small circles. Alternatively, if one is not so concerned to see the individual samples, but more interested in the precision of the group means, Exhibit E.2 shows 95% confidence ellipses for the means, obtained by bootstrapping the samples within each group (see Chapter 29). The groups T.cW and C.gW have only one and two samples respectively, so confidence regions are not well defined, but the other confidence regions show that there is a significant difference between the winter samples of T.lW and T.aW and their corresponding summer samples as well as the summer samples of the other two species. The clear difference between the summer samples T.lS and T.aS contrasts with the proximity of these species in the winter. The variables that are responsible for the separation of the sample means can be clearly seen in the contribution biplot, for example the group of five fatty acids at the top of the ordination are the ones higher in the winter samples, as opposed to the variables across the lower area of the ordination that are higher in the summer samples.

This section is a bit technical but will demonstrate to the statistically minded reader that the chi-square distance, apart from being the key to all the properties of CA, can also be defended on theoretical grounds as an appropriate statistical distance measure. A bit of matrix notation is needed for the weighted Euclidean distance function in (5.1), which can be written as:

$$\text{weighted Euclidean distance} = \sqrt{(\mathbf{x} - \mathbf{y})^{\mathsf{T}} \mathbf{D}_w (\mathbf{x} - \mathbf{y})} \qquad (E.1)$$

where $\mathbf{x}$ and $\mathbf{y}$ are vectors with elements x_j and y_j, $j=1,\ldots,J$, ^{T} indicates transposition of a vector or matrix, and $\mathbf{D}_w$ is the diagonal matrix of the dimension weighting factors w_j's. The rows, say, of a contingency table can be assumed to be realizations of a *multinomial* random variable. The multinomial

Exhibit E.1:
CA centroid discriminant analysis, showing the eight group centroids, each linked to their respective sample points, and the variables in contribution coordinates. Only those variables with contributions to the solution higher than the average are shown (larger font) as well as two more variables that have lower than average contribution but above average correlation with the solution.

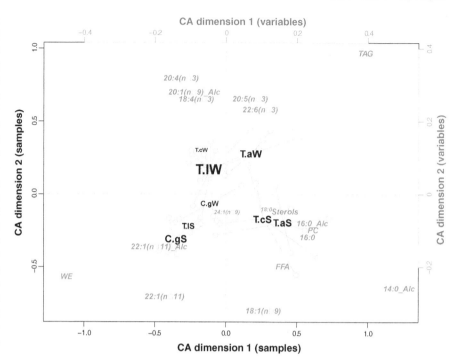

Exhibit E.2:
An alternative version of Exhibit E.1, showing 95% confidence regions for the group average points, rather than the individual sample points. Group C.gW has only two sample points, and group Tc.W only one. Size of the group labels in both these exhibits is related to sample size. Also notice in both exhibits the different scales for samples and variables.

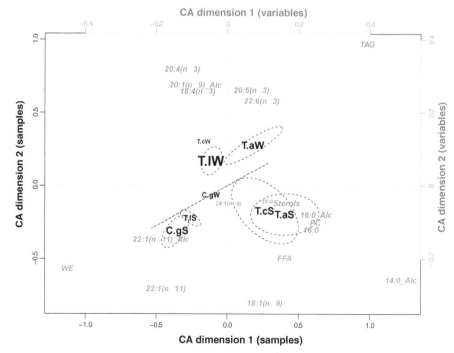

distribution is a generalization of the binomial distribution, and is a model to describe the behaviour of data sampled from a population where there are probabilities p_j, $j = 1, \ldots, J$, of observing a sampling unit in one of J groups, e.g. the three readership groups in Chapter 3 (see Exhibit 3.1 on page 18). Under the null hypothesis that the data are sampled from the same population, the five education groups in this data set would be multinomial samples from a population with probabilities p_1, p_2, p_3, where the estimates of p_j for the three groups are the elements of the average profile $\hat{p}_1 = c_1 = 0.183$, $\hat{p}_2 = c_2 = 0.413$ and $\hat{p}_3 = c_3 = 0.404$ (see last row of Exhibit 3.1). The classic distance function for grouped multivariate data is the so-called *Mahalanobis distance*, based on the inverse of the covariance matrix of the variables:

$$\text{Mahalanobis distance} = \sqrt{(\mathbf{x} - \mathbf{y})^{\mathsf{T}} \boldsymbol{\Sigma}^{-1} (\mathbf{x} - \mathbf{y})} \qquad (E.2)$$

which looks like the weighted Euclidean distance (E.1), except that it involves a full square matrix of weights $\boldsymbol{\Sigma}^{-1}$, not a diagonal matrix. The covariance matrix $\boldsymbol{\Sigma}$ for the multinomial distribution has a simple form, for example for our trinomial case $J = 3$ (the results are similar for any number of groups):

$$\boldsymbol{\Sigma} = \begin{bmatrix} p_1(1 - p_1) & -p_1 p_2 & -p_1 p_3 \\ -p_2 p_1 & p_2(1 - p_2) & -p_2 p_3 \\ -p_3 p_1 & -p_3 p_2 & p_3(1 - p_3) \end{bmatrix} = \mathbf{D}_p - \mathbf{p}\mathbf{p}^{\mathsf{T}} \qquad (E.3)$$

where $\mathbf{p}$ is the vector of the p_j's and $\mathbf{D}_p$ is the corresponding diagonal matrix. (E.3) is estimated by substituting the probabilities p_j by their estimates c_j. To invert the covariance matrix $\boldsymbol{\Sigma}$ in the usual way is not possible since it is a singular matrix, so we cannot find a matrix $\boldsymbol{\Sigma}^{-1}$ such that $\boldsymbol{\Sigma}\boldsymbol{\Sigma}^{-1} = \mathbf{I}$. One way to get around this is to drop one of the categories and use just $J - 1$ categories throughout. Whichever category is omitted, the Mahalanobis distance will be the same. An alternative more elegant approach, which is entirely equivalent but uses all J categories, is to use a so-called *generalized inverse*, denoted by $\boldsymbol{\Sigma}^{-}$, which has the property that $\boldsymbol{\Sigma}\boldsymbol{\Sigma}^{-}\boldsymbol{\Sigma} = \boldsymbol{\Sigma}$ (this is also known specifically as the *Moore–Penrose inverse*). It turns out that the Moore–Penrose generalized inverse of (E.3) is equal to

$$\boldsymbol{\Sigma}^{-} = \begin{bmatrix} 1/p_1 & 0 & 0 \\ 0 & 1/p_2 & 0 \\ 0 & 0 & 1/p_3 \end{bmatrix} = \mathbf{D}_p^{-1} \qquad (E.4)$$

which means that the Mahalanobis distance in (E.2) is estimated exactly by the χ^2-distance. The situation here is similar to that in linear discriminant analysis: to discriminate maximally between groups, the groups are assumed to have equal covariance matrices, which in the multinomial case translates to our assuming the independence model, and then the data vectors are embedded in Mahalanobis space, which translates to chi-square space.

The issue of rotations has not been treated in this book because rotations are seldom justified or needed in CA. On the one hand, the profile space is not unbounded real vector space but a space delimited by the unit points,

Rotation of solutions

or vertices, defining a simplex in multidimensional space. The idea of lining up the category points along specific axes at right angles does not have the same meaning as in factor analysis where right-angledness really means zero correlation between variables (remember that one category point in CA is always determined by all the others, because the elements of a profile add up to 1). Rotations can be appropriate in some contexts in MCA and nonlinear PCA (not treated in this book) where several variables are analysed simultaneously. For example, it frequently occurs that all non-response points in MCA lie together in a bunch owing to high association in the data set but not coinciding with a principal axis, in which case it would be good to be able to rotate the solution as a way to "partial out" the non-response points. But this problem can be better solved by doing a subset analysis (Chapter 21) which completely ignores the non-response points and focuses totally on the substantive responses. If rotation of a solution is required, then the masses of the category points should be taken into account in the rotation. For example, a weighted version of the usual varimax rotation in factor analysis would be to maximize the criterion (assuming rotation of the column points is required):

$$\sum_j \sum_k c_j^2 (\tilde{y}_{jk}^2 - \frac{1}{J} \sum_{j'} \tilde{y}_{j'k}^2)^2 \qquad (E.5)$$

where $\tilde{y}_{jk}$ is the rotated standard coordinate, that is the (j,k)-th element of $\tilde{\mathbf{Y}} = \mathbf{YQ}$, for $\mathbf{Q}$ an orthogonal rotation matrix. Notice that the mass c_j is squared because the objective function involves the fourth powers of the coordinates. Since $c_j \tilde{y}_{jk}^2 = (c_j^{\frac{1}{2}} \tilde{y}_{jk})^2$, an almost identical alternative is suggested, which is a small modification of the usual varimax criterion: perform a rotation (unweighted) on the rescaled standard coordinates $c_j^{\frac{1}{2}} y_{jk}$, which are exactly those used in the contribution biplot (see Chapter 13). In other words, rotate the solution to concentrate (or *reify* in factor analysis terminology) the contributions of the categories on the rotated axes.

CA and   In Chapter 13 CA in K^* dimensions was shown to be a decomposition which
modelling   can be written as follows (see (13.4), also (A.14) in the Theoretical Appendix):

$$p_{ij} = r_i c_j + r_i c_j \left(\sum_{k=1}^{K^*} \sqrt{\lambda_k} \phi_{ik} \gamma_{jk} \right) + e_{ij} \quad i = 1, \dots, I; \; j = 1, \dots, J \quad (E.6)$$

The CA solution is obtained by minimizing the weighted sum of squares of the residuals e_{ij}. The first part of the decomposition, $r_i c_j$, is the expected value under the model of independence, so that the second part is explaining the deviations from the independence model as the sum of K^* bilinear terms (this bilinear part has a geometric interpretation in K^* dimensions which is the subject of most of this book). Any other model of the user's choice can be substituted for the independence model. For example, in the following article, the authors consider log-linear models for a contingency table, and then use CA as a way of exploring the structure, if any, in the deviations from the log-linear model:

— van der Heijden, P.G.M., de Falguerolles, A. and de Leeuw, J. (1989) A combined approach to contingency table analysis and log-linear analysis (with discussion). *Applied Statistics*, 38, 249–292.

This strategy can be used for multiway tables as well, using a contingency table modelling approach to account for main effects and chosen interactions in a first step, then calculating the residuals from the model and analysing these by CA. But note that this is not a straightforward application of CA, since the data have already been centred with respect to the model. The centring step in CA must not be performed and the original margins of the table must be used in the weighted least-squares fitting.

CA has a close affinity to *spectral mapping*, a method developed originally by Paul Lewi in the 1970s and used extensively in the analysis of biological activity spectra in the development of new drugs. Some later references are:

CA and spectral mapping

— Lewi, P.J. (1998) Analysis of contingency tables. In *Handbook of Chemometrics and Qualimetrics: Part B* (eds. B.G.M. Vandeginste, D.L. Massart, L.M.C. Buydens, S. de Jong, P.J. Lewi, J. Smeyers-Verbeke), Chapter 32, pp. 161–206. Amsterdam: Elsevier.

— Wouters, L., Göhlmann, H.W., Bijnens, L., Kass, S.U., Molenberghs, G. and Lewi, P.J. (2003) Graphical exploration of gene expression data: a comparative study of three multivariate methods. *Biometrics*, 59, 1131–1139.

Spectral mapping operates on the logarithms of the table, but incorporates the same weighting of rows and columns as in CA, i.e. by the row and column masses computed on the original table. The log-transformed table is double-centred with respect to the weighted row and column averages before applying the SVD as in CA. Hence, spectral mapping is exactly what was called weighted log-ratio analysis in Chapter 22. If the inertia in the data is low, then spectral mapping and CA are almost identical. The difference between the two methods is more pronounced when the inertia is high. Spectral mapping involves mapping the logarithms of ratios of the data, and has very interesting model-diagnostic properties, demonstrated in Exhibit 22.5. It also obeys the principle of distributional equivalence (see pages 37–38) and, in addition, has the property of *subcompositional coherence*, which is the property that underpins the analysis of compositional data: since the ratio between two data values remains the same whether or not other rows or columns are excluded from the table, subsets of rows or columns can be analysed with impunity. In CA on the other hand, profiles and distances are affected when analysing subsets; that is, CA is not subcompositionally coherent, hence the special adaptation called subset CA described in Chapter 21. For more details and further references, consult the following article:

— Greenacre, M. and Lewi, P.J. (2005) Distributional equivalence and subcompositional coherence in the analysis of compositional data, contingency tables and ratio-scale measurements. *Journal of Classification*, 26, 29–54.

The rather surprising relationship between weighted LRA, alias spectral mapping, and CA (see Chapter 22, page 175), is explained in more detail in the following article:

— Greenacre, M. (2009) Power transformations in correspondence analysis. *Computational Statistics and Data Analysis*, 52, 3269–3281.

The dimensionality of a multivariate categorical data set

To conclude this epilogue, here is an interesting problem, still unsolved since I stated it as a conjecture in the second edition! It is well known that in simple CA the dimensionality of an $I \times J$ table is $\min(I - 1, J - 1)$. For a $J \times J$ Burt matrix based on Q categorical variables, the CA dimensionality is $J - Q$, but we know that $J - Q$ dimensions are many more than are needed to reproduce the off-diagonal tables exactly. We propose that the dimensionality of a Q-variable data set be defined as the number of dimensions required to reproduce the $\frac{1}{2}Q(Q - 1)$ cross-tabulations exactly. In other words, the dimensionality is the number of dimensions required in a joint CA to explain 100% of the inertia. The question is: Can this dimensionality be determined beforehand or does it need to be discovered empirically? The rule in adjusted MCA is to consider only the K^* dimensions for which $\sqrt{\lambda_k} > 1/Q$ (see, for example, the adjusted inertias in Equation (19.6) on page 149, as well as the last paragraph of Chapter 20 on page 160). It would be convenient if this provided the clue to the dimensionality. In empirical studies the inertia explained using this number (K^*) of dimensions is usually very close to 100%, but this is no proof, of course, that the dimensionality is K^*. Perhaps, by the time a fourth edition of this book is published, this problem will have been finally solved!

Here's to correspondence analysis: Cheers!

Index

T - #0216 - 071024 - C0 - 229/178/17 - PB - 9780367782511 - Gloss Lamination